AF814112

STUDIES IN MILITARY GEOGRAPHY AND GEOLOGY

Studies in Military Geography and Geology

Edited by

Douglas R. Caldwell

US Army Topographic Engineering Center, Alexandria, VA, USA

and

Judy Ehlen

US Army Topographic Engineering Center, Alexandria, VA, USA

and

Russell S. Harmon

Army Research Office, Research Triangle Park, NC, USA

KLUWER ACADEMIC PUBLISHERS

DORDRECHT / BOSTON / LONDON

A C.I.P. Catalogue record for this book is available from the Library of Congress.

ISBN 1-4020-3104-1 (HB)
ISBN 1-4020-3105-X (eBook)

Published by Kluwer Academic Publishers,
P.O. Box 17, 3300 AA Dordrecht, The Netherlands.

Sold and distributed in North, Central and South America
by Kluwer Academic Publishers,
101 Philip Drive, Norwell, MA 02061, U.S.A.

In all other countries, sold and distributed
by Kluwer Academic Publishers,
P.O. Box 322, 3300 AH Dordrecht, The Netherlands.

Printed on acid-free paper

All Rights Reserved
© 2004 Kluwer Academic Publishers
No part of this work may be reproduced, stored in a retrieval system, or transmitted
in any form or by any means, electronic, mechanical, photocopying, microfilming, recording
or otherwise, without written permission from the Publisher, with the exception
of any material supplied specifically for the purpose of being entered
and executed on a computer system, for exclusive use by the purchaser of the work.

Contents

Contributing Authors

Robert J. Abrahart
School of Geography, University of Nottingham, Nottingham NG7 2RD UK (bob.abrahart@nottingham.ac.uk)

Charles A. Boland
630 Cannon Drive, Rock Hill, SC 29730 USA

Irene B. Boland
Winthrop University, Rock Hill, SC 29733 USA (bolandi@winthrop.edu)

Tad Britt
US Army Engineer Research and Development Center, Construction Engineering Research Laboratory, PO Box 9005, Champaign, IL 61826 USA (John.T.Britt@erdc.usace.army.mil)

Thomas Bullard
Division of Earth and Ecosystem Science, Desert Research Institute, 2215 Raggio Parkway, Reno, NV 89506 USA (tbullard@dri.edu)

Michael. J. Day
Department of Geography, University of Wisconsin, Milwaukee, WI 53201 USA (mickday@uwm.edu)

LTC James B. Dalton, Jr.
Department of Geography and Environmental Engineering, US Military Academy, West Point, NY 10096 USA (bj7130@usma.edu)

Francis H. Dillon, III
Geography Department, George Mason University, 4400 University Dr., Fairfax, VA 22030 USA (fdillon@gmu.edu)

Richard W. Dixon
Department of Geography, Texas State University, 601 University Drive, San Marcos, TX 78666 USA (RD11@txstate.edu)

Thomas E. Eastler
Department of Natural Sciences, University of Maine at Farmington, Farmington, ME 04938 USA (eastler@maine.edu)

Judy Ehlen
USA Engineer Research and Development Center, Topographic Engineering Center, 7701 Telegraph Road, Alexandria, VA 22315 USA (judyehlen@hotmail.com)

LTC Francis A. Galgano
Department of Geography and Environmental Engineering, US Military Academy, West Point, New York 10996 USA (frank.galgano@usma.edu)

BG Gerald E. Galloway (US Army, retired)
1267 S. Oakcrest Road, Arlington, VA 22202 USA (gerald.e.galloway@us.army.mil)

COL John B. Garver, Jr. (US Army, retired)
6777 Surreywood Lane, Bethesda, MD 20817 USA (jbgarver@aol.com)

MAJ Christopher A. Gellasch*
Department of Geography and Environmental Engineering, US Military Academy, West Point, NY 10996 USA (christopher.gellasch@us.army.mil)
*current address: 71st Medical Detachment (Preventive Medicine), Unit 28130, APO AE 09114

Kajari Ghosh*
Department of Earth Sciences, Florida International University, Miami, FL 33199 USA (kajari@ou.edu)
*current address: School of Geology and Geophysics, The University of Oklahoma, Norman, OK 73019 USA

LTC Daniel A. Gilewitch
Department of Geography and Environmental Engineering, US Military Academy, West Point, NY 10096 USA (bd0930@usma.edu)

Michael R. Gross
Department of Earth Sciences, Florida International University, Miami, FL 33199 USA (grossm@fiu.edu)

Peter L. Guth
Department of Oceanography, US Naval Academy, Annapolis, MD 21402 USA (plguth@usna.edu)

Russell S. Harmon
US Army Research Office, P.O. Box 12211, Research Triangle Park, NC, 27709 USA (harmon@aro.arl.army.mil)

Stephen W. Henderson
Department of Geology, Oxford College of Emory University, Oxford, GA 30054 USA (henderson@emory.edu)

CPT Robert A. Lemire*
Mississippi Army National Guard, Jackson, MS 39296 USA
*current address: 168th Engineer Group, APO AE 09302 USA

William C. Mahaney
Quaternary Surveys, 26 Thornhill Ave., Thornhill, Ontario M3J1P3 Canada (bmahaney@yorku.ca)

Alex K. Manda
Department of Earth Sciences, Florida International University, Miami, FL 33199 USA (amand001@fiu.edu)

Jean Martin
Directorate of History and Heritage, National Defence Headquarters, Ottawa, Ontario K1A 0K2 Canada (martin.js@forces.gc.ca)

Eric McDonald
Division of Earth and Ecosystem Science, Desert Research Institute, 2215 Raggio Parkway, Reno, NV 89506 USA (emcdonald@dri.edu)

LTC Kenneth W. McDonald
Department of Geography and Environmental Engineering, US Military Academy, West Point, NY 10096 USA (bk6124@usma.edu)

Marilyn O'Ruiz
College of Veterinary Medicine, University of Illinois, Urbana, IL 61802 USA (moruiz@uiuc.edu)

COL Eugene J. Palka
Department of Geography and Environmental Engineering, US Military Academy, West Point, NY 10996-1695 USA (gene.palka@usma.edu)

David M. Patrick
Department of Geology, The University of Southern Mississippi, Hattiesburg, MS 39046 USA (david.patrick@usm.edu)

J. David Rogers
Department of Geological Engineering, 129 McNutt Hall, 1870 Miner Circle, University of Missouri-Rolla, Rolla, MO 65409 USA (rogersda@umr.edu)

Edward P.F. Rose
Department of Geology, Royal Holloway, University of London, Egham, Surrey TW20 0EX UK (ted.rose@virgin.net)

Kai M. Roth*

Department of Geology, The University of Southern Mississippi, Hattiesburg, MS 39046 USA

*current address: U.S. Naval Oceanographic Office, Stennis Space Center, MS 39522 USA

Kurt A. Schroeder

Department of Geography and Environmental Planning, Plymouth State University, 17 High Street, MSC 39, Plymouth, NH 03264 USA (kschroed@mail.plymouth.edu

Burl E. Self

Department of Geography, Southwest Missouri State University, 901 South National, Springfield Missouri 65804 USA (bes723f@smsu.edu)

Robert C. Whisonant

Department of Geology, Radford University, Radford, VA 24141 USA (rwhisona@radford.edu)

Dean Whitman

Department of Earth Sciences, Florida International University, Miami, FL 33199 USA (whitmand@fiu.edu)

Dierk Willig

AGeoBw- GeoInfoSt Erfurt, Löberfeld- Kaserne, Zeppelin- Str. 18, 95096 Erfurt, Germany

Preface

This book contains selected papers presented at the International Military Geology and Geography Conference that was held at the US Military Academy, West Point, NY in June 2003. The conference organizers, led by LTC Francis A. Galgano, successfully brought the two military geo-disciplines - military geology and military geography - together for the first time. *Studies in Military Geography and Geology* expands a series of recent publications originating from conferences and symposia on military geology that began at the Geological Society of America annual meeting in Seattle, WA in 1994. The first publication, *Military Geology in War and Peace* (Underwood and Guth, 1998), contains papers from this symposium and emphasizes current research as well as applications of engineering geology principles and practices to military problems. The papers in *Geology and Warfare* (Rose and Nathanail, 2000) were presented at a 1996 conference at the University of Warwick in England. This collection focuses on the work of military geologists and military applications of geology around the world and across time. Military activities and their impact on terrain are examined from an environmental perspective in *The Environmental Legacy of Military Operations* (Ehlen and Harmon, 2001), which is based on a second Geological Society of America symposium in Toronto, Canada in 1998. *Fields of Battle - Terrain in Military History* (Doyle and Bennett, 2002), was developed from research presented at the University of Greenwich, England in 2000. *Fields of Battle* takes a multi-disciplinary approach to understanding the effects of terrain on the outcome of warfare from medieval times to the present. The current volume, *Studies in Military Geography and Geology*, introduces military geography to the series and addresses a broad range of military topics ranging from the strategic perspective, through analyses of historical battles at the operational and tactical levels, to the use of advanced technologies applied to present-day military problems.

We thank all those who reviewed papers in this volume, but especially Dr. Allen Hatheway, Colonel, US Army (Ret.) and University of Missouri-Rolla (Ret.); Danny C. Champion, US Army Training and Doctrine Command Analysis Center (TRADOC), White Sands Missile Range, NM; D.G. Christie, Major, Canadian Forces Liaison Officer, National Geospatial-Intelligence Agency (NGA); and Lubomyr Luciuk, Department of Politics and Economics, Royal Military College of Canada.

Douglas R. Caldwell

Judy Ehlen

Russell S. Harmon

REFERENCES

Doyle, P. and Bennett, M.R., eds. 2002. *Fields of Battle - Terrain in Military History*. Dordrecht, The Netherlands: Kluwer Academic Publishers.

Ehlen, J. and Harmon, R.S., eds. 2001. *The Environmental Legacy of Military Operations*. Boulder, CO: Geological Society of America Reviews in Engineering Geology XIV.

Rose, E.P.F. and Nathanail, C.P., eds. 2000. *Geology and Warfare: Examples of the Influence of Terrain and Geologists on Military 0perations*. London: The Geological Society.

Underwood, J.R., Jr. and Guth, P.L., eds. 1998. *Military Geology in War and Peace*. Boulder, CO: Geological Society of America Reviews in Engineering Geology XIII.

Chapter 1

INTRODUCTION

Gerald E. Galloway
Brigadier General, US Army (retired)

Weather and terrain have more impact on battle than any other physical factor, including weapons, equipment, or supplies.

US Army Field Manual 100-5, *Operations* (1982, 3-1)

Both Napoleon and Hitler learned the impact of fighting the Russian winter in their abortive attempts to conquer the heartland of Europe. The selection of Normandy as the site for allied landings in June 1944 was based on a thorough analysis of the geographic characteristics of the continent's coastline. Montgomery's attempt to seize critical crossings over the Rhine in Operation Market Garden failed in large part because he did not recognize that the alluvial delta terrain would hold his advancing forces to a single highway that the German forces could easily block. MacArthur's decision to land forces at Inchon during the Korean War reflected his and his staff's thorough understanding of the conditions in that harbor. US forces operating near Cu Chi during the Vietnam War had to struggle with the Viet Cong's use of tunnels as effective hiding places and weapons storage areas near US positions. The 1980 special operations effort to rescue US hostages in Iran was aborted because of a dust storm that surprised the helicopters flying from the Arabian Sea to a desert rendezvous, but was known by those who lived on the land to frequently occur. History is replete with other examples of the impact of terrain and weather on military operations.

Knowledge of military geology and geography assists commanders in shaping the conduct of battles. At the strategic level, this same knowledge provides a context to better understand the implications of landforms and climate, and the cultural landscape in which forces must operate. It also assists senior national leaders in understanding the overarching impact of distance in determining the ability of military forces to wage wars thousands of miles from the United States or their overseas home bases.

D.R. Caldwell et al. (eds.), Studies in Military Geography and Geology, 1–4.
© 2004 *Kluwer Academic Publishers.*

At the same time as 21st century leaders prepare their forces to engage in combat, they are expected to act as sensitive stewards of the worldwide installations from which their forces are launched. They must not only support the maintenance and repair of the facilities at these installations but also provide for, under US and foreign environmental laws and policies, the protection and enhancement of the thousands of acres of natural environment at these locations and the diverse species that inhabit them. They must know and understand the non-battle landscape that surrounds them.

As we enter the 21st century, military and civilian leaders are focusing their attention on transforming ponderous Cold War armies into forces that are highly mobile on the ground, transportable by air, and more logistically self-sufficient. The task of leading this transformation, employing technology-enhanced forces and serving as land stewards, demands that military leaders possess an even greater understanding of military geology and geography than was required in the past. More than ever before, the employment of modern forces places heavy responsibility on commanders to understand the natural and cultural environments in which they will operate.

Consider the following:

- In Afghanistan the war against the Taliban and Al Qaeda has been heavily influenced by the terrain and climate of the region. Coalition forces have had to deal with mountainous terrain seldom encountered in their units' histories and where cold, snow, and altitude became the other enemies. Helicopters designed to provide battlefield mobility were not able to carry loads to high altitude battlefield positions. High winds restricted the activity of the helicopters that could fly. Pinpoint accuracy in bombing and use of artillery was offset by the protection found in deep caves. The large number of caves made tracking the enemy difficult. At one point, US forces even turned to geographers to identify, through analysis of earth formations, the location of caves seen in Osama bin Laden video tapes.
- The 2003 war in Iraq and the 1991 war in Kuwait and Iraq pointed out that terrain continues to dictate the location of battles. In the desert, where armored vehicles move rapidly over long distances, battles take place where the terrain affords the defender the most advantage. Lessons learned in the Western Desert in Africa during World War II applied equally to the races across the open spaces in Desert Storm and Iraqi Freedom even though the equipment had been modernized and their lethality increased. Perhaps the most significant technological advance was the availability of global positioning systems that enabled the advancing forces operating in these vast deserts to know where both they and the enemy were located.

- Experiences in Iraq have also pointed out the vulnerability of aircraft flying in the relatively open terrain of the desert to both enemy action and the weather. While experiences in Vietnam and Europe indicated that jungle and forested terrain permit helicopters to fly at treetop level and avoid early detection and ground fire, in the desert helicopters are seen at a distance and, at fly-over, can be brought under fire by those with small arms or other individual weapons. Blinding dust storms, similar to those encountered in the Iran rescue effort, grounded helicopters during the early stages of the US movement into Iraq.

The location and availability of water also occupies the interest of the military today at both the strategic and the tactical level. Finding sufficient clean water for drinking is a challenge on the battlefield. Fighting in the desert increases the need for water at a time when the availability is the lowest. Logisticians supporting the US troops in Iraq had to truck large quantities of water, most of it bottled, long distances through frequently hostile territory. At one point the forces were consuming 45 million 1.5 l bottles a month (Wood, 2003). Then, as the battle stabilized they had to supply not only the troops, but also the noncombatants whose access to water had been disrupted by the battles. Although modern technology can speed the treatment of raw water and move it in bulk to the troops, sources for the water must be located and the logistics movement challenges overcome.

At the strategic level, water plays a critical role. Water can be the source of potential conflict among nations that share a given river or aquifer unless agreements can be reached among the countries as to the equitable use of the water. Over 300 countries share rivers and many downstream nations are at the mercy of the upstream nations that control their water sources. In the Middle East, Turkey's control over the upper reaches of the Tigris and Euphrates Rivers in their multi-dam Southeastern Anatolia water project is of constant concern to Syria and Iraq.

Conflict in the Balkans and the Middle East has illustrated the importance of understanding the culture as well as the historical background of the combatants. Those not familiar with this background fail to realize the animosities that exist and the potential explosiveness of situations that allow one ethnic group to dominate another. Failure to recognize this situation may have led to genocide in the Balkans and pointed out the need for troops on the ground to control contacts in multi-cultural communities, a lesson learned once again in Iraq. US forces in Iraq were frequently surprised by the unfriendly reception and even hostility they faced when they moved into that country. In some cases, this hostility evolved into urban guerilla warfare. Understanding the importance of religion and the sanctity of religious structures, such as mosques, was and remains critical to the success of US activities in the Islamic world.

The 21st century will provide new and better tools for the military analysts to collect and analyze data about the landscape and the people who inhabit it. Modern remote sensing technologies are increasing the resolution and accuracy of the information as well as its timeliness. Massive computing power permits analysts to deal with large amounts of data depicting hundreds of square miles of terrain and to present this information in highly usable formats to the decision makers.

Lessons of the past are means of understanding the future. *Studies in Military Geography and Geology* reminds us of the lessons that have been learned and the ever present need to maintain our skills and knowledge in these fields. The editors and the authors provide us with a perspective on what military geography and geology represent and how broad and important this perspective really is. They have identified and described battles in history, from Hannibal to Hitler, that school us in the advantages gained and the opportunities lost through a grasp of or a failure to understand the local conditions. They have also challenged us with insights into how new technologies may influence the conduct of military geography and geology in support of future military operations. They skillfully remind us that those who know and understand the physical and cultural landscape of the battle area will bring significant advantage to their conduct of military operations in the 21st century.

As Santayana (1905, 284) noted "Those who cannot remember the past are condemned to repeat it." Military leaders must understand that, in warfare, the costs of a failure to understand the role of military geography and geology are too high to repeat.

REFERENCES

Santayana G. 1905. *The Life of Reason: Introduction and Reason in Common Sense.* New York: Scribner's.

US Army. 1982. Operations. Washington, DC: US Government Printing Office. Field Manual 100-5.

Wood, D. 2003. "Designer Water Becomes an Undesigned Logistics Problem for the Army." Newhouse News Service, July 28, 2003.

PART I

GEOPERSPECTIVES

The principles of war are not, in the final analysis, limited to any one type of warfare, or even limited exclusively to warfare itself … but, principles as such can rarely be studied in a vacuum; military operations are drastically affected by many considerations, one of the most important of which is the geography of the region.

Dwight D. Eisenhower
The White House
22 April 1959

The ground, yes, the ground. The ultimate proof of military capability lies in control and advantageous use of the ground. The ground is geology and that geology is military geology.

Allen W. Hatheway
The Environmental Legacy of Military Operations
2001

Chapter 2

PERSPECTIVES ON MILITARY GEOGRAPHY

The Military Operating Environment

Russell S. Harmon,[1] Francis H. Dillon III,[2] and John B. Garver, Jr.[3]
[1] *US Army Research Office*
[2] *George Mason University*
[3] *Bethesda, MD*

Abstract: Military geography is a broad and dynamic subject with boundaries that can be difficult to delineate clearly and that change with time and circumstance. The two key fundamental aspects of military geography at all spatial scales are the physical and cultural landscapes. This paper considers the military operating environment as the foundation of military geography and illustrates the importance of the physical and cultural landscapes at different spatial scales to military operations in war.

Key words: military geography, levels of war, military operating environment, environmental matrix, cultural landscape, physical landscape

1. INTRODUCTION

The value of geography to the warfighter has been long recognized. Field Marshall Lord Bernard Montgomery, British commander of the Allied military campaign in Europe during World War II, made this point clearly at a 1948 Royal Geographical Society meeting in London:

> I feel the making of war resolves itself into very simple issues and the simplest in my view is what is possible and what is not possible. ... What is possible will depend firstly on geography (Falls, 1948, 16).

Military activities occur on landscapes with distinct physical and cultural character. Understanding how the spatial distribution of landscape elements affects the military operating environment at the tactical and operational scales, and how this applies to military concerns at the strategic scale, is the substance of military geography. As such, military geography is the

D.R. Caldwell et al. (eds.), Studies in Military Geography and Geology, 7–20.
© 2004 *Kluwer Academic Publishers.*

application of geospatial concepts, approaches, and tools to military problems in war; military operations other than war, such as peacekeeping; and military activities during peace-time, such as disaster relief (e.g., Peltier and Pearcy, 1966; O'Sullivan and Miller, 1983; Palka, 1995, 2002; Palka and Galgano, 2000). Thus, military geography is a broad subject, whose boundaries with closely related areas such as military science and history, military geology, cultural geography, political geography, and geopolitics are diffuse, not clearly delineated, and change with time and circumstance.

As with all geography, the value of military geography lies in its unique spatial perspective and methodology. The mere listing of information and data concerning an area of operations does not in itself constitute military geography. Instead, the contribution of military geography lies in identification and analysis of the significant elements creating the environmental matrix in areas of potential or actual military operations, i.e. the von Clausewitz "*sense of place*" (Greene, 1943), which assists formulation, preparation, and execution of military plans. In conflict, military geography provides the foundation for, and the means to develop, a coherent and selective mission-oriented assessment of the environmental matrix at the tactical, operational, and strategic levels.

This paper focuses on the military operating environment in war and discusses the military importance of environmental factors at different spatial scales. It presents a set of examples reflecting the different, but related, geographic considerations that are important at the strategic, operational, and tactical levels of war.

2. THE MILITARY OPERATING ENVIRONMENT

The problem of war is rooted in geography because military conflict is primarily undertaken to gain control of land and influence people. The conduct of warfare is conditioned by the physical and cultural characteristics of the conflict area, defined as the military operating environment. Military plans and operational doctrine, mission force structure, weapons, and equipment requirements vary depending on the physical and cultural conditions in which they are deployed. Some military operating environments contain critical military or psychological objectives, whereas others constitute maneuver corridors or essential communication and supply lines. Some environments and conditions favor an attacker, others a defender; some areas present severe operating constraints requiring employment of dismounted forces, whereas others permit extensive maneuver and employment of mechanized formations. Therefore, in conflict, the outcome of battles, campaigns, and wars rests in large part on how well military leaders at each level of command understand, and take advantage of, opportunities for success provided by geographic factors of the military operating environment.

The military operating environment can be described in terms of an "environmental matrix" (Fig. 1) that defines the physical and cultural elements of a landscape. Peltier and Pearcy (1966) view the environmental matrix as the sum of factors and forces operating at a place which have the potential to affect the performance of any military function there, with its specific expression resulting from interactions among a host of different coexisting factors. It is the environmental matrix that provides the unique character of each military engagement or campaign. Understanding how significant elements of the matrix affect a particular military activity is critical to planning and executing military operations. It is this unique combination of factors that defines a particular environmental matrix and determines the most appropriate training, equipment, organization, force structure and deployment, tactics, and strategy for a military operation (Peltier, 1961).

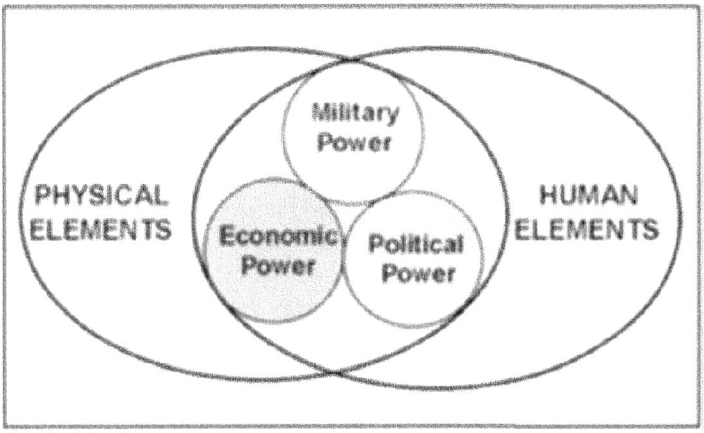

Figure 1. The environmental matrix. Modified from Peltier and Pearcy (1966).

Primary geographic elements of the environmental matrix are the physical and cultural landscapes (i.e., the physical and human elements of Fig. 1). The physical landscape includes the diversity of terrain features, resulting from dynamic interaction of underlying geology and geomorphic processes. Additionally, climate determines soil and vegetation types at a regional scale. Together, climate and terrain define physical landscapes around the globe as different regional military operating environments. The broad range of environmental conditions found in different climates can present significant opportunities to an army that understands and appropriately utilizes the environment, or formidable barriers to the army that does not. The cultural or human landscape consists of political structures, population distribution, settlement and landuse patterns, economic activity, transportation networks, and cultural groupings. It is the interaction of these elements upon which the social, economic, and political fabric - and military power base - of a state is derived. Every place is constantly in a state of flux. Natural processes continually modify the physical landscape, while humans change the physical landscape and their use of it. Therefore, analysis of the implications of changes in the environmental matrix of a country or region should also be a continuing process.

3. LEVELS OF WAR

The geographic elements discussed here vary according to the spatial scale of the military activity. The scale effect is most evident when viewed through the prism of the three levels of war - strategic, operational, and tactical. Although the geographic considerations differ in detail at each level, they are related to one another.

The strategic level of war deals with integrated plans and operations on the largest spatial scale - the global, national, and theater levels. It is concerned with political, economic, social, and psychological factors in addition to military considerations. The focus is on territory and space, together with the total capabilities, resources, and culture of both adversaries and allies in a geospatial context. Geography at this level involves the use of military force to achieve national objectives and asks questions related to *when*, *where*, and *how* wars are fought. Historically, strategic analysis has involved geopolitics and presentation of global strategic views (e.g., von Clausewitz, 1832; Jomini, 1862; Mahan, 1890; Mackinder, 1904; Haushofer, 1941; Spykman, 1944; Keegan and Wheatcroft, 1986; Huntington, 1997; Gray and Sloan, 1999; Martin, 2004, this volume). Strategic planning is the broadest form of military planning and flows from national political policy and strategy (Shalikashvili, 1995) and, therefore, has a strong affiliation with political geography. Military considerations are only one of the many elements that enter into formulation of strategic estimates and plans at this level.

The operational level of war relates to larger military units (e.g., army, corps, and brigade) allocated by the strategic planning process. It is dominated by campaign planning, logistics, and administration over large areas of operation, while connecting tactical operations to the achievement of strategic objectives. Knowledge of general enemy resources and military capabilities, together with analysis of the physical character of the operational area, longer-term climatic conditions, and cultural factors, are important considerations. Operational planning is broad military planning concerned with synchronization of joint and combined forces and movement of troops and resources to and within the theater of operations, enabling battles and campaigns to be fought on favorable terms.

The tactical level of war deals with battlefield geography and involves those factors related to specific engagements. It includes small unit (i.e., division and lower) plans and operations plus the direct concerns of combat. Primary interests are enemy forces, resources, and maneuverability in restricted areas of operation. Tactical planning is military planning in the narrowest sense, dealing with how to fight battles with respect to the effects of terrain and short-term environmental and weather conditions on the deployment of troops, weapons, and other resources on the battlefield.

The paragraphs below illustrate the importance of physical and cultural aspects of the environmental matrix at each level of war. The examples cited are but a few of the many possibilities from ancient times to the present. O'Sullivan (1991), Collins (1998), Winters et al. (1998), Palka and Galgano (2000), and Stephenson (2003) provide additional examples that illustrate the importance of these geographic considerations in individual military battles or campaigns, as either impediments or facilitating factors.

3.1 The Cultural Landscape

The more obvious effects of the physical landscape frequently overshadow the influence of the cultural landscape. However, warfare is intrinsically a cultural activity and, therefore, frequently is governed by cultural considerations. Because most warfare is conducted to control territory or population, such control usually focuses on culturally defined resources or access to resources. The decision of *when* and, for our purposes most importantly, *where* war will be conducted is largely driven by the cultural landscape. All commanders must be cognizant of the cultural or human elements of the military operating environment, particularly settlement and landuse patterns, economic activity, and transportation networks.

In the American Civil War, a series of low hills determined the course of the Battle of Gettysburg and ultimate Union victory (Brown, 1961). The topographically lower hill, Little Round Top, not the adjacent higher and dominating hill, Big Round Top, was the key terrain during the battle. Critical sloping ground around Little Round Top had been cleared by logging, a cultural landuse. As a result, Little Round Top provided excellent observation and unobstructed fields of fire to whichever side held it.

Similarly, the Bocage region of Normandy, a terrain of ancient hedgerows 100-200 m apart with 3-4 m high banks surrounding fields, dictated how this area had to be attacked by Allied forces after the D-Day invasion of 1944 (Keegan, 1990). Such agricultural land use is one of the most persistent factors shaping battlefields and controlling tactical military operations.

Control of transportation networks is essential for movement of troops and their logistical support, making them critical objectives at all levels of war. This is illustrated by the 1940 German glider assault onto the roof of Fort Eben Emael in Belgium, considered to be the most impregnable fortification in Europe (Shirer, 1959), to secure local bridge crossings and control the Muese River-Albert Canal junction, which permitted the rapid and unhindered advance of German armored units across Belgium.

Distribution of cultural resources often dominates military planning. The selection of an amphibious landing site for Operation Overlord, the Allied invasion of France in 1944, was largely dictated by the need to insure air cover over the landing beaches (Churchill, 1954a). Operational ranges of fighter aircraft and the locations of airfields in England limited possible locations to the Pas de Calais or Normandy coasts.

Places of high cultural value, political, religious, or historical, are often objectives of military operations. The importance of Richmond during the American Civil War, Paris and Berlin during World War II, or Baghdad during the Iraq invasion of 2003 cannot be overstated, although the cities themselves may not have contributed in a material way to defeat of adversary forces.

The cultural characteristics of a population can shape a battle area or military campaign. This is most acutely the case in guerrilla wars and counter-insurgency movements (Galvla, 1964; Asprey, 1975; Beckett, 1998), as in the Maroon Wars of Jamaica during the 17th century (see Day, 2004, this volume). Similarly, recent operational experience of US military forces illustrates the importance of developing greater cultural awareness sensitivity within the military (Konorow, 2004). It has been argued that the negative outcome for the US in Vietnam resulted from a failure to appreciate the complexity of the conflict and an ignorance of the country, its history, and culture (Record, 1996-7). The US military experience in Somalia during the 1980s, and the present difficulties in stabilizing Afghanistan and Iraq, likewise, can be attributed, in part, to lack of cultural understanding (Peters, 1994; Hansen, 1997; Terrill, 2003). It is incumbent that military leaders at all levels understand the cultures of both adversaries and allies as a basis for developing sound strategies and more insightful operational plans.

3.2 The Physical Landscape

The geographic position and distribution of many cultural features results from the nature of terrain and climate. Therefore, understanding the physical terrain is one of the first factors to consider in tactical planning, and also is of importance in operational and strategic planning (Collins, 1998). Recognizing and taking advantage of high ground and controlling natural barriers, choke points, and movement corridors has always been among the central tenets of military thinking (O'Sullivan, 1991). The North African Campaign during World War II was restricted to a narrow coastal region by the constraining topography of the Saharan Ergs and the Qattara Depression (Toppe, 1952). The Mitla Pass, a strategic choke point in the Gebel al Raha Mountains of western Sinai, historically has been a focal point because it provides the most direct route through the rugged desert terrain of Sinai Peninsula to the Suez Canal crossing at Suez (Winters et al., 1998).

Natural vegetation is a terrain characteristic that always has influenced the course of military operations. At various places and under different circumstances, vegetation has offered advantages to both defender and attacker. Vegetation certainly affects battlefield visibility at the tactical level, but the obscuration value of vegetation also impacts military planning at the operational and strategic levels.

In 9 AD, Varus cost the Roman Empire all territory between the Rhine and the Elbe Rivers in Germany (Bentley, 1907; Winters et al., 1998). Having conducted previous, open-area campaigns in Mediterranean regions during the dry season, Varus had no understanding of the climate of Central Europe and was unprepared for the cold, wet, swampy conditions of a mixed-deciduous forest during the rainy season. The Germans used the dense woods for tactical advantage, first for attrition of the superior Roman force and later to destroy the Roman legions at the Battle of the Teutoburg Forest, while Varus' cavalry and wagons were immobilized in mud and water.

The last major German counteroffensive of World War II was launched in winter through the lightly defended Ardennes region between the Rhine River and Belgian-Luxembourg-German border, with Antwerp as its objective (Churchill, 1954b). The densely forested uplands of the Schnee-Eifel area east of the Ardennes provided ideal cover and concealment for the logistic buildup and massing of forces for this unexpected German operation (Elstob, 1971; Winters et al., 1998), which, although it failed, prolonged the Allied breakthrough into Germany (Keegan, 1990).

The physical environment of the Hudson Highlands in New York determined the conduct of the American War of Independence at all three levels of war (see Dalton, 2004; Galgano, 2004; and Palka, 2004; this volume). Similarly, regional physiography was arguably the most important factor in determining Union and Confederate strategies and operational planning in the Eastern Theater during the American Civil War. The different topographic domains (see e.g., Thornbury, 1965, Fig. 5) largely dictated the strategy of the two adversaries and determined the styles of military engagements (e.g., the Confederate Shenandoah Valley Campaigns of 1862 and 1863 vs. the Union Peninsular Campaigns of 1862 and 1864-65; Henderson, 1949; Deaderick, 1951; Winters et al., 1998). This was particularly true in the Valley and Ridge province, where ridges compartmentalized the region and provided barriers to observation, valleys provided corridors of rapid movement, and cross-compartment movement was only possible where streams formed cross-cutting river gaps (Henderson, 2000). By contrast, the Piedmont and Coastal Plain provinces were amenable to expansive troop movements and amphibious operations. The effects of topography and local geology on the Chickamauga campaign of 1863 are clearly shown by Henderson (2004, this volume).

The Western Front in World War I provides another example of how terrain affected strategic planning and operational campaigns. Long, arcuate cuestas, separated by major rivers, determined corridors and barriers for military action, but strongly favored the defender, who could occupy forested high ground and continuously threaten adversaries on adjacent, open, poorly drained lowlands. A common view of the Western Front is that it was a progression of battles of attrition to control cuesta-ridge high ground (Johnson, 1917; Brooks, 1920). Topography and local geology frequently determined the location of trench systems and greatly influenced citing of defensive positions, movement of men and materiel, and logistical supply. Throughout the war, the German army resorted to an operational strategy of retreat to pre-planned defensible positions, taking advantage of terrain, whereas the Allies, thinking in terms of offensive action and not disposed to yield ground to obtain a strategic advantage, frequently had to contend with disadvantageous terrain (Doyle and Bennett, 1997).

Climate is another aspect of the physical landscape of military concern, focusing on long-term weather patterns and extremes. First, regions of extreme climate are fundamentally difficult areas in which to conduct military campaigns. For example, off-road maneuver is generally difficult in desert regions and dust generated by vehicles commonly discloses movement and interferes with mechanical functions of vehicles, aircraft, and weapons. Also, sparse desert vegetation requires different camouflage techniques, and the increased demand for water combined with the paucity of dependable local water sources seriously strains logistical systems designed to carry fuel, ammunition, and other warfighting materiel (see e.g., Gellasch, 2004, this volume; King et al., 2004).

Extreme climates can be the greater adversary for the soldier. Tropical illnesses inflicted more casualties on the US Army during its 1943 Burma campaign than battle injuries (Stone, 1949). As a consequence of drinking non-potable water and the lack of bathing water, a majority of the first brigade of Germans deployed to North Africa during World War II was affected by dysentery (Toppe, 1952). The Soviet army encountered a similar problem during its ten years in Afghanistan (Grau, 1998).

Climatic factors have played important roles in military campaigns throughout history and, in some instances, provided military advantage to one side in a conflict. The extremely harsh winters and vast expanse of northeastern Europe three times protected Russia from western armies by halting major invasions, first by Charles XII in 1708-09 (Scott, 1996), then Napoleon in 1812 (Tarle, 1942), and finally, Hitler in 1941-42 (see Self, 2004, this volume). Particularly illustrative is Napoleon's futile attempt to invade Russia with an army of some 400,000 during 1812. In June, rainy weather slowed the advance and heavy supply wagons, stuck in deep mud, were abandoned (van Creveld, 1977). Unable to decisively engage the Russian army, which tactically withdrew eastward, the French army arrived at Moscow after the city was evacuated, burned, and abandoned. Having failed to achieve his ultimate military objective, Napoleon ordered a retreat in mid-October. Departing Moscow with 100,000 soldiers, Napoleon's army was reduced by 40,000 by the end of October, by almost 20,000 in early November, and then by another 20,000 at the end of the month (Minard, 1861; Winters et al., 1998).

Both long- and short-term climate change can have military implications, particularly in the environmental security context, because such perturbations can cause displacement of formerly settled populations and generate regional instabilities that affect military alliances and shift military power balances between states (Wirth, 1989; King, 2000). A shift in prevailing climate within Europe from 1200-1400 ushered in the "Little Ice Age." This regional cooling event caused shifts in agricultural patterns, loss of the herring industry in the Baltic region, and halted the Viking expansion into southern Europe (Brand, 2002). It has been argued (e.g., Clairborne, 1970) that this climate perturbation initiated social unrest and disunity across northern Europe and generated changes in social structures that ultimately led to the socio-political situations that produced World Wars I and II centuries later. More recently, severe, persistent drought and the resulting desertification across the Sahel region of sub-Saharan Africa in the 1970s and 1980s, the consequence of short-term climate perturbation, caused instability and conflict. The pastoral lifestyle of the nomadic Tuareg peoples of the Sahelian states was undermined and stable populations were displaced into refugee camps, which led to the Tuareg rebellion of the 1990s in Mali (Kieta, 1998).

Weather frequently is a critical, short-term, but generally unpredictable, component of war. On many occasions, weather has determined success or failure for individual battles or had an unexpected impact on a military operation.

Wind can be a significant problem, particularly in extreme environments. Winds blowing across unconsolidated sands often expose or bury minefields, limiting their effectiveness. Shifting sands close roads vital to logistical support or military maneuver (King et al., 2004). Dust storms limit visibility and can cause operations to stop altogether as routinely happened in North Africa during World War II (Toppe, 1952; Perrett, 1988) or during US military operations in Kuwait and Iraq in 1990-91 and 2003 (DOD, 1992; Espo, 2003). The attempt by the US military in April 1980 to deploy an elite helicopter force to rescue American hostages in Tehran, Iran was prematurely terminated because of a sudden, unexpected dust storm that severely impacted military personnel and equipment. (Greeley, 2001; Huchthausen, 2003).

Twice in the 13th century, typhoons prevented Mongol invasions of Japan (Newmann, 1975). Kublai Khan first invaded with an armada of 800 vessels and 40,000 soldiers in 1274, but a typhoon approached just after the Mongol army landed. The invasion was halted because of concern about grounding the ships during the storm and the soldiers returned to their ships. The ships set sail but encountered the typhoon, incurring a loss of some 13,000 men. A second attempt seven years later with a larger force of 140,000 met a similar fate. This time, a typhoon occurred as troops were disembarking. Most returned to the ships, which put out to sea. A large part of the fleet was destroyed, and some 50,000 lives lost, whereas the small force remaining ashore was defeated and executed by the Japanese defenders.

Perhaps the most striking and important example of weather effects at the operational and strategic levels is the Allied invasion of Western Europe in June 1944. Operation Overload was the most complex and completely planned military operation in history (Ambrose, 1998). The time for the amphibious landing along the Normandy coast was selected based on favorable tides and adequate moonlight. The need for air cover over the amphibious landing and the ability of landing craft to reach shore required at least a minimum ceiling and visibility and no more than a moderate sea state. A few days before the planned landing date, a storm moved into the North Sea. Overcast, stormy weather was forecast by German meteorologists, causing the German commander, Field Marshal Erwin Rommel, to leave the region. High winds and waves caused General Dwight Eisenhower to postpone the landing for 4 and 5 June, as these conditions persisted. On the evening of 4 June, meteorologists at Allied headquarters forecast a break in the weather for the evening of 5 June and the following day (Stagg, 1971). The order for invasion was given and the Normandy landings subsequently succeeded as planned, partly because the German commanders deemed conditions unfavorable for the invasion (Keegan, 1990) and the stormy conditions prevented deployment of German reconnaissance aircraft (Brand, 1981).

4. CONCLUSION

Military forces operate across a variety of environmental conditions that, by necessity, affect virtually all facets of military activity, from training to warfighting and from peacekeeping to disaster relief. Military doctrine has long dictated that commanders know their terrain and weather. Current US Army doctrine envisages a mobile force that, when deployed, can operate effectively and efficiently in a distributed and largely self-sufficient manner across the full variety of terrain and environmental conditions that might be encountered during deployment. Intelligence preparation of the battlefield includes assessments of the effects of the cultural and physical landscape on military operations. Commanders at all levels must be able to read the cultural and physical landscapes and understand how these conditions will affect their actions, as well as those of their adversary, and a successful commander will use this knowledge for military advantage. The same is true during the deployment of the military for non-combat purposes. More than two millennia ago, the Chinese General Sun Tsu recognized the need to know about one's enemy and cited the importance of terrain and weather in the conduct of military campaigns (Sun Tzu, 2003). Today, the two fundamental aspects of military geography at all spatial scales remain essentially the same, physical elements and human elements.

ACKNOWLEDGEMENTS

We are deeply indebted to E.J. Palka, D.R. Caldwell, and particularly J. Ehlen for constructive and helpful reviews of the various drafts of this paper.

REFERENCES

Ambrose, S. 1998. *The Victors*. New York: Simon and Schuster, Inc.

Asprey, R.B. 1975. *War in the Shadows: The Guerrilla in History*, 2 volumes. New York: Doubleday, Inc.
Beckett, J.F. 1998. *The Roots of Counterinsurgency: Armies and Guerrilla Warfare, 1900-1945*. London: Blanford, Ltd.
Bentley, R. 1907. Weather in wartime. Quarterly Review of the Royal Meteorological Society 33:82-83.
Brand, S. 1981. Weather and the military: A historical perspective. National Weather Digest 6:8-10.
Brinkerhoff, J.R. 1963. The nature of modern military geography (unpublished Master's thesis). New York: Columbia University.
Brooks, A.H. 1920. The use of geology on the Western Front. US Geological Survey Professional Paper 128-D, 85-124.
Brown, A. 1961. Geology and the Gettysburg campaign: Geotimes 6:8-12, 40-41.
Clairborne, R. 1970. *Climate, Man, and History*. New York: W.W. Norton Company.
Churchill, W.S. 1954a. *The Second World War, V: Closing the Ring*. London: Cassell Company, Ltd.
Churchill, W.S. 1954b. *The Second World War, VI: Triumph and Tragedy*. London: Cassell Company, Ltd.
von Clausewitz, K. 1832. *On War* (translated in 1943 by O.S. Matthijs Jolles). New York: Modern Library.
Collins J.M. 1998. *Military Geography for Professionals and the Public*. Washington, DC: Brassey's, Inc.
van Creveld, M. 1977. *Supplying War: Logistics from Wallenstein to Patton*. Cambridge: Cambridge University Press.
Dalton, J.B., Jr. 2004. Saratoga - A military geographic analysis. In *Studies in Military Geography and Geology*, D.R. Caldwell, J. Ehlen, and R.S. Harmon, eds., Dordrecht, The Netherlands: Kluwer Academic Publishers, 121-131.
Day, M. 2004. Military campaigns in tropical karst terrain - The Maroon Wars of Jamaica. In *Studies in Military Geography and Geology*, D.R. Caldwell, J. Ehlen, and R.S. Harmon, eds., Dordrecht, The Netherlands: Kluwer Academic Publishers, 79-88.
Deaderick, B. 1951. *Strategy in the Civil War*. Harrisburg, PA: Military Service Publishing Company.
Doyle, P. and Bennett, M.R. 1997. Military geography - terrain evaluation and the British Western Front 1914-1918. Geographical Journal 163:283-311.
Doyle, P. and Bennett, M.R., eds. 2002. *Fields of Battle - Terrain in Military History*. Dordrecht, The Netherlands: Kluwer Academic Publishers.
Elstob, P. 1971. *Hitler's Last Offensive*. New York: Macmillan, Inc.
Espo, D. 2003. "Storm, battles stall U.S. Push," Phoenix, AZ, East Valley Tribune, 26 March 2003.
Falls, C. 1948. Geography and war strategy. Geographical Journal 62:4-18.
Galgano, F.A. 2004. Decisive terrain - A military geography of Fortress West Point. In *Studies in Military Geography and Geology*, D.R. Caldwell, J. Ehlen, and R.S. Harmon, eds., Dordrecht, The Netherlands: Kluwer Academic Publishers, 105-120.
Galvla, D. 1964. *Counter-Insurgency Warfare: Theory and Practice*. New York: Preager, Inc.
Gellasch, C. 2004. Groundwater - Past, present, and future uses in military operations. In *Studies in Military Geography and Geology*, D.R. Caldwell, J. Ehlen, and R.S. Harmon, eds., Dordrecht, The Netherlands: Kluwer Academic Publishers, 307-319.
Greene, J.I. 1943. *The Living Thoughts of Clausewitz*. New York: Longman Greene.

Grau, L.W. 1998. *The Bear Went Over the Mountain: Soviet Combat Tactics in Afghanistan.* Portland, OR: Frank Cass Publishing Company.

Gray, C.S. and Sloan, G. 1999. *Geopolitics, Geography, and Strategy.* London: Frank Cass Publishers.

Greeley, J. 2001. Desert One: A mission of hope turned tragic. A case of what could've been. Airman Magazine. (Online Version: http://www.af.mil/news/airman/0401/hostage.html)

Hansen, D.G. 1997. The immutable importance of geography. Parameters 27:55-64.

Haushofer, K. 1941. *Wehr-Geopolitik, Geographische Grundlagen einer Wehrkunde.* Berlin, Germany: Junker and Dunnhaupt.

Henderson, G. 1949. *Stonewall Jackson and the American Civil War.* New York: Longmans Green and Company.

Henderson, J.P. 2000. Military geography of the Civil War - The Blue Ridge and Valley and Ridge Province. In *The Scope of Military Geography: Across the Spectrum from Peacetime to War,* E.J. Palka and F.A. Galgano, eds. New York: McGraw Hill, Inc., 53-73.

Henderson, S.W. 2004. The geology of the Chickamauga Campaign, American Civil War. In *Studies in Military Geography and Geology*, D.R Caldwell, J. Ehlen, and R.S. Harmon, eds., Dordrecht, The Netherlands: Kluwer Academic Publishers, 173-184.

Huchthausen, P. 2003. *America's Splendid Little Wars: A Short History of U. S. Military Engagements: 1975-2000.* New York: Viking Press.

Huntington, S.P. 1997. *The Clash of Civilizations and the Remaking of the World Order.* New York: Simon and Schuster, Inc.

de Jomini, Baron A.H. 1862. *The Art of War* (English translation). Philadelphia, PA: Lippincott and Company (reprinted 1992, London: Greenhill Books).

Keegan, J. 1990. *The Second World War.* New York: Penguin Books.

Keegan, J. and Wheatcroft, A. 1986. *Zones of Conflict: An Atlas of Future Wars.* New York: Simon and Schuster.

Kieta, K. 1998. Conflict and conflict resolution in the Sahel: The Tuareg insurgency in Mali. Carlisle Barracks, PA: Strategic Studies Institute.

King, W.C. 2000. *Understanding Environmental Security: A Strategic Military Perspective.* Atlanta, GA: US Army Environmental Policy Institute.

King, W.C., Harmon, R.S., McDonald, E., Redmond, K., Gilies, J., Doe, W.W., Warren, S., Gilewitch, D., Morrill, V., Stullenbarger, G., and Havrilo, L. 2004. Scientific characterization of desert environments for military testing, training and operations: Unpublished report for US Army Yuma Proving Ground, Yuma, AZ.

Komarow, S. 2004. "U.S. forces train in Arab culture," USA Today, 17 February 2004.

Mackinder, H. 1904. The geopolitical pivot of history. Geographical Journal 23:421-437.

Mahan, A.T. 1890. *The Influence of Seapower on History, 1660-1783.* Boston, MA: Little, Brown and Co.

Martin, J. 2004. Canadian military geography, 1867-2002 - From empire to alliance: Guarding the south or watching the north. In *Studies in Military Geography and Geology*, D.R. Caldwell, J. Ehlen, and R.S. Harmon, eds., Dordrecht, The Netherlands: Kluwer Academic Publishers, 53-64.

Minard, J. 1861. *Tableux Graphics et Cartes Figuratives de M. Minard, 1845-1869.* Paris: Bibliotheque de l'Ecole Nationale des Ponts et Chaussees.

Newmann, J. 1975. Great historical events that were significantly affected by the weather - I. The Mongol invasion of Japan. Bulletin of the American Meteorological Society 61:1167-1171.

O'Sullivan, P. 1991. *Terrain and Tactics.* New York: Greenwood Press.

O'Sullivan, P. and Miller, J. 1986. *The Geography of Warfare*. London: Croom Helm, Ltd.
Palka, E.J. 1995. The US Army in operations other than war: A time to revive military geography. GeoJournal 37:201-208.
Palka, E.J. 2002. Perspectives on military geography. Geographical Bulletin 44:5-9.
Palka, E.J. 2004. A military geography of the Hudson Highlands - Focal point in the American War of Independence. In *Studies in Military Geography and Geology*, D.R. Caldwell, J. Ehlen, and R.S. Harmon, eds., Dordrecht, The Netherlands: Kluwer Academic Publishers, 89-103.
Palka, E.J. and Galgano, F.A. Jr., eds. 2000. *The Scope of Military Geography: Across the Spectrum from Peacetime to War*. New York: McGraw Hill, Inc.
Peltier, L. 1961. The potential of military geography. Professional Geographer 13:3-4.
Peltier L.C. and Pearcy, G.E. 1966. *Military Geography*. Princeton, NJ: Van Nostrand, Inc.
Perrett, B. 1988. *Desert Warfare: From its Roman Origins to the Gulf Conflict*. Wellingborough, UK: Stephens, Ltd.
Peters, R. 1994. The new warrior class. Parameters 24:16-26.
Self, B.E. 2004. War in the heartland - The role of geography in Operation Barbarossa 1941-1942. In *Studies in Military Geography and Geology*, D.R. Caldwell, J. Ehlen, and R.S. Harmon, eds., Dordrecht, The Netherlands: Kluwer Academic Publishers, 215-226.
Record, J. 1996-97.Vietnam in retrospect: Could we have won? Parameters 26:55-61.
Scott, H.M. 1996. Great Northern War. In *Reader's Companion to Military History*. R.. Crowley and G. Parker, eds. New York: Houghton Mifflin Company, 187-188.
Shalikashvili, J.M. 1995. *National Military Strategy of the United States of America*. Washington, DC: US Government Printing Office.
Shirer, W. 1959. *The Rise and Fall of the Third Reich*. New York: Simon and Shuster, Inc.
Spykman, N. 1944. *The Geography of Peace*. New York: Harcourt and Brace, Inc.
Stagg, J.M. 1971. *Forecast for Overlord*. London: Ian Allen, Ltd.
Stone, J.H. 1949. The marauders and the microbes. Infantry Journal March 1949: 4-11.
Sun Tzu. 2003. *On the Art of War* (translation in 1910 by L. Giles with introduction by D. Gavin). New York: Barnes and Noble Classics.
Tarle, E. 1942. *Napoleon's Invasion of Russia, 1812*. New York: Oxford University Press.
Toppe, A. 1952. Desert warfare: German experiences in World War II. Fort Leavenworth, KS: Combat Studies Institute.
Terrill, W.A. 2003. Nationalism, sectarianism, and the future of the U.S. presence in post-Saddam Iraq. Carlisle Barracks, PA: Strategic Studies Institute.
US Department of Defense (DOD). 1992. Conduct of the Persian Gulf War. Final report to Congress. Washington, DC: US Government Printing Office.
Winters H.A., Galloway, G.E., Jr., Reynolds, W.J., and Rhyne, D.W. 1998. *Battling the Elements: Weather and Terrain in the Conduct of War*. Baltimore, MD: Johns Hopkins University Press.
Wirth, D. 1989. Climate chaos. Foreign Policy 74:3-22.

Chapter 3

MILITARY USE OF UNDERGROUND TERRAIN

A Brief Historical Perspective

Thomas E. Eastler

University of Maine at Farmington

Abstract: The use of man-made underground environments for military operations is reported back to about 5500 BP, although the use of natural caves for defensive purposes must date back much further. Regardless of precisely when mankind first used underground terrain for military purposes, this use has continued, and has had a remarkably successful and noteworthy history. It is the contention of this author that military use of naturally occurring or man-made space underground has been underrated in its contribution to sustained military operations. Retaliatory strikes in Afghanistan subsequent to the September 11, 2001 terrorist attacks against the US have demonstrated the strategic as well as the tactical importance of underground terrain, and as such, have shown that its military use is now more important than it has ever been on a global scale, and may soon challenge outer space as the pivotal battlefield of the future.

Key words: military underground terrain, tactical strategic battlefield, caves, Underground Facilities (UGF), fortress

1. INTRODUCTION

Literature concerned with military use of terrain throughout history deals almost entirely with the elements of geomorphology and physical geography - the landscape. It is an unwritten but nearly universal given that terrain is the land surface characterized by such geologic, geographic, and biologic variables as are needed to define the nature and configuration of surface features. These surface features include shape, size, vertical relief, slope, erodability, trafficability, viewscape, and a host of other specialized parameters that are used to classify the land surface for military purposes.

D.R. Caldwell et al. (eds.), Studies in Military Geography and Geology, 21–37.
© 2004 *Kluwer Academic Publishers.*

fortress, a well-defined excavated shaft in bedrock intersects a secure ground water supply for use during times of siege. Many fortresses had such shafts leading to groundwater or deeply incised cisterns. Herod's spectacular, although much younger (c. 2040 BP), fortress at Masada had a network of 12 cisterns in the rock base of the mesa, outside the fortress, that held nearly 1.5 million cubic feet of water. Water was transported by men and mules to cisterns within the fortress (Bourbon and Lavagno, 2001).

Greek colonists constructed complex underground aqueducts, called Quanats (also Qanat, Qaraz, Karaz, Fogarra), and complex grottos (Fig. 1) around 2500 BP (Papnek, 1994), about the same time that the Medes (an early Iranian tribe) were developing the art of constructing underground aqueducts. Thousands of these Quanats are still in use today in the Middle East (Liaqati, 1997; Qassemi, 2001). Although neither the grottos nor the Quanats were originally constructed for military purposes, both have been used as conduits for evasive activities in warfare, such as for temporarily hiding equipment and personnel, and or for escape and evasion. Evasive activities also included the burial of the dead to keep them from falling into the hands of the enemy. Whereas not always related to military evasion, the protection of the dead played a very important role in early cultures, even as it does today. The oldest well-documented use of man-made underground terrain for both military and civilian purposes dates back to about 2400 BP in the Cappadocia region of Turkey (Fig. 2).

Figure 1. Sybil's Grotto at Cumae in southern Italy, hewn into rock by Greek colonists in the 5th century BC. The chamber is 427 ft long and was excavated as a place for the prophetess of the God Apollo to dispense her oracles. Modified from Papnek (1994).

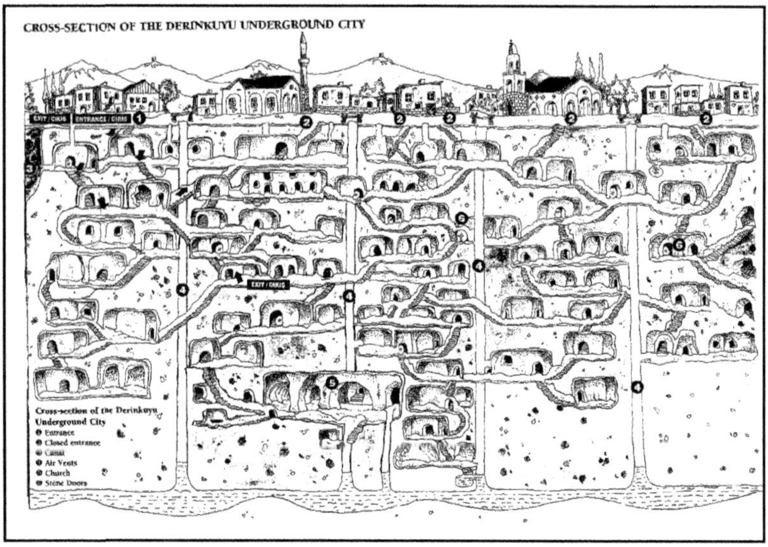

Figure 2. Cross-section of Derinkuyu City in the Cappadocia region of Turkey. Turkey has 450-500 similar cities, some up to 20 stories deep. Adapted from Demir (1997).

The Hittites are credited with construction of these underground cities and secret tunnels to ambush attackers and to hide and defend the population of their more vulnerable above-ground cities. Use of the Cappadocian UGF by the neo-Hittites, the Byzantines, and finally, the newly converted Christians continued well past the 10th century AD, and some of the underground space is still in use today for tourism. Some of the underground cities go as deep as 20 stories (http://www.cappadociaonline.com; Demir, 1997; Aydan and Ulusay, 2003).

About the same time as the development of Cappadocia (2433-2431 BP), the besieged Plataeans dug a counter-siege tunnel under their fortress to try to defeat the wooden/earthen siege ramp constructed by attacking Peloponnesians. The plan was to tunnel under the ramp and set the wooden structure afire. The ramp was set afire, but, as fate would have it, a torrential rain storm quickly doused the flames and the defense failed (Wary, 1980).

Perhaps the most spectacular bedrock excavations are at 2100 year old Petra in Jordan (Fig. 3), where the sandstone cliffs are carved to appear as multi-story buildings (Maqsood, 1994). Petra was the site of many military confrontations throughout its history (Hammond, 2003), and one can speculate that the myriad rock-cut funerary monuments (over 800) scattered nearby were occasionally used for escape and evasion or for offensive military activities. Entombment in rock-cut chambers and construction of a vast underground necropolis may not seem relevant to military use of UGF,

Figure 3. Carved sandstone facade known as the Obelisk Tomb, Petra, Jordan. Modified from Maqsood (1994).

but such catacombs are very well suited for purposes of concealment and evasion and cannot be overlooked. The literature is replete with numerous examples of underground catacombs that were used primarily for entombment (e.g., Greenhut, 1992; van der Horst, 1992; Maqsood, 1994; Papnek, 1994; Rose, 1994; Lemonick, 1995; Bourbon, 1999; Bourbon and Lavagno, 2001). Catacombs that were originally underground quarries in France became burial grounds but also served as bomb shelters and locations for French resistance forces in World War II (Linhardt, 1996). Catacombs in other European countries and in the Pacific theatre also served military functions in World War II.

2.1 The Medieval Chateau Gaillard (801 BP)

The Chateau Gaillard (Fig. 4), a massive walled and moated fortress that occupies high ground above the great meander in the River Seine in Andelys, France is an excellent example of early use of underground terrain in conjunction with classic fortification. The fortress was constructed by Richard the Lionheart, King of England and Duke of Normandy, over a two year period ending in 1198, and was "state-of-the-art" in fortification. The multi-layered defensive construction allowed defenders to fire down on attackers while being protected from enemy projectiles. The design was such as to thwart both battering rams and sappers (military tunneling engineers). King Philip Augustus II of France put the fortress under siege in the autumn of 1203.

Figure 4. Chateau Gaillard. Forces of England's King Richard the Lionheart defend against beseiging forces of French King Philip Augustus in 1203. Both sides dig tunnels as tactical elements of battle. Modified from Brice (1990).

As depicted in Fig. 4, King Richard's forces are using catapulted projectiles from inside the fortress to try to defeat a massive multistory siege tower constructed by King Philip's army. The high tower allowed French archers to fire directly into the Chateau defenders. The illustration also shows the defenders have tunneled beneath the outer wall of the fortress and moat to a position under the tower. There they have excavated a large chamber directly under the tower and have erected vertical-support timbers. Their plan is to surround the timbers with flammable materials and then to set them afire, weakening the earth above and causing the tower to collapse into the large chamber. In the meantime, the besiegers, assuming that defensive sappers are at work trying to undermine the siege tower, have begun tunneling to counter attack the sappers in hopes of preventing the conflagration; they were successful. This battle for underground terrain did not prove to be pivotal, because as with most medieval sieges, the combination of hunger, disease, and death for those inside the fortress led to eventual defeat in early March 1204.

3. NINETEENTH CENTURY UGF

Napoleonic France conceived an attack over, on, and under a "moat" against the British in 1803 as depicted in Fig. 5 (Anon, 1803). In this case, though, the moat was the English Channel separating Britain from France. The plan was to launch a three-pronged attack using hot air balloons, naval vessels, and an excavated tunnel under the English Channel. It is evident that military thinking was involved in trying to use underground terrain to the attacking force's best advantage.

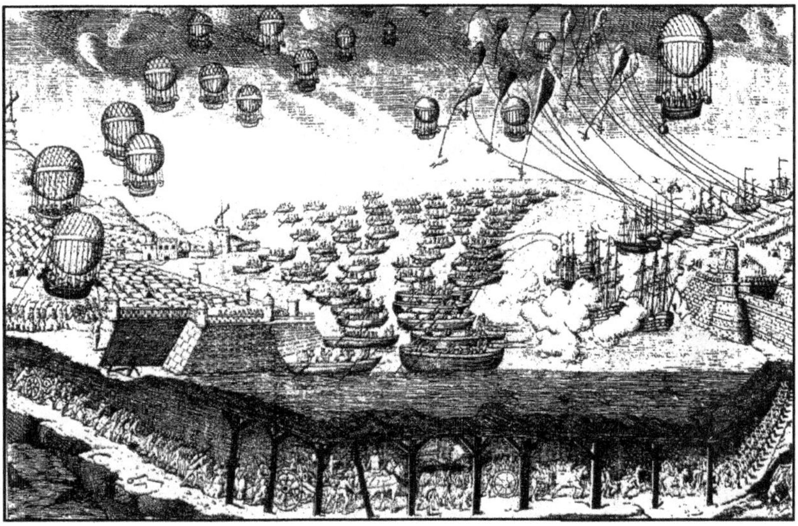

Figure 5. Napoleonic France. Artist's rendition of three-pronged attack of Britain. From Anon (1803).

Another 19th century example of using underground terrain for defensive and offensive purposes occurred during the siege of Vicksburg, MS (19 May to 4 July 1863) during the American Civil War (1861-65). From a defensive point of view both Confederate civilians and military made effective use of man-made caves in the thick clay soils near Vicksburg. Describing Grant's siege, Lossing (1868, 621) states:

> For a month General Grant closely invested Vicksburg. Day after day he drew his lines nearer and nearer, crowning hill after hill with batteries, and mining assiduously in the direction of the stronger works of his foe, with the intention of blowing them high in air. Day and night, with only slight intermissions, his heavy guns and those of (Admiral) Porter were hurling shot and shell with fearful effect into the city, and its suburbs within the lines, making it hell for the inhabitants, and the soldiers too, who sought shelter for limb and life in caves dug in the steep banks where streets passed through the hills [Fig.6].

> In these the woman and children of whole families, free and bond, found protection from the iron hail that perforated the houses, plowed the streets, and even penetrated to these subterranean habitations, where gentle women were waiting and praying for deliverance, and where children were born. It was a terrible ordeal, and yet during that long siege very few persons, not in the army, lost their lives.

Figure 6. Caves dug in steep banks where streets passed through the hills in Vicksburg. From Lossing (1868).

Lossing (1868, 625) additionally describes Union use of underground terrain:

> Toward the close of June the most important of Grant's mines was completed. It extended under Fort Hill Bastion...Between four and five o'clock in the afternoon of the 25th it was fired. The explosion was terrific. . . The garrison, expecting the event, were partly removed, and but a few were injured. But a great breach was made.

The use of offensive trenching and tunneling gave Grant a powerful tool with which to destroy forts and breach lines and was instrumental in his ultimate victory at Vicksburg. From an offensive perspective, mining was conducted in many campaigns throughout the American Civil War, e.g., during the sieges of Petersburg and Port Hudson (Lossing, 1868; Davis, 1986). Counter mining was also used extensively in the same battles.

In the period between the Civil War and World War I improvements to rifled barrels (artillery and small arms; see Schroeder, 2004, this volume) and the development of armor-piercing rounds signaled the beginning of the end of the classic fortress and an evolution in warfare tactics.

4. TWENTIETH CENTURY UGF

In World War I (1914-17) fortress construction continued and France, Belgium, and Germany all built fortresses, in spite of the development of new penetrating weapons. In Belgium there were 12 concrete fortresses built ringing the city of Liège. The Germans used their 420-mm howitzer (Big Bertha) very effectively and its 2000-lb projectiles destroyed these fortresses in short order (Mallor and Ottar, 1973). Although no tunneling was involved in their destruction, the fortresses had numerous underground chambers used for protection of personnel and ammunition; neither was immune to the penetrating rounds from Big Bertha.

Another ubiquitous underground warfare tactic used during World War I was tunneling as an offensive weapon, much like at Vicksburg. No place along the Western Front personifies this type of warfare better than at the Durand mine and Vimy Ridge. Here the opposing forces using clay spades (a foot-powered digging tool used in soft ground) dug lengthy tunnels across "no man's land" towards each other's lines, loaded the tunnels with ammonal, and detonated the whole works (Robinson, 1989; Bostyn, 2002). The British alone had 31 tunneling or mining companies of 500-600 men each; the equivalent of 14 British and French mining engineer regiments were involved in mining and counter mining, in excess of 20,000 troops (Trounce, 1918; Barrie, 1961). Combat engineering was a very important component of World War I tactics. Mining was usually done in winter when surface battles were at a minimum. Charges as large as 96,500 lbs were used to blast massive craters under or behind enemy lines. In specific instances the mining tactic was perceived to have saved thousands of lives that would otherwise have been lost in frontal attacks. In other instances, the perception was that mining did not materially aid either side (Barrie, 1961).

As weapons of war and their delivery methods improved by World War II (1939-45), so too did defensive fortifications (O'Brien, 1955). Some countries, such as Germany, developed highly effective underground utilization techniques (AFM, 1954; Rose et al., 2002) and other countries, such as France and its Maginot line (Fig. 7), had the right ideas but were less adept at making them effective (Gonzalez and Gonzalez, 1993; Chelminski, 1997). Due primarily to the ability of Allied intelligence to find and target dispersed above-ground industrial factories, the Germans quickly and effectively developed a UGF conversion and new construction program which

> ...by January 1945 had converted mines, caves, highway tunnels, railway tunnels, beer storage tunnels, and cellars into first class industrial facilities. By the end of the war there were more than 108 underground aircraft factories planned or under construction of which 78 reached some stage of production... A grand total of 96,502,900 square ft. of underground floor space was planned...14,085,000 square ft. were completed and 73,337,000 square ft. were under construction when the war ended (AFM, 1954, 1).

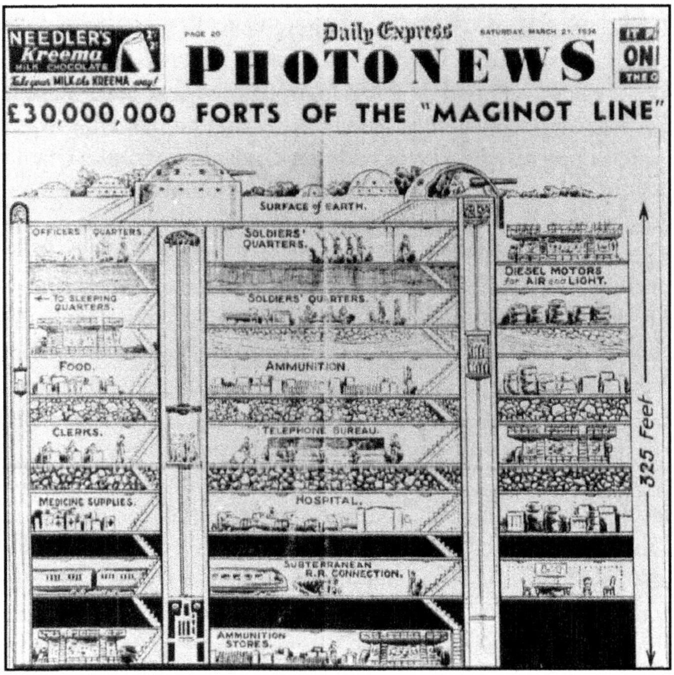

Figure 7. Artist's rendition of the forts of the Maginot Line showing the complexity of eight layers of support for the surface artillery pieces. Modified from London Daily Express (1936).

All in all some 227 underground factories were in production not to mention underground hospitals, air raid shelters, and various military headquarters and warehouses (AFM, 1954). Everything from ball bearings to aircraft (Fig. 8) to V1 and V2 rockets was being produced underground where the factories were well protected from allied bombing (AFM, 1954; Neufield, 1995). Germany's bunkered U-Boat bases were also noteworthy for their survivability (Pallud, 1987; Blair, 1996).

France built a fabulous underground support system for its above-ground defensive line against the Germans. The construction of the multi-level Maginot Line was an impressive engineering feat (Fig. 7). The failure of the Maginot Line to repel attacking German forces came not from poor design or construction but from Germany's ability to develop tactics to defeat the fortifications (Mallory and Ottar, 1973; Pallud, 1988; Gonzalez and Gonzalez, 1993). Germany was willing to exploit neutral countries in order to circumvent what would have been a highly effective line of defense had it stretched into bordering countries.

British troops were able to make the Rock of Gibraltar into a nearly impenetrable fortress (Ramsey, 1978; Elliott, 1978; Rose, 2001) and in London, the underground system and rubblized buildings on the surface

Figure 8. He 162 fusclage assembly line in the Junkers underground factory located 1600 ft below the surface in a salt mine in Tarthun, Soviet Zone, Germany. From AFM (1954).

afforded safe conditions for civilians and war planners. The rubblized buildings acted the same as a present-day burster slab or rock rubble layer which defeats an incoming round by causing it to turn sharply because of edge effects of building material slabs. Such rapid change in trajectory produces forces against the munition that exceed the strength of the casing and the weapon fails to detonate.

On Iwo Jima the US Marines suffered their greatest losses in the World War II because the Japanese were able to take defensive advantage from natural cavities in the honeycombed volcanic rocks of the island (Thomey, 1995). The Japanese caves were interconnected by hand-dug tunnels to increase their effectiveness. The battle for Iwo Jima pitted 75,000 Marines above ground against only 23,000 Japanese soldiers and sailors underground. It was the only World War II battle where Marine amphibious force suffered more casualties then the enemy (Thomey, 1995).

Although the name Vietnam conjures up visions of jungles, rivers, deltas and the like, it also reminds some of terms like "tunnel rats" and "snake eaters." The tunnels of Cu Chi (Fig. 9) exemplify how effective using underground space can be in both defensive and offensive modes (Mangold and Penycate, 1986; Duong, 1994; Mydans, 1999). These hand-dug tunnels played a pivotal role in the defeat of US troops because North Vietnamese and Viet Cong forces were able to disappear below ground and reappear at will to strike fatal blows against US forces. This effective defense and offense came from the makeshift UGF, not from sophisticated, heavily fortified facilities. History repeated itself in Vietnam because, in 1953, the French succumbed to the very tunnel tactics that ultimately defeated the US. French forces

> ...made the fatal error of trying for set-piece confrontation with the Viet Minh in a remote northern valley, Dien Bien Phu...(General) Giap's men besieged and finally overran the French fortress...they approached the perimeter in tunnels, and burrowed underneath the French defenses. (Mangold and Penycate, 1986, 22)

The Viet Cong used their tunnels for intelligence gathering, for offensive and defensive operations, as hospitals, kitchens, and burial grounds, and for tactics that were ultimately successful against all of their enemies.

Figure 9. American GI exiting one of the tunnels at Cu Chi, Vietnam. Modified from Duong (1994).

The Democratic People's Republic of Korea (North Korea) has been building so-called invasion tunnels (more appropriately infiltration tunnels because of their small cross-section) beneath the demilitarized zone that separates the two Koreas since the late 1940s, with a number of them being detected in the early- to mid-1970s. Four such tunnels have been found and neutralized by the Republic of Korea (South Korea; Anon., 1985; Cameron,

1998). How many more such tunnels have been completed and await activation in time of war? In addition and according to the Library of Congress Country Study for North Korea (1993, section 1 of 1, Weapons and Equipment):

> The army has an extensive facility-hardening program. Almost all the forward deployed artillery can be stored in well-protected underground emplacements. The passive defenses in the forward corps include a large bunker complex to conceal and protect infantry forces, mechanized units, and war material stockpiles... (the Air Force has) a system of well-dispersed and well-protected air facilities... Many North Korean navy bases have hardened berths and other passive defenses... In 1990 North Korea had some 134 arms factories, many of them completely or partially concealed underground. These facilities produce ground service arms, ammunition, armored vehicles, naval craft, aircraft (spares and subassemblies), missiles, electronics, and possibly chemical-related materials. In addition, some 115 nonmilitary factories have a dedicated wartime matériel production mission.

Numerous military actions around the world subsequent to the fall of the Soviet Union in 1991 have continued the use of underground terrain for defensive purposes and whence to stage offensive activities. UGFs have played pivotal roles in many of these campaigns. Lessons from Vietnam should have taught that no matter how strong a force you might have, if you can't find your enemy or can't flush him out or pin him down, you will not easily or readily defeat him. This lesson should have been learned many times throughout history, and it wasn't. But it appears that some are now learning from those lessons of history. Afghanistan, for instance, and its vast subterranean world of limestone caves has proved the point more than once (see Gross, 2004; Day, 2004, both this volume). The vast deserts of the Middle East with their drifting sands and harsh environment offer new challenges and opportunities for use of underground terrain to impact the course of military operations.

4.1 Rogue Nations and Terrorism

Newly constructed UGFs have mushroomed globally in the last few decades, particularly in what are considered by the Western World to be nations that support terrorism. The recent military actions in Iraq, for instance, center on the US perception of that country's ability to hide the implements of war. Hiding the implements of war, of course, is not a new tactic, but the scale on which many countries are constructing UGFs to hide things seems to be new. Not all UGF projects are high-tech. Mangold and Penycate (1986) describe several very successful low-tech human-powered UGF excavation projects undertaken during the Vietnam War. For example, in the 1960s Viet Cong soldiers were able to strip down two 105-mm field howitzers, take them underground, reassemble them, service them, disassemble them, and bring them outside to reassemble and use again. One must keep in mind that commercially available underground construction technology is very well developed and that the technology is available to any country that has the money or resources to pay or trade for it. Weapons of mass destruction can be hidden almost anywhere, and those countries that have chosen to go underground have an unparalleled opportunity to hide personnel, resources, and most importantly, intent and capability, from the rest of the world (Eastler et al., 1998).

Keep in mind that any UGF, civilian or military in nature, can be used for military purposes. That broadens the potential underground military theater to virtually all underground mines in the world, and all naturally occurring cavities, including those discussed at the beginning of this paper. Also consider that the evolution of more effective attack weapons is what drives the design of more survivable UGF, a topic that has not been considered here.

5. CONCLUSIONS: THE PAST IS PROLOGUE

Perhaps hundreds of thousands of years ago our ancestors occupied natural caves and used them for escape from the elements and protection from attacking groups. With time and increasing sophistication, their descendents built increasingly more complex fortifications to defend against increasingly more effective siegecraft, and a continuing cycle of improving fortifications followed by better siegecraft ensued. Perhaps the ultimate in siegecraft was the development of nuclear weapons that could pulverize even the most robust UGF, but whose use has been politically incorrect in recent decades. With the knowledge that collateral effects of nuclear weapons pose a significant threat even to the attacking force, the development of more and varied types of UGF have again leveraged the art of fortification above that of effective siegecraft, playing out the inevitable battle between the two.

Since the September 11, 2001 attack on the United States, it has become painfully clear that something as simple as a naturally occurring limestone cave can effectively carry out the same military functions as a multi-billion dollar hardened and deeply-buried command post. No longer can strategic targets be considered as only those well-defined man-made structures and locations, such as missile silos that have ICBMs aimed at our country. We now know that a lowly cave harboring an effective leadership can bring military might against even the most powerful of nations, and that such a cave must therefore now be considered a strategic target.

With that in mind, how will we locate all such caves globally (see Gross, 2004, this volume), and how will we determine if caves are occupied and armed? The classic shell game writ large translates into the question of which cave, or which UGF as the case may be, has the weapons?

It appears even now that circumstances on our finite globe have shown that natural caves continue to be an effective way to escape and evade (see Day, 2004, this volume), and to stage ambush attacks even continents away. Be it natural caves, simple hand-dug tunnels, or very sophisticated hard and deeply buried UGF, it now appears that the underground terrain may very well be the pivotal battlefield of the not so distant future.

REFERENCES

Air Force Manual (AFM). 1954. Underground installations, photographic interpretation keys. Washington, DC: Department of the Air Force AFM 200-35.

Anonymous (Anon.). 1803. Engraving of Napoleonic three-level attack on Britain. Source unknown.

Anonymous (Anon.). 1985. Tunnels of aggression, North Korea catacombs the DMZ. Seoul, Republic of Korea: Korean Overseas Information Service, January, 18-19, 1985.

Anonymous (Anon.). 2003. Cappadocia. (Online Version: http://www.capadociaonline.com)

Aydan, O. and Ulusay, R. 2003. Geotechnical and geoenvironmental characteristics of man-made underground structures in Cappadocia, Turkey. Engineering Geology 69:245-272.

Barrie, A. 1961. *War Underground: The Tunnellers of the Great War.* London: Tom Donovan Publishing.

Blair, C. 1996. *Hitler's U-boat.* New York: Random House.

Bostyn, F. 2002. Zero hour: Historical note on the British underground war in Flanders, 1915-1917. In *Fields of Battle - Terrain in Military History*, P. Doyle and M.R. Bennett, eds., Dordrecht, The Netherlands: Kluwer Academic Publishers, 225-236.

Bourbon, F. 2000. *Petra: Jordan's Extraordinary Ancient City.* New York: Barnes and Noble, Inc.

Bourbon, F. and Lavagno, E. 2001. *The Holy Land: Guide to Archaeological Sites and Historic Monuments.* New York: Barnes and Noble, Inc.

Brice, M. 1990. *Forts and Fortresses.* New York: Quatro Publishing, PLC.

Brunet, M., Guy, F., Pilbeam, D.R., Mackaye, H.T., Likius, A., Ahounta, D., Beauvilain, A., Blondel, C., Bocherens, H., Boisserie, J.-R., de Bonis, L., Coppens, Y., Dejax, J., Denys, C., Duringer, P., Eisenmann, V., Fanone, G., Fronty, P., Geraads, D., Lehman, T.M., Lihoreau, F., Louchart, A., Mahamat, A., Merceron, G., Mouchelin, G., Otero, O., Peláez-Campomanes, P., Ponca de León, M., Rage, J.-C., Sapanet, M., Schuster, M., Sudre, J., Tassy, P., Valentin, X., Vignaud, P., Viriot, L., Zazzo, A. and Zollikofer, C.P. E. 2002. A new Hominid from the Upper Miocene of Chad, Central Africa. Nature 418:145-151.

Cameron, C.P. 1998. Clandestine Tunnel-4, northern Punchbowl, Korean Demilitarized Zone. In *Military Geology in War and Peace*, J.R. Underwood, Jr. and P.L. Guth, eds., Boulder, CO: Geological Society of America Reviews in Engineering Geology XIII:99-110.

Chelminski, R. 1997. "The Maginot Line," Smithsonian Magazine, June 1997, 91-99.

Davies, P. 1993. "Middle-east Tunneling from Abraham to Today," Tunnels and Tunneling, November 1993, 36-37.

Davis, W.C. 1986. *Death in the trenches: Grant at Petersburg*. Alexandria, VA: Time-Life Books.

Day, M.J. 2004. Military campaigns in tropical karst terrain - The Maroon Wars of Jamaica. In *Studies in Military Geography and Geology*. D.R. Caldwell, J. Ehlen, and R.S. Harmon, eds., Dordrecht, The Netherlands: Kluwer Academic Publishers, 79-88.

Dehan, D. 1993. *Megiddo Armageddon*. Tel Aviv, Israel: Palphot Press.

Demir, O. 1977. Cappadocia: Cradle of History. Demir Color Kartpostal ve Turistik Yayincilik; Derinkuyu - Nevsehir, Turkey.

Duong, T.P. 1994. The Document Album of CU CHI 1960-1975. Publisher No. 7 (no place): Hoang Anh and Liksin Publishers.

Eastler, T.E., Percious, D.J., and Fisher, P.R. 1998. Role of geology in assessing the vulnerability of underground fortifications to conventional weapons attack. In *Military Geology in War and Peace*, J.R. Underwood, Jr. and P.L. Guth, eds., Boulder, CO: Geological Society of America Reviews in Engineering Geology XIII:173-187.

Elliott, H.M. 1978. The war in Gibraltar. After the Battle 21:1-14.

Gore, R. 2002. The face of the first human to leave Africa. National Geographic 202-2: New Find (no page numbers).

Gonzalez, M. and Gonzalez, A. 1993. "Maginot's Folly," The Retired Officer Magazine, August 1993, 30-32.

Greenhut, Z. 1992. Burial cave of the Caiaphas family. Biblical Archaeology Review 18:46-57.

Gross, M.R., Ghosh, K., Manda, A.K., and Whitman, D. 2004. A GIS-based spatial analysis of caves and solution cavities - Application to predicting cave occurrence in limestone terrain. In *Studies in Military Geography and Geology*, D.R. Caldwell, J. Ehlen, and R.S. Harmon, eds., Dordrecht, The Netherlands: Kluwer Academic Publishers, 287-306.

Hammond, P. 2003. Petra. (Online Version: http://www.bibleinterp.com/articles/petra)

Landberg, H. and Davis, L.K. 1993. Attack of deep underground target entrances with penetrating warheads - literature analysis. Vicksburg, MS: US Army Corps of Engineers Waterways Experiment Station, Technical Report SL-93-18.

Leaky, M.G., Feible, C.S., McDougall, I., and Walker, A. 1995. New four-million-year-old Hominid species from Kanapoi and Allia Bay, Kenya. Nature 376:565-571.

Lemonick, M.D. 1995. "Secrets of the Lost Tomb," Time Magazine, 29 May 1995, 46-53.

Liaquati, M.B. 1997. "A passing glance at the appearance and extension of Quanats (aqueducts)," The Construction Jihad Quarterly, Spring/Summer 1997, 1-3.

Linhardt, P. 1996. The catacombs of Paris. (Online Version: http://members.aol.com/_ht_a/plinhardt/ Linhardt_Paul.htm)

Library of Congress. 1993. North Korea Country Study. (Online Version: http://memory.loc.gov/frd/cs/kptoc.html)

London Daily Express. 21 March, 1936.

Lossing, B.J. 1868. *The Civil War*, Vol. II. Hartford, CT: T. Belknap Publisher.

Maqsood, R. 1994. *Petra: A Traveller's Guide*. Reading, UK: Garnet Publishing, Ltd.

Mallory, K. and Ottar, A. 1973. *The Architecture of War*. New York: Random House, Inc.

Mangold, T. and Penycate, J. 1986. *The Tunnels of Cuchi*. London: Pan Books.

Mydans, S. 1999. "Visit the Vietcong's world: Americans welcome," The New York Times, 7 July 1999.

Neufeld, M.J. 1995. *The Rocket and the Reich*. Cambridge, MA: Harvard University Press.

O'Brien, T. 1955. *History of the Second World War: Civil Defense*. London: Her Majesty's Stationery Office.
Pallud, J.P. 1987. U-boat bases in France. After the Battle 55:1-53.
Pallud, J.P. 1988. The Maginot Line. After the Battle 60:1-40.
Papnek, J.L., ed. 1994. *Greece: Temples, Tombs, and Treasures*. Virginia: Time-Life Books.
Piggot, S. 1961. *The Dawn of Civilization*. London: Thames and Hudson, Ltd.
Qassemi, A. 2001. 3000 year old Zarch aqueduct. Mosaferan, Iran: Iran's Domestic and International Tourism Monthly 7:17.
Ramsey, W.G. 1978. *Gibraltar*. London: Plaistow Press Magazines, Ltd., After the battle 21.
Robinson, G.P.G. 1989. The Durand mine and the First World War tunnel system in the Grange area of the Canadian Memorial Park, Vimy. Chatham, UK: Royal School of Military Engineering RE ADP 10/84.
Rose, M. 1994. The tombs of Silwan. Biblical Archaeology Review 20:39-51.
Rose, E. P. F. 2001. Military Engineering on the Rock of Gibraltar and its geoenvironmental legacy. In *The Environmental Legacy of Military Operations*, J. Ehlen and R.S. Harmon, eds., Boulder, CO: Geological Society of America Reviews in Engineering Geology XIV:95-121
Rose, E. P. F., Ginns, W.M., and Renouf, J.T. 2002. Fortification of island terrain: Second World War German military engineering on the Channel Island of Jersey, a classic area of British geology. In *Fields of Battle, Terrain in Military History*, P. Doyle and M. R. Bennett, eds., Dordrecht, The Netherlands: Kluwer Academic Publishers, 265-309
Schroeder, K.A. 2004. The development of tactical geography in the nineteenth century. In *Studies in Military Geography and Geology*, D.R. Caldwell, J. Ehlen, and R.S. Harmon, eds., Dordrecht, The Netherlands: Kluwer Academic Publishers, 39-51.
Thomey, T. 1995. "Hill 362A," The Retired Officer Magazine, March 1995, 33-39.
Trounce, H.D. 1918. Notes on military mining. Washington, DC: US Army Washington Barracks, press of the US Army Engineer School, Occasional papers 57.
van der Horst, P.W. 1992. Jewish funerary inscriptions. Biblical Archaeology Review 18:46-57.
Wary, J. 1980. *Warfare in the Classical World*. New York: St. Martin's Press.

Underground terrain exists in the form of openings and cavities that have resulted from either natural geologic processes or human excavation. Natural features include, but are not limited to, caves in karst landscapes, lava tubes and other igneous voids, and fold/fault/dissolution-generated voids. Man-made underground structures can be classified as tunneled bedrock or, if excavated from the surface, cut and cover facilities. Since all such underground spaces large enough to house individuals or groups of individuals and their implements of war can be considered potential military strongholds, then it is clear that such structures, herein referred to as Underground Facilities (UGF), are all integral to the study of past, present, and future military geology and military geography, in the realm of combat engineering.

This paper presents a brief history of the little-studied and even less-referenced field of military use of underground terrain along with an analysis of the role underground terrain is increasingly playing in military matters.

2. EARLY UNDERGROUND FACILITIES

The first instance of mankind organized into groups engaged in invading the territories of other groups underground, or in avoiding the attack of other groups underground, marks the beginning of military use of underground terrain. Paleoanthropologists cannot yet pinpoint such military activities, but they can put a lower age limit on them by acknowledging the date assigned to the first appearance of hominids at either four million BP (Leaky et al., 1995) or seven million BP (Brunet et al., 2002). By 1.8 million BP hominids were traveling outside of Africa in groups and using stone tools (Gore, 2002). One might speculate that such groups used caves and engaged in offensive and defensive activities in those caves, but there is no definitive research on such use. It is logical, however, to assume that military activities took place at some time prior to the first man-constructed UGF, since natural caves have always been present.

Excavated rock and soil aqueducts have been reported in the Middle East as far back as 5500 BP (Piggot, 1961). Megiddo Tell (artificial mountain), near Haifa, Israel is possibly the biblical Armageddon. The Tell is made up of 20 layers of strata each of which contains settlement ruins; the city of Megiddo lies on the uppermost layer of ruins. At the very base is a 400 ft long excavated tunnel in which flows a spring-fed canal that was constructed to supply the city with water during any siege. The tunnel is accessed by a vertical shaft nearly 200 ft deep dug from within the protected city (Dehan, 1993; Bourbon and Lavagno, 2001). Agamemnon's 4000 year old "Bronze Age" citadel at Mycenae contains a similar man-made UGF. Within the

Chapter 4

DEVELOPMENT OF TACTICAL GEOGRAPHY IN THE NINETEENTH CENTURY

Kurt A. Schroeder
Plymouth State University

Abstract: The geographic patterns of warfare have changed greatly in the past 200 years. The development of the technology of warfare and the influence of technology on tactics in the 19th century resulted in a more extensive geography of tactical warfare. Pre-Napoleonic armies were largely constrained by relatively short-range weapons and limited command control to be tactically linear. This began to change at about the time of the Napoleonic Wars. By the end of the 19th century, increases in army size, weapons range, rate of fire, and accuracy led to both a lengthening and thickening of linear tactical forms and also an increasing use of terrain features and field entrenchments.

Key words: military geography, technology, tactics

1. INTRODUCTION

War, like any field of human activity, has geographic form. These forms can be defined in terms of basic geometric entities such as the point, the line, and the area. The geographic pattern of tactical warfare changed greatly during the 19th century. The development of weapons technology resulted in a more spatially extensive tactical geography. The relatively small linear tactical forms of pre-Napoleonic warfare, limited in length and depth to a few kilometers, were gradually extended, until by World War I, they extended for hundreds of kilometers in length. These tactical lines were also thickened from several hundred meters to dozens of kilometers. By World War I, the result of this process was a more extensive geography of tactical warfare that corresponded to the more comprehensive effects of modern warfare on the economy, technology, and society.

D.R. Caldwell et al. (eds.), Studies in Military Geography and Geology, 39–51.
© 2004 *Kluwer Academic Publishers.*

This paper traces the major technological changes that created the extensive and thick linear tactical forms of World War I from the limited linear patterns of pre-Napoleonic warfare. It first examines the heritage of tactics received from the pre-Napoleonic era, since these formed the basis for the development of Napoleonic and post-Napoleonic tactics. It then examines the major technological developments of the 19th century, emphasizing the effect of technological change on the geography of tactical warfare.

2. THE MATURATION OF PRE-NAPOLEONIC TACTICS

The 18th century saw the development of large national armies in Europe (for France, see Lynn (1990a); for the United Kingdom, see Brewer (1990) and Addington (1994)). Before the end of the 18th century, most warfare was conducted for limited territorial objectives by armies that could not easily operate far from pre-established bases and magazines (van Creveld, 1977; Rothenberg, 1978).

At the beginning of a battle the marching columns would deploy into tactical line formations beyond the range of small arms fire from the enemy forces, at 250 m or more (Fig. 1; see also Rothenberg, 1978; Chandler, 1994). The linear form was necessary because it maximized the effect of the extremely inaccurate and unreliable musket fire, it allowed officers to see and more effectively control their troops, and it limited the damage artillery could inflict (Lynn, 1990b). Lines were typically three to five soldiers deep; some armies, such as the British, preferred a two-deep line (Addington, 1990; for the development of firearms tactics in the pre-Napoleonic era, see Chandler, 1994). Ideally, the troops would deploy from a road that was perpendicular to the tactical line desired. This would best guarantee the security of the baggage train and the lines of communications and retreat. Open, flat country without obstacles or structures was viewed as best for battles, because the fragile linear formations would become disorganized in irregular terrain; however, obstacles such as rivers or swamps were desirable at either end of the line to prevent out-flanking by the enemy (Addington, 1990; Muir, 1998). Tactical manuals of the time emphasized the desirability of a flat and unobstructed battleground, saying that such a site was superior for linear tactical forms; the only activity that could take advantage of irregularities of terrain was skirmishing (see, for example, Hardee (1855) which was heavily influenced by French training manuals of the Napoleonic era). Troops were trained to continuously monitor and adjust the straightness of their lines (Hardee, 1855).

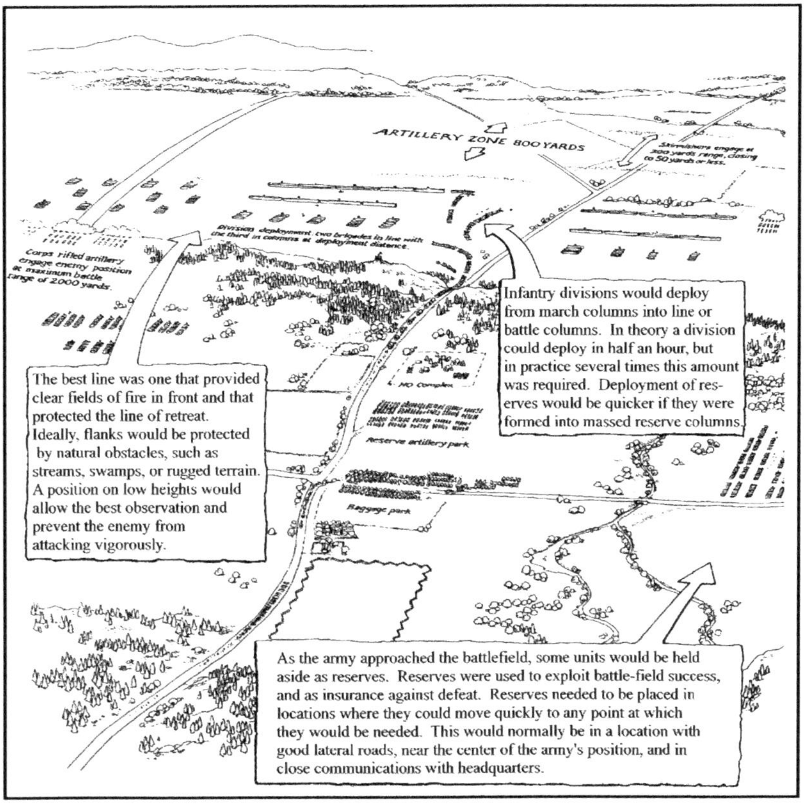

Figure 1. Deploying into line for a battle. Modified and simplified from Griffith (1986).

Deployment into line usually took hours as battalions, which marched as columns, moved off the road into linear formations perpendicular to that road (Addington, 1994; for discussion of methods of deployment, see Hardee (1855) and Chandler (1994)). Battle was joined as opposing lines moved toward one another; fire often began at distances as short as 60 m (Chandler, 1994). However, firing often began prematurely, at ranges that were ineffective (Muir, 1998). After the battle, pursuit was often limited to the cavalry, since it would take many hours for the infantry to redeploy into column format for a pursuit, whereas cavalry units were normally deployed in column and could be used for shock value during the battle and for pursuit afterwards (Chandler, 1994). These linear tactical forms dominated in open battles that usually lasted one or rarely two days. Typical examples of these linear battles from the Seven Years' War (1756–63) include Minden (August, 1759), Rossbach (November, 1757), and Leuthen (December, 1757). During this last battle, Frederick the Great of Prussia demonstrated what was considered unusual form by deploying a wing of his army at an angle to the rest (Britt et al., 1984).

3. NAPOLEONIC TACTICS

Napoleonic operational maneuver had two closely related distinguishing characteristics that were designed to lead to decisive battle. These were the corps system and distributed maneuver (Addington, 1994). The basic unit of Napoleonic operational maneuver was the corps, an all-arms body of troops that could maneuver independently and could sustain itself in battle for at least one day. In distributed maneuver, these corps marched parallel to each other and were distributed spatially across the theatre of operations, such that if one of the corps were attacked, the other corps could move to its assistance, and outflank the opposing forces within a day or two (Rothenberg, 1978; Fig. 2). Distributed maneuver was largely made possible in Napoleon's time because of improvements to the European road network in the late 18th and early 19th centuries; the feasibility of distributed maneuver increased with the development of railroad networks and the telegraph (van Creveld, 1989).

In Fig. 2, the maneuvers of the corps before the battle of Ulm in October 1805 are shown; numbers indicate days of the month. As the Corps advanced through Germany north of the Danube River, they moved quickly and then, from the 5th through the 7th, drew rapidly closer to each other on the north bank of the Danube River. As they crossed the Danube, beginning on the 7^{th,} and carefully advanced south of the river, they cut the main line of communications of the Austrian army (Chandler, 1966). Note that the rate of maneuver slowed considerably once the Corps crossed the Danube and needed to carefully reconnoiter in the potential presence of enemy forces.

As the corps approached the enemy army, they would commit the final maneuvers before battle; these final maneuvers would typically position the French corps on the lines of communication of the enemy force (Strachan, 1983). These created the necessity for battle, or perhaps surrender, in the mind of the enemy commander. Distributed maneuver had a major impact on the tactical situation. When it worked well, the corps would arrive nearly simultaneously on the battlefield, and were used to outflank enemy forces (Strachan, 1983). However, opposing generals eventually became aware of the pattern and took steps to prevent this, such as increased reconnaissance and improved communications (Muir, 1998).

Napoleonic battle tactics were characterized by flexibility of formation between line and column, the use of skirmishers, and the rapid movement

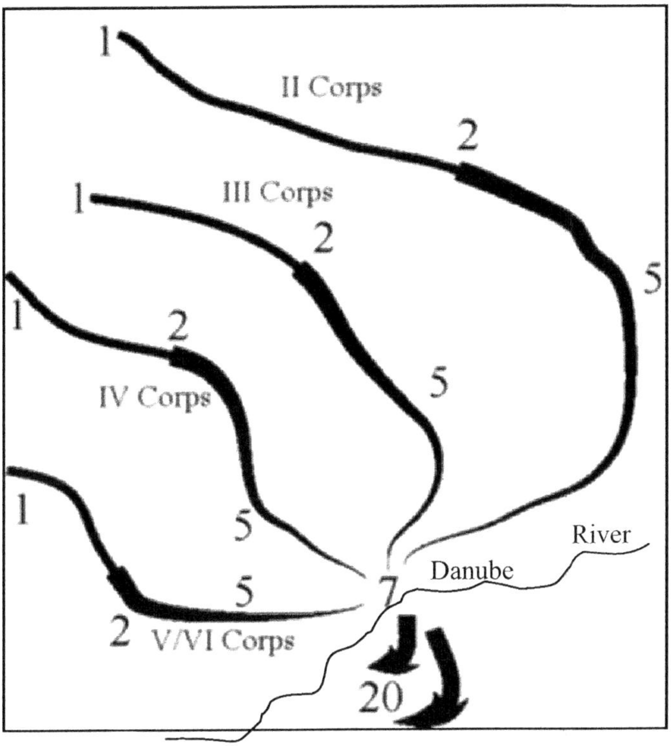

Figure 2. Napoleonic Distributed Maneuver before the Battle of Ulm. Numbers indicate dates in October, 1805. Modified and simplified from Chandler (1966).

and concentration of artillery (Chandler, 1966; van Creveld, 1989; Lynn, 1990b; Muir, 1998). The French and other European armies had experimented with the development of non-linear tactical forms such as columns-of-divisions for shock attacks, and skirmishers to absorb the force of advancing enemy lines (Addington, 1994). Officers in the old French royal army had pioneered all these innovations; Napoleon familiarized himself with them during and after his training as an artillery officer. Although these innovations had been tried separately during the Wars of the French Revolution, Napoleon saw the potential of integrating them into a new system of battle, which he employed from his first campaign in Italy in 1796 (Chandler, 1966; McElwee, 1974; Rothenberg, 1978; Addington, 1994). The Napoleonic Wars also saw increased tactical use of terrain features. The British general, the Duke of Wellington, usually used terrain features such as ridges to protect his line until the last possible moment before the impact of French troops was felt; this violated the general pattern of warfare, and was done to protect scarce British manpower (Addington, 1994). Occasionally, earthworks were used to protect an army or city from the enemy (Strachan, 1990; Muir, 1998). Sometimes a house, farm, or chateau would be used as an advanced strongpoint to retard the advance of the enemy, as with the chateau at Hougemont and the farm of La Haie Sainte in the battle of Waterloo (Chandler, 1966). Villages also were often used as strongpoints (Muir, 1998).

4. DEVELOPMENT OF TACTICAL GEOGRAPHY DURING THE NINETEENTH CENTURY

After the Napoleonic wars, the nations of Europe entered a century of rapid technological change. These changes led to further development of the linear tactical geography of Napoleon and his predecessors. Probably the most significant technological change during this era, which heralded the beginning of truly modern warfare, was the increase in small arms range, rate of fire, and accuracy during the 19th century (van Creveld, 1989; O'Sullivan, 1991). Some of the major small arms innovations are summarized in Table 1, which lists the approximate date of each innovation, the type of innovation, the effect on range and rate of fire, and a brief description of other weapons effects.

Increased range and accuracy will each have a more than proportional effect on the efficiency of fire. Assuming open terrain and no covering artillery fire, doubling the effective range of small arms will more than double the time spent in the zone of fire by advancing troops, since the rate of advance is inversely related to the distance to be covered. Doubling the depth of the zone of fire from 400 m to 800 m will mean that advancing troops will spend more than twice as much time in the zone of effective fire.

Unless they are trained to moderate their pace at the beginning of a rush toward enemy lines, troops that are tiring during a long run may move most slowly at the end of their run towards the enemy lines, precisely when enemy fire is most lethal (O'Reilly, 2003). Much training of the era was devoted to regulating the pace of marches so that this would not happen (Hardee, 1855). On the other hand, many observers noticed a natural tendency for advancing lines and columns to accelerate, due to both the stress of being under more effective fire and anticipation of hand-to-hand combat, as they approached enemy positions (Muir, 1998).

As accuracy of fire increases, the number of wasted rounds declines dramatically. The accuracy of fire from small arms will depend on three factors: (1) accuracy of aiming of the weapon at an aiming point; (2) mechanical inaccuracies in the weapon which cause the path of projectile to depart from the path to the aiming point; and (3) external factors such as wind. All these factors combine to create ineffective fire, that is, fire that

Table 1. Some small arms innovations before and during the 19th century. From Strachan (1983) and Addington (1994).

Date	Weapon	Range and rate of fire	Notes
1650	Smoothbore, muzzle-loading matchlock	100 m, 0.5/min	Accuracy poor due to tendency to flinch, stiffness of firing mechanism; susceptible to poor weather
1700	Smoothbore, muzzle-loading flintlock	50-200 m, 1-2/min	Only took 26 movements to ready, as opposed to 44 for matchlock; principle small arm of the Napoleonic era
1750	Muzzle-loading flintlock rifle	200-500 m, 1/min	Accuracy doubled, range increased; rate of fire low due to more difficult ramming process; accuracy sometimes suffered as a result of fatigue
1830s	Percussion lock rifle with copper cap	200-500 m, 2/min	Much less susceptible to weather, less flash and smoke
1840s	Breech-loading rifle	800 m, 7/min	Breech brought center of gravity back for more accurate firing; could be loaded and fired from kneeling or prone position
1849	Cylindro-conoidal bullet	1000 m, 8/min	Improved accuracy and range, decreased mechanical problems
1880s	Magazine-loaded rifle	1200 m, 20/m	Allowed more rapid fire
1883	Machine gun	2000 m, 400/min	Primary limitation on use was the supply of ammunition

does not cause casualties. The theoretical effect of a change in accuracy can be calculated (Fig. 3).

Assume that rounds will be distributed randomly within a circular area of radius r (the radius of error) at any distance from the firing line. Assume also that the line of approaching soldiers is of height h. The fraction of the rounds that strikes the line of soldiers is equal to the sum of the area of sector ABC and the area of triangle ABD, divided by the area of one quarter of the circle with radius r. From the formula for the area of a sector of a circle:

$$Area(ABC) = \frac{r^2}{2} \cdot \arcsin\left(\frac{h}{2r}\right)$$

From the formula for the area of a triangle and the Pythagorean Theorem:

$$Area(ABD) = \frac{h}{4} \cdot \sqrt{r^2 - \frac{h^2}{4}}$$

Thus, the fraction P of rounds that strikes the line is given by:

$$P = \frac{2r^2 \cdot \arcsin\left(\frac{h}{2r}\right) + h \cdot \sqrt{r^2 + \frac{h^2}{4}}}{\pi r^2}$$

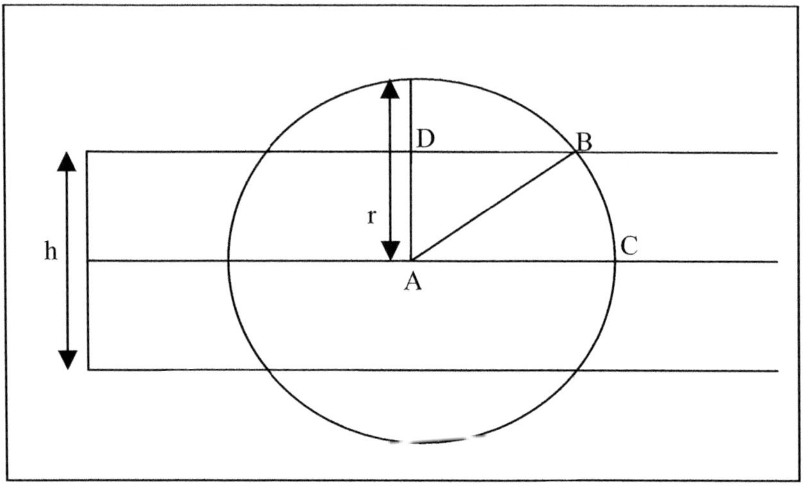

Figure 3. Effect of weapons radius of error on successful fire, where *h* is the height of a line of soldiers, and *r* is the radius of error for small arms fire.

The percent of rounds striking the line of soldiers declines rapidly as the diameter of the circular error of probability increases (Table 2). If the radius is one-and-one-half times the height of the line of troops, or 9 ft, about 42% of the rounds will be effective; if the radius of error is doubled to 18 ft, only about 20% of the rounds will strike the horizontal area covered by the approaching line of troops.

Table 2. Effect of accuracy on effectiveness of fire. The radius of error is *r*; *P* is the percent of rounds fired which strikes a line of advancing troops 6 ft high.

r	Ratio of *r* to *h*	*P*
3	0.5	100.00
6	1.0	60.90
9	1.5	41.64
12	2.0	31.50
15	2.5	25.29
18	3.0	21.12
21	3.5	18.13
24	4.0	15.87
27	4.5	14.12
30	5.0	12.71
33	5.5	11.56
36	6.0	10.60

During the Napoleonic era, it probably took between 200 and 600 rounds to create one casualty (Muir, 1998). This fire was so ineffective not only because it was inaccurate, but also because ". . . much ammunition was fired at ludicrously long ranges, much was wasted in bickering between skirmishers, and much was probably thrown away by broken troops or deliberately squandered" (Muir, 1998, 83). The rapid increase in small arms effectiveness meant that frontal assaults were increasingly futile, as demonstrated in the American Civil War (1861-65) and the Danish War of 1864 (Addington, 1994). Prussian doctrine after the war with Denmark stressed the avoidance of frontal attacks, the turning of enemy positions by outflanking, followed by the assumption of defensive positions that would force the enemy to leave their defensive positions. This tactical doctrine reached its apex at Sedan during the Franco-Prussian War of 1870-71, when the largest French field army was surrounded and forced to surrender (Strachan, 1983; Addington, 1994).

During the last half of the century, the development of small arms was complemented by the rapid development of artillery (Table 3; see also Ehlen and Abrahart, 2004, this volume). This development of artillery reinforced the superiority of the defensive created by the development of small arms fire, making defensive positions even more difficult to take by direct assault. By the time of World War I, the main restriction on the use of small arms and artillery was the supply of munitions (Addington, 1994).

Table 3. Artillery advances in the late 19th century, and the characteristics of some typical artillery weapons of the period. From Strachan (1983) and Addington (1994).

Date	Weapon	Range and rate of fire	Notes
1850	Smooth-bore, muzzle-loading field artillery	1500 m 2/min	Eclipsed by small arms in range and rate of fire by 1880s
1852	Fused case shot	-	Effective against dispersed formations
1864	Rifled breech-loading field artillery	5000 m, 5/min	Use of percussion fuses increased lethality of shells
1890	Smokeless powder	-	Allowed higher velocity, reduced visual obstruction on battlefield
1897	French 75 mm	8000 m, 10/min	Typical World War I artillery piece; could be used for direct and indirect fire; recoil mechanism increased rate of fire

The adjustments made to increases in the range, accuracy, and rate of fire of small arms and artillery included the use of terrain to protect troops, the adoption of entrenchments, which had previously been primarily used in sieges, and the increased use of skirmishers (Addington, 1994). The practice of using terrain features, such as ridges, or cultural features, such as houses, as defensive positions became more prevalent during the 19th century. This practice became more effective as a small group of soldiers in a house or other strongpoint could produce increased fire (Addington, 1994). Trenches and other earthen field fortifications were used in conflicts such as the Crimean War (1854-55; Strachan, 1990; Royle, 2000), the American Civil War and the Franco-Prussian War (van Creveld, 1989; Hess, 1997). A good example of this from the American Civil War was the Battle of Fredericksburg: Confederate entrenchments on Marye's Heights and the use of terrain such as the Sunken Road foiled repeated Union attacks (O'Reilly, 2003). At the battle of Gettysburg, as at many battles during this era, the linear form typical of Napoleonic and pre-Napoleonic warfare was replaced by adaption to the local terrain, in this case a "fish-hook" pattern that followed the shape of the hills south and east of the town (Fig. 4).

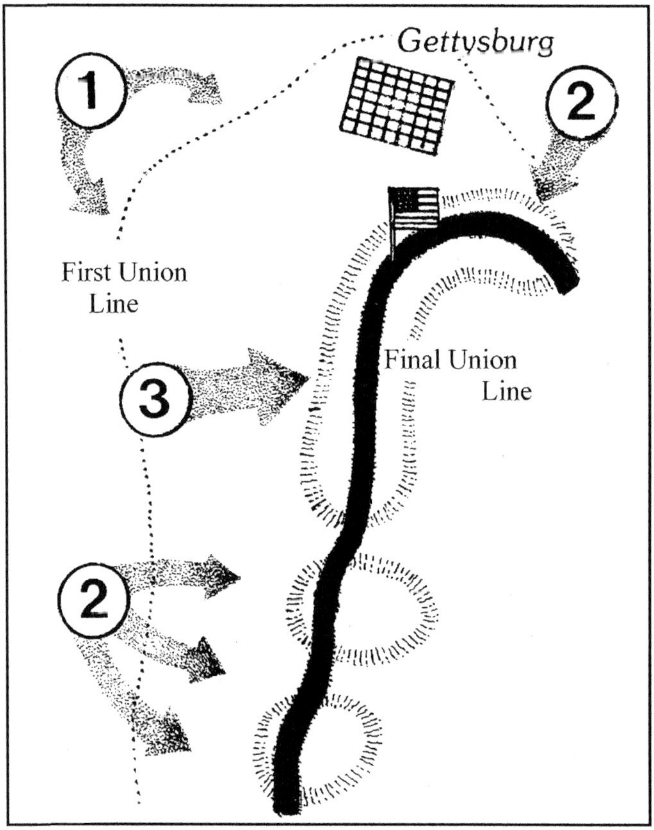

Figure 4. Union positions at Gettysburg. Confederate attacks are shown against the first Union line on Day 1 of the battle, and against the final Union line on Days 2 and 3. Modified and simplified from Griffith (1986).

Skirmishers were increasingly used during this period, and skirmisher tactics became increasingly elaborate. For example, by the end of the

American Civil War, some units specializing in skirmishing developed tactics of coordinated small group action that allowed them to attack linear positions much more effectively; these tactics included short rushes, open (skirmisher) order, and suppressive fire (Luvaas, 1988; Addington, 1994). The use of skirmishers increased the depth of the line.

Another major innovation in the 19th century was the use of continuous warfare during the last year of the American Civil War (Hess, 1997). In previous Civil War campaigns, battles had been relatively infrequent and surrounded by extended periods of recuperation and maneuver. The Union generals Ulysses S. Grant and William T. Sherman planned to apply continuous pressure to an over-stretched Confederate army; this was to be accomplished by a rapid succession of high intensity battles with short periods of rapid maneuver and active reconnaissance (Hess, 1997). In the Eastern Theater of the war in Virginia in late 1864 and early 1865, the series of battles summarized in Table 4 placed continuous tactical pressure on the Confederate Army in Virginia. These operations also posed a strategic threat, since they were aimed at Richmond, the Confederate capital and a major manufacturing center.

Table 4. Union operations during the period of continuous warfare. From Addington (1994) and Hess (1997).

Battle	Date
The Wilderness	5-7 May, 1864
Spotsylvania	8-21 May, 1864
North Anna	23-26 May, 1864
Cold Harbor	31 May - 12 June, 1864
Siege of Petersburg	9 June, 1864 - 2 April, 1865
Pursuit to Appomattox	3-9 April, 1865
Ancillary operations:	
The left operational flank: Bermuda Hundred	Continuous operations
The right operational flank: Shenandoah Valley	Continuous operations
The strategic flank: Sherman's March to the Sea	22 November - 13 December, 1864

Ancillary Union operations also listed in Table 4 combined to create continuous pressure on the other Confederate armies so that they could not use their interior lines to stave off eventual defeat (Addington, 1994; Hess, 1997). Battles during this phase of the war were characterized by the use of field entrenchments (McElwee, 1974). The Sieges of Sevastopol during the Crimean War and Port Arthur during the Russo-Japanese War (1904-05) also extended over the winter, in large part because of the use of entrenchments by the defender (McElwee, 1974; Westwood, 1986;

Addington, 1994; Royle, 2000). At the same time, the increased sizes of armies and the extended ranges of weapons led to an extension of linear forms to distances of tens of kilometers in length and depth (Strachan, 1983; Westwood, 1986; Bellamy, 1990).

5. CONCLUSION: THE CULMINATION OF THE NINETEENTH CENTURY

Before the Napoleonic era, when armies made contact, they would assume linear tactical formations for battles which lasted one or two days and extended in length and depth for a few kilometers at most. Technical changes and the growth of armies in the 19th century led to a lengthening and widening of the linear tactical form, and increased dispersal of force. Van Creveld (1989) estimates that the dispersion ratio of soldiers to space, which stood at about 1:10 in the 18th century, declined to approximately 1:25 during the American Civil War, and continued to decline during the latter half of the 19th century until it reached approximately 1:250 during World War I. World War I marked the culmination of the contribution of technological advances during the previous century to the dominance of the tactical defensive, and the extension of tactical geography to such a length and depth that it assumed a strategic significance; the entire theatre of operations became one continuous battle (Gilbert, 1994).

REFERENCES

Addington, L.H. 1990. *The Patterns of War through the Eighteenth Century*. Bloomington, IN: Indiana University Press.

Addington, L.H. 1994. *The Patterns of War since the Eighteenth Century*. Bloomington, IN: Indiana University Press.

Bellamy, C. 1990. *The Evolution of Modern Land Warfare: Theory and Practice*. London: Routledge.

Brewer, J. 1990. *The Sinews of Power: War, Money, and the English State, 1688-1783*. Cambridge, MA: Harvard University Press.

Britt, A.S., O'Connell, J.A., Palmer, D.R., and Stadler, G.P. 1984. *The Dawn of Modern Warfare*. Wayne, NJ: Avery Publishing Company.

Chandler, D.G. 1966. *The Campaigns of Napoleon: The Mind and Method of History's Greatest Soldier*. New York: Macmillan.

Chandler, D.G. 1994. *The Art of Warfare in the Age of Marlborough*. New York: Sarpedon.

van Creveld, M. 1977. *Supplying War: Logistics from Wallenstein to Patton*. Cambridge: Cambridge University Press.

van Creveld, M. 1989. *Technology and War: From 2000 B.C. to the Present*. New York: The Free Press.

Ehlen, J. and Abrahart, R.J. 2004. Terrain and its affect on the use of artillery in the American Civil War - The Battle of Perryville 8 October 1862. In *Studies in Military Geography and Geology*, D.R. Caldwell, J. Ehlen, and R.S. Harmon, eds., Dordrecht, The Netherlands: Kluwer Academic Publishers, 155-172.

Gilbert, M. 1955. *The First World War: A Complete History*. New York: H. Holt.

Griffith, P. 1986. *Battle in the Civil War*. Surry, UK: Field Books.

Hardee, W.J. 1855. *Rifle and Light Infantry Tactics*. Philadelphia, PA: Lippincott, Grambo and Company.

Hess, E.J. 1997. Tactics, trenches, and men in the Civil War. In *On the Road to Total War: The American Civil War and the German Wars of Unification, 1861-1871*, S. Förster and J. Nagler, eds., Cambridge: Cambridge University Press.

Luvaas, J. 1988. *The Military Legacy of the Civil War: The European Inheritance*. Lawrence, KS: University Press of Kansas.

Lynn, J.A. 1990a. The pattern of army growth, 1445-1945. In *Tools of War: Instruments, Ideas, and Institutions of Warfare, 1445-1871*, J.A. Lynn, ed., Urbana, IL: University of Illinois Press.

Lynn, J.A. 1990b. En avant! The origins of the revolutionary attack. In *Tools of War: Instruments, Ideas, and Institutions of Warfare, 1445-1871*, J.A. Lynn, ed., Urbana, IL: University of Illinois Press.

McElwee, W. 1974. *The Art of War: Waterloo to Mons*. Bloomington, IN: Indiana University Press.

Muir, R. 1998. *Tactics and the Experience of Battle in the Age of Napoleon*. New Haven, CT: Yale University Press.

O'Reilly, F.A. 2003. *The Fredericksburg Campaign: Winter War on the Rappahannock*. Baton Rouge, LA: Louisiana State University Press.

O'Sullivan, P. 1991. *Terrain and Tactics*. New York: Greenwood Press.

Rothenberg, G.E. 1978. *The Art of Warfare in the Age of Napoleon*. Bloomington, IN: Indiana University Press.

Royle, T. 1983. *Crimea: the Great Crimean War, 1854-1856*. New York: St. Martin's Press.

Strachan, H. 1983. *European Armies and the Conduct of War*. London: George Allen and Unwin.

Strachan, H. 1990. The British army and 'modern' war: The experience of the Peninsula and of the Crimea. In *Tools of War: Instruments, Ideas, and Institutions of Warfare, 1445-1871*, J.A. Lynn, ed., Urbana, IL: University of Illinois Press.

Westwood, J. 1986. *Russia against Japan, 1904-1905: a new look at the Russo-Japanese War*. Albany, NY: State University of New York Press.

Chapter 5

CANADIAN MILITARY GEOGRAPHY 1867-2002

From Empire to Alliance - Guarding the South or Watching the North

Jean Martin

Department of National Defence, Canada

Abstract: Military geography enjoyed some level of popularity in Europe during the late 19th and early 20th centuries. In Canada, though, no systematic and deliberate attempt was ever made in that direction. This is largely because most of the thinking in that field was already done by others: Great Britain until World War II, and the US later on. The Canadian Army and Air Force took a large part in the surveying and mapping of the country, particularly in the vast northern areas. The Canadian Forces are still very good at mapping today, but geography is nearly absent from the curriculum at the Royal Military College, and trained geographers are hard to find at the Department of National Defence. This paper's first objective is to present a survey of the relationship between military and geography in Canada since the Confederation of 1867, and then to try to explain the Canadians' disregard for military geography.

Key words: history, Canada, imperialism, mapping, Great Britain, academic training, United States

1. INTRODUCTION

It is often suggested that Canada is dominated by its geography. The second largest country in the world, Canada's total surface area is approximately ten million square kilometers of mountain, flatland, ice, and water. A few small, densely populated areas alternate with farmlands and unbounded, uninhabited expanses of forest and tundra. Several hydrographic basins and nine great "natural regions" comprise this gigantic country (National Atlas of Canada, 1989).

One would expect this kind of environmental dominance to resonate heavily on Canadian studies. Geography departments were not established in

D.R. Caldwell et al. (eds.), Studies in Military Geography and Geology, 53–64.
© 2004 *Kluwer Academic Publishers.*

Canadian universities until the late 1940s, although they rapidly proliferated thereafter and geographers can now be found in every institution throughout the country. The Geography Department at the University of Toronto was the only one established during the 1930s, although geography was included within departments of geology at the Universities of Western Ontario and British Columbia. Hundreds of atlases, monographs, and geographic studies were produced during the past 50 years in sharp contrast to the near absence of geography publications before World War II.

Canada is not only subject to its geography. For most of its history Canada was a colony, first of France, then of Great Britain and, as some think nowadays, of the United States (e.g., Newman, 1983; Hellyer, 1997). The study of Canadian geography was consequently oriented towards fulfillment of French and then British expectations, before it began to reflect more Canadian, or at least North American, concerns. As for Canadian military geography, I argue that it still awaits development. As part of an empire or as a minor player in international alliances, Canada has always chosen not to involve herself too deeply in matters of world strategy, and has left the dominant power of the day cope with these concerns.

In this paper I suggest that, although seriously involved in surveying and mapping activities, the Canadian military has never shown much interest in formal geographical thinking. Indeed, Canadian Forces have acquired a remarkable expertise in the development and use of geographical techniques and instruments, maps, air photographs, remote sensing, Geographic Information Systems (GIS), etc., over the last century. Canadians were among the first to issue field maps to every infantry section during World War I, for example, and the young Royal Canadian Air Force (RCAF) was a leader in the development of cartographic techniques based on aerial photography in the 1920s (Winterbotham, 1926). The Canadian Forces Mapping and Charting Establishment is actively involved today in surveying and map making in several theatres throughout the world.

It is very difficult to find an academically trained geographer working as such in the Canadian Forces or in the Department of National Defence (DND). The officers who helped to develop the excellent mapping and surveying services were engineers who specialized in surveying. They were very proficient, but generally showed no interest in venturing outside this strictly practical area of activity, otherwise some form of military geography would surely have developed. Of course, spatial analysis has always been performed by the military, and the Canadian Forces Mapping and Charting Establishment is at the leading edge of modern GIS and remote sensing technologies. However, just as historians, chemists and biologists are needed to study history, chemistry, and biology, one would expect to find geographers practicing geography. I believe that without geographers, there can be no consequential geography; and where there is no geography, there can be no military geography other than on an incidental basis.

Geographers are present in large numbers in Canadian universities and if they were to show interest in military geography, it is likely that a Canadian military geography would develop. I have found only one case of a professional geographer who broached the subject of Canadian military geography (see Kimble, 1949, below). He was not, in fact, even Canadian, but freshly embarked from England, and he did the job only because he was commissioned to do so by the DND. The assessment made fourteen years ago by the French geographer, Paul Claval (1990), rings true today: Military considerations have been totally neglected by Canadian geographers at least since the end of World War II. It is likely that military geography is too closely associated with geopolitics to find much support in geography departments (Chauprade and Thual, 1998). At present, I know of no Canadian universities that put any kind of emphasis on military geography in their programs.

2. WHAT IS CANADIAN MILITARY GEOGRAPHY?

Canada has never had ambitions for foreign conquest, nor has she had any "natural enemy" since the normalization of relations with her closest and only immediate neighbor, the United States, in the latter part of the 19th century. Moreover, her foreign affairs were, at least until the 1930s, largely conditioned by British imperial policies. Consequently, the need to develop an independent geopolitical strategy was not as strong as it was among the closely intermingled and often hostile European nations. Canada's relative geographical isolation bestowed upon the country a certain natural security and national efforts in matters of defense have essentially been devoted to strengthening this security. Apart from the occasional, but always wholehearted, support for Britain's bellicose undertakings, Canada has aspired to be left alone on this vast land situated between three large oceans and the one great neighbor that we like to consider benevolent. (e.g., Cuthbertson, 1977; Bothwell, 1998-99; and Rioux, 1998-99).

However, the United States has not always been well disposed toward Canada, and Canada has not always responded to US overtures with the expected friendship. For well over a century, tension existed between the young United States and the northern colonies that had remained faithful to the British Crown. This era, punctuated with frequent scuffles, became deeply imprinted on the Canadian psyche. The American foe remained for a long time, in the mind of many Canadians, the only menace openly directed at their territory and population. It is this threat that was, to some extent, the driving force behind the first Canadian approach to a continental geostrategy, if not military geography. At least until the early 1920s, the Canadian military had contingency plans in case of an American invasion (Preston, 1977, 1990).

Until World War II, Canadian military geography remained part of a wider British imperial geography. One book stood in good place on the shelves of the Royal Military College (RMC) Library, in Kingston, Ontario, most Canadian officers' *alma mater*: D.H. Cole's *Imperial Military Geography*, first published in London in 1924. In the 1937 edition, Canada was allotted Chapter XIII's 20 pages, a fair share compared to 38 pages devoted to India and Burma; 32, to the Pacific; 29, to the Middle East; 26, to the ensemble Australia-New Zealand; and 15, to South Africa. As for other parts of the British Empire, Canada's chapter (which also included Newfoundland, the last province to join the Canadian confederation in 1949) dealt with natural resources, industrial capacity, and transportation infrastructures, with a particular emphasis on railways and inland navigation. The military situation was only mentioned in relation to coastal defense and the safety of sea trade. The US threat was no longer regarded as credible: "Ties of race, languages, sport, finance, industry and commerce bind Canada and her great neighbour" (Cole, 1937, 208). The real threat, explained Cole, ". . . is not military attack but peaceful absorption by a powerful and wealthy neighbour" (Cole, 1937, 211).

A similar book was published in 1936 by A.G. Boycott of the Royal Air Force Educational Service. Although containing more detailed information than Cole's, Boycott's book was obviously intended for the education of students and his analysis was thus less developed than Cole's. The assessment of the Canadian military situation covered only one page, in which a possible attack by the United States was still noted. Should this occur, the author concluded, ". . . little could be done for the Defence of the Dominion" (Boycott, 1936, 169). Another book, more concise than the two aforementioned, was published by J. Fitzgerald Lee in 1908. Still older British books dealing with military geography can be found at the RMC Library: *Outlines of Military Geography* (Maguire, 1900), *Geography in Relation to War* (May, 1907), *An Introduction to Military Geography* (May, 1909), and *The Outlines of Military Geography* (Macdonnell, 1911) were all produced in the years between the South African War and World War I by senior British officers, with the exception of May's 1907 book. May was a London lawyer who served as a lieutenant in the British Territorial Army. A series of papers by geographer Vaughan Cornish was also read in 1917 to British officers serving in France. These papers were published under the title *The Strategic Geography of the Great Powers* (Cornish, 1918).

The first book purportedly addressing the military geography of Canada was not published until 1949 after World War II. George H.T. Kimble, head of the newly established Department of Geography at McGill University, was commissioned by the Directorate of Military Training of the DND to prepare a handbook for the education of officers in the Canadian army. Over 200 pages long, this book consists of a series of lectures on broad topics of Canadian geography, many of which are only loosely related to military matters. Despite its title (*Canadian Military Geography*), the book was little more than an introductory manual to Canadian geography with little or no relation to military geography. Each chapter included a relatively brief description (less than 15 pages) on a general subject, and was usually completed with one or two small scale maps and several statistical tables. Five questions were proposed to the student at the end of each of the ten chapters for further consideration and discussion.

After a brief introductory chapter describing Canada's relative position in the world, Kimble devotes two chapters to the climate and wildlife in different parts of the country. Human activities are dealt with beginning in Chapter 5, which is devoted to agriculture. Mineral and power resources (Ch. 6), manufacturing industries (Ch. 7), external trade (Ch.8), and transportation (Ch. 9) follow. The last chapter provides an overall picture of population and settlement problems throughout the country. Notwithstanding the "Confidential" tag on the document, this book neither shows a particularly progressive approach nor provides an exceptionally thorough analysis; the only sources mentioned are Currie's *Economic Geography of Canada* (1945) and Taylor's *Canada* (1947).

Kimble was obviously not a military geographer and it is doubtful that he ever planned to become one. He had studied and taught in Great Britain before World War II, during which he served with the Royal Navy Meteorological Department. He came to Canada in 1945 and became the first teacher of geography at McGill University in Montreal. After receiving his Ph.D. from the French-speaking Université de Montréal in 1948 and publishing his Canadian Military Geography in 1949, he left for the United States, where he remained for the rest of his career. He retired first to France and then finally to Britain. He seems not to have maintained any special connection with Canada and, apart from his transitory Montreal experience in the late 1940s, the rest of his career was devoted primarily to other geographical issues, such as ancient cartography and explorations, climatology, and African geography.

3. SHIFTING EMPIRE

Several books were published after World War II dealing with military topics from a geographical perspective. Most of those that can be found on the shelves of the RMC Library originated in the United States and pay little attention to the already forgotten British Empire. Some were published by the office of the US Secretary of Defense or other US government agencies (e.g., Air Force Reserve Training Corps, 1959; Peltier and Percy, 1966; Collins, 1998). Even Kimble's book had adopted a very continental approach when published in 1949. That a British geographer came to Canada after World War II and produced the only book ever published on the military geography of Canada, just before he left for the US, is perhaps the most revealing symbol of the shift in influence that occurred in Canada about this time.

Military geography was no more popular outside official circles than within them. A survey of a comprehensive bibliography published in 1993 (Conzen et al., 1933) includes a few studies on early military cartography and explorations in North America, as well as other publications on military construction, and some military atlases, but not one single attempt at geographical analysis of the Canadian military situation. Only one master's thesis that can truly be labeled a study in Canadian military geography has been discovered: *A Military Geography of the Great Lakes Area* (McDaniel, 1954) focuses on one of the most important areas in Canadian military history.

The author, R. McDaniel, was an officer in the Princess Patricia's Canadian Light Infantry when he did his research at the University of Western Ontario in London. About one third of the thesis' 180 pages are specifically devoted to defense matters, although as the title indicates, the economic importance of the region also receives a good deal of attention. Surprisingly, McDaniel says little about the military history of the area, a traditional zone of conflict between France and Britain and, later, between the US and Canada. He could naturally not avoid the War of 1812, which was for a large part fought around Lakes Erie and Ontario. McDaniel also spends a few pages on the strategic importance of the commercial and communication axis formed by the Great Lakes and the St. Lawrence River.

But McDaniel's thesis was produced in 1954, at the height of the Cold War, and it is strongly influenced by the general feeling prevalent in Canada during the aftermath of the Korean War. The fear of a third world war is clearly expressed, and McDaniel focuses on the vulnerability of the industrial core of the Great Lakes. An attack from the south was no longer considered; the emphasis was now placed on the new threat from the north over the Arctic. McDaniel's study thus followed a trend that was initiated during the interwar period. The fast and sustained improvement in aircraft performance following World War I led many Canadians to believe that they could no longer rely entirely on their relative seclusion for protection against possible threats from overseas (Fig. 1). Aircraft, and soon missiles from the Soviet Union, could reach any part of Canada, a prospect that obviously had deep impact on McDaniel's approach. His thesis abstract explains: "This area is the 'heartland' of the free world and its defence is a major and necessary undertaking."

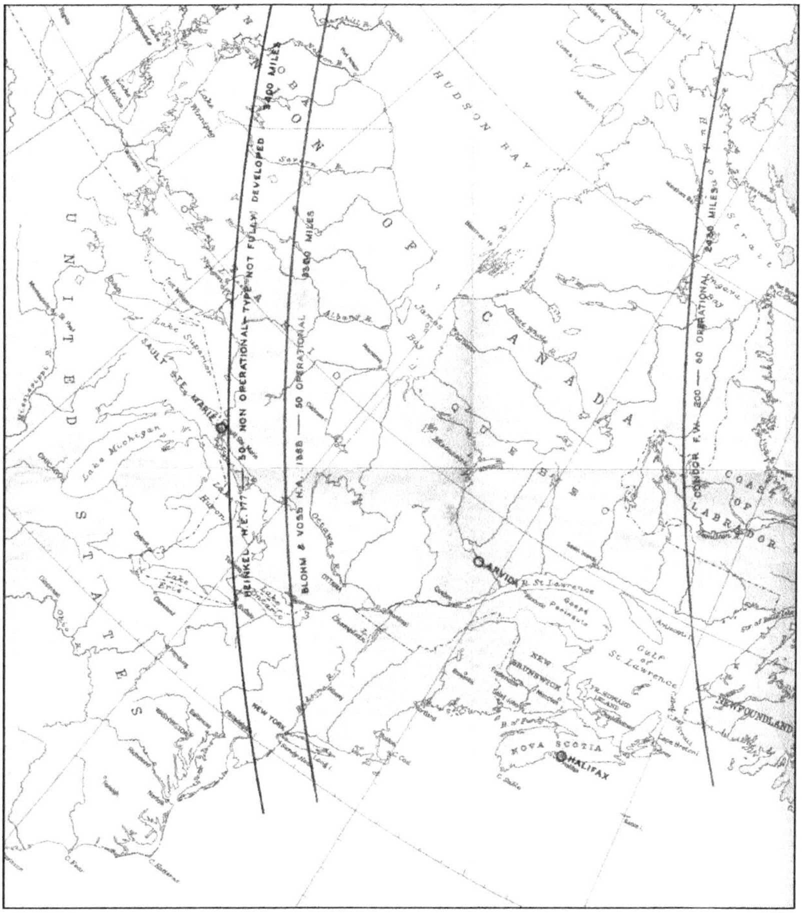

Figure 1. The threat from the north (1). The reach of German long-range bombers as estimated by Canadian intelligence. Section of a map contained in DHH, 193.009 (D23) from JIC (1943).

As early as the late 1930s, RCAF Captain A. Carter had warned: "It can be done" (Carter, 1938). This was a reference to the new Canadian vulnerability to possible attacks or infiltrations from the north. Aircraft of that era could hardly have launched an attack directly from Europe to the Canadian heartland, but Carter surmised that an enemy force could well find its way through the northern wilderness and secretly set up bases from which cities and industries in the south could be reached (Fig. 2). The north thus became the new frontier to be secured, and this developed into an increasing concern in the military.

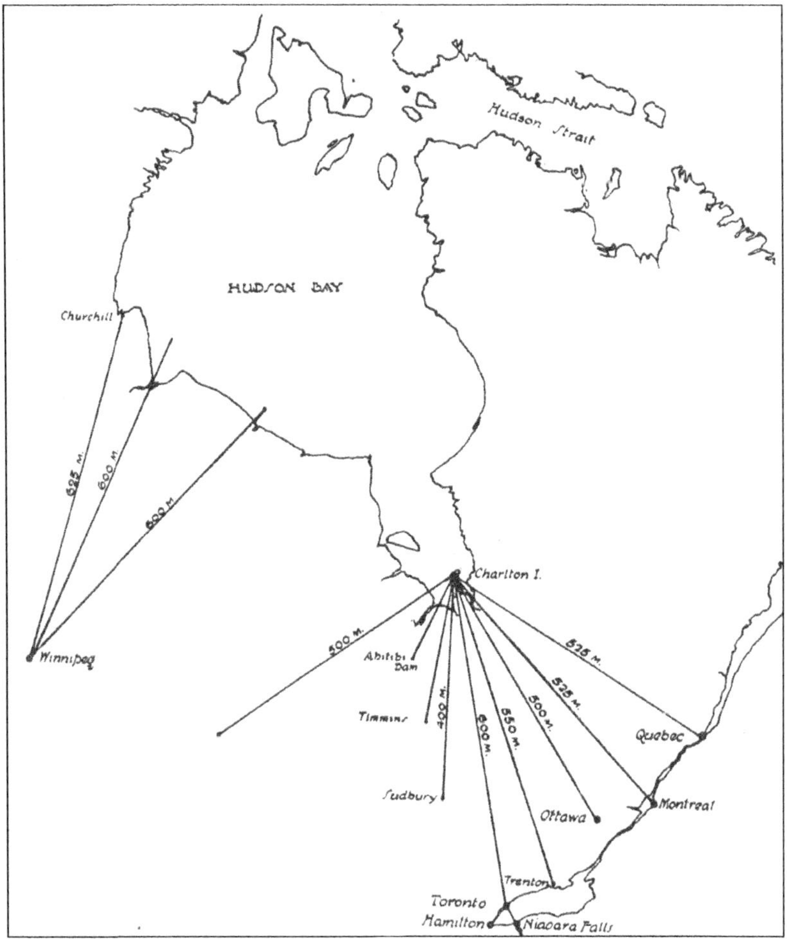

Figure 2. The threat from the north (2). Targets within range of aircraft launched from enemy bases secretly set up in the Canadian north. From Carter (1938).

As far as the United States was concerned, this was only a matter of pushing the first line of defense further to the north, without any change to its orientation: In times of tension between the US and Great Britain, the

Canadian border had to be guarded. This new northern threat simply meant that the protection had to be extended in order to include Canada into a new continental area of defense. The creation of the Joint Board on Defence in 1940, followed by the establishment of a line of radar stations (the Distant Early Warning system, or DEW line), was the first manifestation of this new reality. The later establishment of the North American Air Defence Command (NORAD) and the newly proposed continental area of defense only confirmed the trend.

Historians have traditionally shown more interest than geographers in spatially related military issues in Canada. Military matters have always been largely ignored by Canadian academics, so it is not really surprising that no university geographer ever thought of making a specialty of Canadian military geography. Historians, however, particularly in English Canada, are a little more interested in military topics and some military historians approached their studies through questions related to space and landscape. The work of Colonel Charles P. Stacey (1940) is a good example, as is that of Richard A. Preston (1977). Stacey, a scholar and career officer who headed the Canadian Army Historical Section from 1945-59 and the new Directorate of History of the newly unified Canadian Forces between 1965-66, has had a deep and durable influence on Canadian military history. *The Military Problems of Canada* (Stacey, 1940), for example, contains a long chapter entitled "Geography and the Canadian Security." Preston's and Stacey's works were published between the 1940s and 1970s, however, and not much ensued in the following decades. There is apparently renewed interest in the subject recently, exemplified by books like *The Canadian Military Atlas* (Zuehlke and Daniel, 2001) and *Forging a Nation, Perspectives on the Canadian Military Experience* (Horn, 2002b).

The work of an historian, Zuehlke's military atlas is more of a military history of Canada, merely presenting campaign maps, than a true geographical study. However, it suggests a return to spatial concerns in the study of Canadian military history. As for *Forging a Nation*, it contains two papers that clearly raise geographic questions going back to two recurrent themes from the last century: The protection of the interminable Canadian coastline and the defense of the north. In "Geography and history: The six fathom line of Canadian naval policy" Lieutenant-Commander Gregg Hannah (2002) re-examines ideas promoted by the founders of the Royal Canadian Navy in the early 20th century. His purpose is to highlight what he regards as deficiencies in Canadian naval policies. These early ideas were inspired by Canada's near-insularity - surrounded by three large oceans - a situation which, according to Hannah, should induce Canadians to regard themselves as a maritime nation and thus prompt them to adopt productive policies, rather than relinquishing their responsibilities, along with the accompanying benefits, to external protecting powers.

In "Gateway to invasion or the curse of geography? The Canadian Arctic and the question of security, 1939-1999," Lieutenant-Colonel Bernd Horn (2002a) explains the evolution of Canadian defense policies in relation to the new focus on the north in our global strategy. Horn, who also taught at RMC, argues that since World War II, Canada has been more concerned in the north with the assertion of her sovereignty against possible US encroachments than with the collective defense of the continent in collaboration with our American ally.

4. CONCLUSIONS

The most significant conclusion that can be drawn from this quick survey is most likely the apparent lack of interest demonstrated by Canadians in their own military geography. I admit that I expected to find much more material when I began this research. I was bewildered to discover that, probably because of her Colonial past, Canada has more or less ignored this important aspect of Canadian studies, relying on others to investigate questions that should normally be a major Canadian concern. The absence of nearby enemies might appear to be a reasonable explanation for this carelessness, but this does not account for the greater interest in the subject displayed by historians than by geographers. There might still be a Canadian military geography, but it would be an orphan discipline, largely subjected to operational needs, and totally deprived of academic guidance, for no one has ever claimed to be a military geographer in this country.

Be that as it may, there appears to be renewed interest worldwide in military matters and, more specifically, in military geography. One might hope that Canada would not miss this opportunity to play a part, but a link must first be established between academic geographers and the military in Canada. At present, mapmaking is the only idea that comes to mind when the word "geographer" is mentioned at DND. There are no trained geographers working as such at DND, as noted previously, and a cadet can go through his program at RMC and become an officer without taking one single course in geography. This is not often the case, but most cadets take only one or two of the few geography courses offered by the Department of Politics and Economics. No doubt the military should develop a wider and more elaborate geographical approach in the future, but Canadian academic geographers should also turn their attention towards a too-long forgotten field of study and do their share to define modern Canadian military geography.

DISCLAIMER

The views expressed in this paper are those of the author and do not necessarily reflect those of either the Government of Canada or the Canadian Department of National Defence.

REFERENCES

Air Force Reserve Officers Training Corps. 1959. Military aspects of world political geography. Maxwell Airbase, AL: Air Force Reserve Officers Training Corps.

Bothwell, R. 1998-99. The Canadian isolationist tradition. International Journal LVI:76-87.

Boycott, A.G. 1936. *The Elements of Imperial Defence. A Study of the Geographical Features, Material Resources, Communications, and Organization of the British Empire.* Aldershot, UK: Gale & Polden, Ltd.

Canadian Joint Intelligence Committee (JIC). 1943. Enemy capabilities in the North Atlantic area. 24 September, 1943.

Carter, A. 1938. It can be done. Canadian Defence Quarterly XVI:54-58.

Chauprade, A. and Thual, F. 1998. *Dictionnaire de géopolitique.* Paris: Ellipses.

Claval, P. 1990. La géographie militaire du Canada: Avant-propos. Les Cahiers de Géographie du Québec 34:279-284.

Collins, J.M. 1998. *Military Geography for Professionals and the Public.* Washington, DC: National Defense University Press.

Conzen, M.P., Rumney, T.A., and Wynn, G. 1993. *A Scholar's Guide to Geographical Writing on the American and Canadian Past.* Chicago, IL: The University of Chicago Press, Geography Research Paper 235.

Cornish, V. 1918. *The Strategic Geography of the Great Powers.* London: George Philips and Son, Ltd.

Currie, A.W. 1945. *Economic Geography of Canada.* Toronto: Macmillan.

Cuthbertson, B. 1977. *Canadian Military Independence in the Age of the Superpowers.* Toronto: Fitzhenry & Whiteside.

Hannah, G. 2002. Geography and history: The six fathom line of Canadian naval policy. In *Forging a Nation, Perspectives on the Canadian Military Experience*, B. Horn, ed., St. Catharines, Ontario: Vanwell, 288-306.

Hellyer, P. 1997. *The Evil Empire: Globalization's Darker Side.* N.P.: Chimo Medias.

Horn, B. 2002a. Gateway to invasion or the curse of geography? The Canadian Arctic and the question of security, 1939-1999. In *Forging a Nation, Perspectives on the Canadian Military Experience*, B. Horn, ed., St. Catharines, Ontario: Vanwell, 307-332.

Horn, B., ed. 2002b. *Forging a Nation, Perspectives on the Canadian Military Experience.* St. Catharines, Ontario: Vanwell.

Kimble, G.H.T. 1949. *Canadian Military Geography.* Ottawa: Department of National Defence, The Directorate of Military Training.

Lee, J.F. 1908/1922/1923. *Imperial Military Geography.* London: William Clowes and Sons.

Macdonnell, A.C. 1911. *The Outlines of Military Geography.* London: Hugh Rees, Ltd.

Maguire, T.M. 1900. *Outlines of Military Geography.* Cambridge: Cambridge University Press.

May, E.S. 1907. *Geography in Relation to War.* London: Hugh Rees, Ltd.

May, E.S. 1909. *An Introduction to Military Geography.* London: Hugh Rees, Ltd.

McDaniel, R. 1954. A military geography of the Great Lakes area with an emphasis on industrial dispersion and defence (unpublished Master's thesis). London, Ontario: University of Western Ontario.

National Atlas of Canada, 5th ed. Ottawa: Energy Mines, and Resources.

Newman, P.C. 1983. *True North. Not Strong and Free*. Toronto: McClelland and Stewart.

Peltier, L.C. and Pearcy, G.E. 1966. *Military Geography*. Princeton, NJ: D. Van Nostrand Company, Inc.

Preston, R.A. 1977. *The Defence of the Undefended Border. Planning for War in North America, 1867-1939*. Montreal and London, Ontario: McGill-Queen's University Press.

Preston, R.A. 1990. Buster Brown was not alone: American plans for the invasion of Canada, 1939-1939. Canadian Defence Quarterly 20:29-37.

Rioux, J.-F. and Hay, R. 1988-99. Canadian foreign policy: from internationalism to isolationism? International Journal LIV:57-75.

Taylor, G. 1947. *Canada*. London: Methuen.

Stacey, C.P. 1940. *The Military Problems of Canada*. Toronto: The Ryerson Press.

Winterbotham, H.S.L. 1926. The Surveys of Canada. The Geographical Journal LXVII:403-421.

Zuehlke, M. and Daniel, C.S. 2001. The Canadian Military Atlas: the Nation's Battlefields from the French Indian Wars to Kosovo. Toronto: Stoddart.

PART II

HISTORICAL VIGNETTES

In peace, soldiers must learn the nature of the land, how steep the mountains are, how the valleys debouch, where the plains lie, and understand the nature of rivers and swamps – then by means of the knowledge and experience gained in one locality, one can easily understand any other.

Niccolo Miachiavelli
Discoursi
1503

The study and intricate knowledge of the terrain is the beginning and the end of tactics. Everything to do with the ground, its shape, contours, texture, and even its color at times of the day, affects everything both you and the enemy do or cannot do. Both adversaries have the terrain of the battlefield in common. Other things being equal, victory goes to the commander who best understands the terrain.

Ralph Ingersoll
The Battle is the Payoff
1943

Chapter 6

A GEOLOGICAL/TOPOGRAPHICAL RECONNAISSANCE OF HANNIBAL'S INVASION ROUTE INTO ITALIA IN 218 BC

William C. Mahaney
Quaternary Surveys

Abstract: During the Second Punic War, in 218 BC, the route Hannibal followed from the Rhône River to the Alps is subject to some controversy. A reconnaissance undertaken in summer 2002 assessed the most likely invasion route as across the Col de Grîmone to the area around Gap in the Durance River Basin, thence east along the Guil River that would take the Carthaginians to the Col de la Traversette, the main alpine pass north of Mt. Viso. Once into the upper Guil Valley, Hannibal encountered snow and firnpack, both formidable physical and psychological barriers. However, on the lee side, a rockfall blocked his passage below the 2800 m contour. Employing the ingenious strategy of ordering trees felled, timber laced around boulders and set alight, it took three days to clear a passage through the rockfall so that his starving army could wind its way along the Po River to the fertile plains below.

Key words: Second Punic War (218-202 BC), Alpine invasion

1. INTRODUCTION

The expanding empires of Carthage and Rome covered most of the Mediterranean world after the First Punic War (264-242 BC). In 218 BC, Carthage and Rome came to loggerheads over economic spheres of influence and political domination of the western Mediterranean. Attempting to avoid the pitfalls of the First Punic War, which led to the defeat of Carthage, Hannibal took on a bold new initiative and planned a land invasion of Italy across the Pyrénées and the Alps (Cavan, 1980). His route (Fig. 1) from New Carthage (now Cartagena) lay across the Col le Perthus in the Pyrénées, along the Mediterranean Coast to the upper Rhône Delta where he encountered his first real opposition from the Gallic Volcae (Ellis, 1990).

D.R. Caldwell et al. (eds.), Studies in Military Geography and Geology, 67–78.
© 2004 *Kluwer Academic Publishers.*

The Volcae had recently ratified treaties with Rome and most probably objected to the ecological damage that would ensue with the passage of the Carthaginians. Living close to Saguntum (Greek city on the Iberian Coast sacked by Hannibal in 219 BC), they may have learned the fate of its citizenry, the survivors sold off into slavery. Following defeat of the Volcae in a flanking action that caught them unawares, Hannibal met with Gallic representatives of the Boii people of northern Italy who advised him he would have their allegiance once his army issued out onto the Po River Plains.

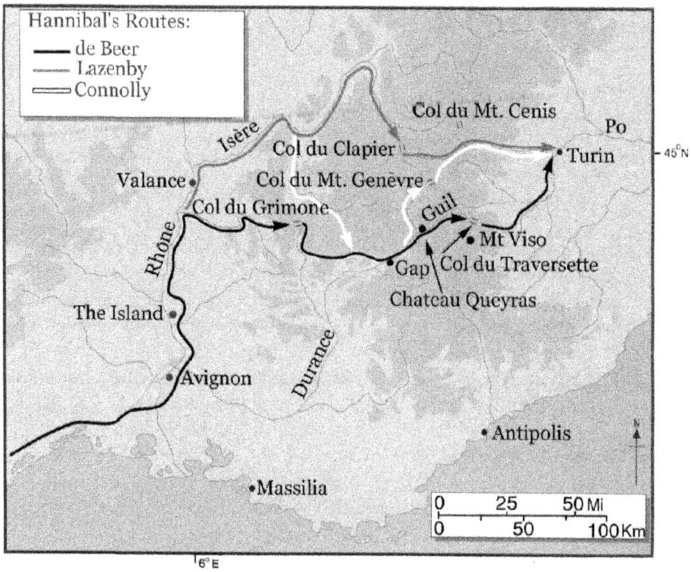

Figure 1. Route followed by Hannibal, north along the Rhône River to the Drôme River, east to the Col de Grîmone, north along the Durance River to the confluence with the Guil River, thence east to the Col de la Traversette, and along the upper Po River into Italy. Map modified from Healy (1994).

Considering his main strategic goal was to invade Italy by land, Hannibal moved his army north along the Rhône towards the "Island," an area fortified with a palisade and inhabited by lowland Gauls. From there, Hannibal followed either the main valley of the Rhône north toward present-day Valance, or the tributary valley of the Drôme east towards present-day Gap (Fig. 1). The northern route favored by Hart (1967) and Lazenby (1978) eventually reaches the Isère, the Little St. Bernard Pass or Mt. Cenis, and continues to Italy close to Turin. Mt. Cenis, also the route considered by Napoleon to be the invasion route, is lower in elevation (1800 m asl), and lacking the snow (firnpack; old snow with a density of 0.4-0.8) and ice documented by Polybius and Livy, is an unlikely choice. An alternate route proposed by Connolly (1981) is a variant of the northern route that winds into Mt. Gen||vre Col from the west. The southern route, east along the Drôme River to the valley of the Durance, is considered the most likely route by a number of historians (Dodge, 1891; de Beer, 1969; Bath, 1981; Cottrell, 1992; Prevas, 1998; Bagnall, 1999; and Baker, 1999). The northern route along the Rhône, north of the Drôme to the Isère, follows a topographically challenging course that makes cavalry operations exceedingly difficult. Moreover, the time allowed for travel along the northern route is insufficient to cover the distance involved, especially since Hannibal would have been lucky to cover 5 km per day in rough country. Considering the southern route would have been the likely one taken by the Carthaginians, a geological reconnaissance of the area from the Rhône to the Po River Plains was carried out in 2002.

2. HISTORICAL SOURCES

What we know of Hannibal (247-183 BC) comes from his Roman enemies, who feared and hated him in the extreme. During the Second Punic War (218-202 BC), hardly a Roman family had not lost a member or members in battle with the Carthaginian invaders. Called the "Father of Strategy," Hannibal, leading a polyglot army formed around his Carthaginian Corps, was a gifted tactician who seemed to outguess his opponents at nearly every juncture.

One of the major sources on Hannibal and Carthage is "The Rise of the Roman Empire" by Polybius (200-118 BC; Scott-Kilvert (trans.), 1979), a former cavalry general, statesman and historian. As a military officer, Polybius had first-hand experience of battle, and as a friend and confidante of Scipio Aemilianus, he was present at Carthage in the Third Punic War (149-146 BC). Polybius drew on primary sources no longer available to us, many having been lost in the great fire in the library of Alexandria in the 3rd century AD. Polybius certainly knew and interviewed survivors of the Hannibalic War (Lazenby, 1978), including men who fought with Hannibal, and he alone followed the invasion route across the Alps into Italy.

A second main source of the Hannibalic War, among classical historians, is Livy (64 BC-12 AD; de Sélincourt (trans.), 1965). Of his 142 books on the history of Rome, only 35 survive in whole including books 21 to 30, which cover the Hannibalic War.

3. FOLLOWING THE ROUTE: DISCUSSION

Following the defeat of the Volcae, Hannibal marched north along the Rhône to an area known as the "Island," inhabited by lowland Gauls as noted previously. The exact location of the "Island" is unknown, but it had to be within eight to ten days march from the upper Rhône Delta, somewhere between Avignon and Valence. A reconnaissance of the lower Rhône Valley during the summer 2002 showed the most likely location to be at or near the confluence of the Drôme and the Rhône rivers south of Valence where even today numerous islands exist, partially controlled by river flood work projects (Fig. 1). When faced with a choice between the northern and southern route, Hannibal must have appreciated the tactical importance of a wide flood plain and low, expansive terraces along the Drôme, nearly flat surfaces that could be easily controlled by his Numidian horsemen.

Hannibal's first major test would come as he approached the Col de Grîmone, the interfluve between the Drôme and Durance valleys (Fig. 1); a tactically important position well garrisoned by unfriendly mountain Gauls, most probably the Allobroges (Lazenby, 1978). The time required for Hannibal to transit the Drôme Valley from the "Island" to the Col was in the range of 10-15 days. Assuming the Col de Grîmone is the area where the mountain Gauls first attempted to stop the Carthaginians, Hannibal made full use of his intelligence by resorting to deception and surprise, two tactical elements he would use over and over again once he came up against Roman legions in the Po Valley.

Relying on reports from scouts (de Sélincourt (trans.), 1965) that the Gauls defending the col abandoned their positions at night, retreating to the safety of a nearby town, Hannibal decided on a night action. Spreading his forces out with orders to light many fires, he hoped to deceive the Gauls into thinking he was bivouacked for the night. With a handpicked force of infantry he took the heights abandoned by the Gauls and waited until first light. When the Gauls appeared climbing up to the col from the east, they were taken by surprise and Hannibal easily defeated them bringing about a general rout. With the col now safely in Carthaginian hands, Hannibal had only to encourage his troops to move his baggage train across into the Durance Valley. The timing of this movement is obscure but it must have been in September of 218 BC, possibly late in the month. The 125 km transit of the Drôme Valley would have taken at least two to three weeks, the last 30 km in the foothills proving as much an obstacle as the Queyras he would encounter in some ten days time.

Once across the col, Hannibal would have found the wide flood plain and terraces of the Durance Basin easy to negotiate. His cavalry could protect his flanks and if the Gauls were massing in strength he would have known about it before they could inflict any damage. Presumably, Hannibal now faced two possibilities: Continuing north to the pass at Mt. Gen||vre (route later taken by Julius Caesar), or turning east into the Guil Valley that would take him to the Col de la Traversette to the north of Mt. Viso (Fig. 1), the latter route crossing the highest pass into Italia.

Polybius (Scott-Kilvert (trans.), 1979) points out that as Hannibal turned into the Guil catchment he encountered two gorges, the first a narrow defile stoutly defended by Gauls intent on stopping him. This first gorge must certainly be the high buttress upon which the Chateau Queyras stands today, a point noted by Cottrell (1992) and Prevas (1998). Built in the early 19th century, this fortress was part of the French defence system erected by Napoleon, but in Hannibal's time it was certainly an imposing geological structure, laced with large outcrops of schist and a source of numerous large boulders that could be easily defended. By every account the Gauls dislodged many large boulders, some of mega tonnage, monoliths that could be rolled in an instant and dropped on soldiers passing along the winding banks of the Guil below (Fig. 2). Sliding boulders on beds of mica, it would be relatively easy to pry large stones loose and at the right moment let loose a barrage of material that in effect would act like a rockfall or rock avalanche tearing loose many smaller rocks on the way down.

Figure 2. Inclined plane of mica schist with unloaded beds along the north flank of the Guil Valley, etched with mass-wasted deposits, mainly debris flows. The presence of debris flows indicates appreciable surface water that would have aided the Gauls in positioning boulders for eventual release. Gentle lower slopes may have been more heavily forested in Hannibal's time.

On the ground along the river at the base of the gorge Hannibal's troops would have been compressed into a narrow line. The Gallic road then, as now, would have been on the south side of the river, crossing over to the north side about 0.5 km upstream. Nevertheless, the cascade of rock falling on the Carthaginians exacted a heavy toll, killing hundreds of troopers by the hour, and no doubt causing considerable pandemonium. According to Polybius (Scott-Kilvert (trans.), 1979), Hannibal seemed content to let the slaughter continue, perhaps thinking that after the initial onslaught of stones, the Gauls would run out of "ammunition" and give up the attack. However, it seems the Gauls had considerable knowledge of local geology and continued dislodging large boulders after sliding them on mica beds, ultimately pitching them over the side. There are large boulders in the gorge still but no present-day evidence of large rockfall debris. In all likelihood, road building and reconstruction on the south side of the gorge have resulted in the removal of most of the debris flung from the heights by bands of Gallic sappers intent on destroying the infidel invaders.

Once through the first gorge, Polybius states that Hannibal faced a second gorge that offered a major obstacle, one that he had to fight his way through (Scott-Kilvert (trans.), 1979). The exact location of the second gorge remains problematic. It might be located at any one of several locations up-valley from the Chateau Queyras through Abries to L'Echalp, even in the flank of the U-shaped valley forming a natural gliding plane along which boulders could be let loose from the ridge above (Fig. 2). It is almost certain that Hannibal would have followed the existing road on the north side of the Guil, and whereas there are several gorges along the river, there are none so steep as at the Chateau Queyras. There is relatively little talus against the valley sides, despite the presence of schist outcrops all the way to the Col de la Traversette, and no identifiable masses of landslide or rockslide material.

The Guil flows on the south side of the valley between the Chateau Queyras and the Col de la Traversette with the road following a wide expanse of bedrock and terraces on the north side, probably little changed from the time Hannibal passed through. There is no sign of the ancient Gallic road that must have wound up valley to the col, but there are signs of present-day hearths (Fig. 3) in the upper valley, situated in forest remnants amid likely camping places, which may in fact have been used temporarily by the Carthaginians. Standing on a prominent bedrock bar at about 2500 m asl in the upper Guil catchment (Fig. 4), it is difficult to imagine what the valley would look like with some 50,000 to 60,000 troops and 37 elephants in it. Even with the army strung out, as it must have been, the sight would have been impressive to Gallic observers perched on the valley sides.

Figure 3. Hearth in upper Guil Valley located in a depression similar to hundreds of similar sites where Hannibal's army may have bivouacked. Excavations of hearths often yield artifacts important in reconstruction of the cultural/ecological use of the land (Mahaney, 1990).

As already mentioned, Polybius relates that Hannibal, once in the upper valley, watched the carnage in the lower valley with a certain amount of detachment. Of the 20,000 Carthaginians lost in the valley, most were ably dispatched in the lower gorge, which would have been a perfectly defensible site for the Gauls. The upper valley sites where the Gauls may have ambushed the Carthaginians are not so defensible. It may have been in the upper valley, with the Gauls raining boulders down on his forces, that Hannibal decided (Scott-Kilvert (trans.), 1979) to send infantry onto the higher slopes in order to drive off the enemy.

Continual harassment by small groups of Gauls probably issuing from forests on the south side of the river presented something of a problem for Hannibal as he could not use his cavalry in thickly wooded areas. Certainly the Gauls knew the lay of the land perfectly, so well in fact that they could fade away into the landscape, singly or in small groups, as documented by Polybius (Scott-Kilvert (trans.), 1979).

According to Polybius (Scott-Kilvert (trans.), 1979), Hannibal rested his army on or near the summit for two days before descending along the Po. Given the rough topography along the crest of the Col de la Traversette, a sharp-crested ridge with scree-strewn slopes, there is no available space to bivouac a squadron of company strength, let alone an army of 25,000 men with horses and elephants. More to the point, available aerial imagery of the area shows landslide debris or rockfall on the French side at 2900 m asl, a deposit that is most certainly of Late Glacial or early Holocene age (12-8 ka)

Figure 4. Upper Guil Valley approximately 6 km from the Col de la Traversette. View to the east with steep, well-forested, north-facing slopes to the right.

judging from weathering characteristics and topographic position (see Mahaney (1991) for weathering data and chronology in the nearby French Alps) below the rock bar that forms the col itself. Hannibal could easily have skirted or cut through the landslide/rockfall, following what must have been a Gallic trail to the col. To the east-southeast of the rockfall (Fig. 5), a rock glacier consisting of several lobes of fresh, boulder-size clasts lacking appreciable lichen cover, is most probably of late Holocene age (Little Ice Age). Had this feature been present when Hannibal traversed the Alps it would have presented an effective barrier to his passage, more effective perhaps than what awaited him on the lee (Italian) side.

Bivouacking troops in large numbers on the col - the tactical situation described by Polybius - is clearly impossible, but resting troops in the lower valley as far as present-day L'Echalp, or lower down, on the French side is more realistic (Fig. 6). The valley of the Guil (Fig. 4), as far as the Chateau Queyras, could probably provide space for the remnants of Hannibal's army, the survivors of Gallic attacks in the lower valley. Numerous hearths in the Queyras, as shown in Fig. 3, were likely camping places in Hannibal's time and may well contain a wealth of artifacts from his passage. But even with two days rest, the entire area would have been left ecologically destroyed by the passage of the Carthaginians, landscape recovery coming with the ingress of vegetation after Hannibal had traversed the entire valley and the Gauls resumed their pastoral occupation of the land.

The present-day trail to the east out of the Col de la Traversette is a switchback (see Fig. 5) along to the north of the Po River, eventually crossing to the south side of the river below the Pian del Re. The exact position of a blocking "landslide" (probably a rockfall) lying below the col is mentioned by Livy (de Sélincourt (trans.), 1965), but no specific distance is given to the timber in the subalpine forest. While the exact location is unknown, it must lie either above the albergo at 2020 m elevation, or below towards the town of Crisollo. In Hannibal's time the trail down from the col must have negotiated the steep slopes above the Pian del Re, which is the most likely source for rockfall material west of the Truc Teston at 2523 m.

Figure 5. Air photo of the Col de la Traversette showing the present-day pass (arrow; shaft points along the upper Po River Valley). The aiguilles de la Traversette, a rock buttress that Hannibal had to march around, is north of the arrow.

If the switchback is indeed the location of the rockfall, Hannibal would have had to forge a path through it, or backtrack up to the col and move his army north about 3 km, descending in the area of present-day Bobbio. Clearly, considering the state of his starving men, the change in temperature with newly fallen snow, and most probably his lack of intelligence regarding the lay of the land to the north, finding a new route was not a viable option. By all accounts the descent from the col was on firnpack now covered with fresh snow, a condition that made walking difficult if not downright treacherous.

According to Livy (de Sélincourt (trans.), 1965), Hannibal had to be summoned from the rear of the column to deal with this new obstacle. His engineers, stumped as to how to remove the boulders and clear a path through to lower elevations, seemed powerless to move the large boulders. To the south, a steep precipice dropped off to the Po and to the north the seeming insurmountable mass of rock rubble extended up the flanks of the mountain blocking their path. Hannibal ordered timber cut and laced around the boulders; the wood was set alight and allowed to burn for a time to

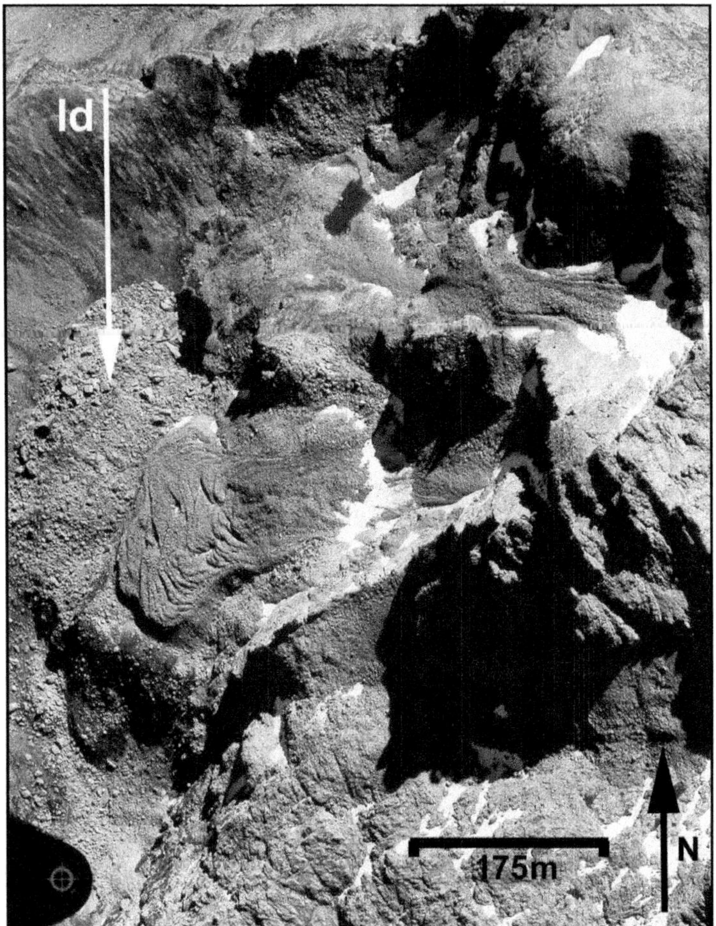

Figure 6. Various lobes of young Neoglacial (Little Ice Age) rock glaciers and protalus ramparts are oriented downvalley just west of the col. Rock glaciers are juxtaposed with an older mass of landslide debris (light gray tone at arrow ld) that is likely of Late Glacial age (~12 ka). A trail is visible through the landside debris (tip of arrow) nearly cutting the landslide in half. The area to the west of the drainage divide is just above the probable location of Hannibal's bivouac area. Map source: Mount Viso, 3637 OT, scale 1:25,000.

insure a high temperature. Sudden dousing with sour wine (water was presumably unavailable) produced the desired effect, splitting and spalling many of the large stones, which allowed their removal. Assuming Livy (de Sélincourt (trans.), 1965) is correct, it would seem the switchback shown on

Fig. 5 (below the arrow) cuts across the rockfall that blocked Hannibal's descent along the Po. If so, there is the problem of where the wood came from to fire the boulders, not to mention the added problem of firing freshly fallen timber, and what means were available to haul wood from lower slopes some 2-3 km away. It is possible the timberline was closer to the switchbacks in Hannibal's time, as logging over the last few centuries and pastoral occupation of the high slopes on the Italian side have probably altered the timberline elevation. In any case, Hannibal must have known that Greek farmers routinely removed large boulders from fields by firing the timber around them, and he used a similar technique to forge a path through the rockfall. To his starving army he must have seemed the leader who could overcome any obstacle. There was no return possible across the Alps, only the certainty of engaging the Consular Army of Publius Cornelius Scipio who was certain to be searching the hills and valleys of Cisalpine Gaul looking for them.

4. CONCLUSIONS

Assessing the topography in the lower Rhône, Hannibal was faced with two choices: Give battle to the Romans who were marching up from Massilia or follow his grand strategy and march up the Rhône to the "Island," where he could resupply his army and march into the Alps. Following his strategic plan, he marched along the Drôme to the Durance Valley, thence north to the Guil where he encountered fierce opposition from the mountain Gauls.

The battle of the first gorge in the Guil catchment was fought near present-day Chateau Queyras in the lower drainage near its confluence with the Durance. Most probably Hannibal's army was strung out down the Guil and along the Durance when the Gauls set upon them, rolling large boulders of schist down upon them. According to both Polybius and Livy, Hannibal at this time was with his rearguard, his elephants and cavalry up front. Even though his force was divided for a time, he managed to drive off the Gauls re-establishing a unified command.

Once through the lower gorge, Hannibal would have made almost unimpeded progress through the Guil except for local harassment from the Gauls. The location of the second gorge mentioned in the ancient literature is still a mystery; there are several possible locations for it, but the most likely sites are hardly the equivalent of the lower gorge.

The Col de la Traversette is clearly impossible as a bivouacking area for the army given the limited amount of camping space. On the one hand, it is not large enough to accommodate between 25,000 and 30,000 troops, assuming this is the number that managed to fight its way through the Guil catchment; for another, it is windswept and in the fall months would provide little to no protection for troops lightly clad, with only blankets to sleep in. However, the area below the col, covered by forest, offered protection against the elements. Most importantly, numerous hearths in the lower valley have probably been used as camping spots for millennia and offer the prospect of recovering an artifact record.

Hannibal's method of dealing with a major mass wasted obstacle on the lee side of the range shows he anticipated Blackwelder (1927) and realized the value of fire as an agent to provide the necessary stress to break rock. His course of action, a human-induced form of physical weathering, certainly shows innovative genius and a talent for dealing effectively with the unexpected, something he was to do over and over again in the coming years.

ACKNOWLEDGMENTS

I thank Quaternary Surveys (Toronto) for funding to complete this project in 2002 and 2003. I am indebted to John Dawson for preparing the figures and Hazel O'Loughlin-Vidal for formatting the manuscript.

REFERENCES

Bagnall, N. 1999. *The Punic Wars.* London: Pimlico.
Baker, G.P. 1999. *Hannibal.* New York: Cooper Square Press.
Bath, T. 1981. *Hannibal's Campaigns.* New York: Barnes and Noble.
de Beer, G. 1969. *Hannibal: Challenging Rome's Supremacy.* New York: The Viking Press.
Blackwelder, E. 1927. Fire as an agent in rock weathering. Journal of Geology 35:134-140.
Caven, B. 1980. *The Punic Wars.* New York: Barnes and Noble.
Connolly, P. 1981. *Greece and Rome at War.* New York: MacDonald and Co.
Cottrell, L. 1992. *Hannibal enemy of Rome.* New York: Da Capo Press.
Dodge, T.A. 1891. *Hannibal.* Boston, MA: Houghton-Mifflin.
Ellis, P.B. 1990. *The Celtic Empire.* Durham, NC: Carolina Academic Press.
Hart, B.H.L. 1967. *Strategy.* London: Faber and Faber, Ltd.
Healy, M. 1994. Cannae 216 BC. In *Campaign Series*, D.G. Chandler, ed., Oxford: Osprey Military 36.
Lazenby, J.F. 1978. *Hannibal's War.* Norman, OK: University of Oklahoma Press.
Mahaney, W.C. 1990. *Ice on the Equator*, Ellison Bay, WI: Wm. Caxton, Ltd.
Mahaney, W.C. 1991. Later Pleistocene and Holocene glacial chronology at Chamonix and Argentière, French Alps. Zeitschrift für Geomorphologie 35:225-237.
Prevas, J. 1998. *Hannibal Crosses the Alps.* Rockville Centre, NY: Sarpedon.
Scott-Kilvert, I. (translator of Polybius). 1979. *The Rise of the Roman Empire*, London: Penguin.
de Sélincourt, A. (translator of Livy). 1965. *The War with Hannibal.* London: Penguin.

Chapter 7

MILITARY CAMPAIGNS IN TROPICAL KARST

The Maroon Wars of Jamaica

Michael J. Day
University of Wisconsin-Milwaukee

Abstract: Topography, restricted surface water supply, and cave use for refuge and ambush afford strategic offensive and defensive advantages to native combatants familiar with the surface and underground terrain, and pose tactical problems for unfamiliar foreign forces. The Maroon Wars of 1690-1796 pitted numerically superior British forces against organized bands of escapee slaves and others in the classic tropical karst of Jamaica's Cockpit Country. In this complex landscape of rugged conical hills surrounding deep sinkholes, British efforts to engage and subdue their antagonists were largely unsuccessful, while they suffered mounting casualties from guerrilla attacks. Ultimately, the British adopted a strategy of attrition and containment, selected deforestation, and bombardment, forcing Maroon withdrawal. The British established a road cordon around the Cockpit Country and limited Maroon access to water supplies. In 1796, weakened by measles, the Maroons agreed to duplicitous peace terms. Vestiges of the conflicts remain, particularly in Cockpit Country place names.

Key words: tropical karst landscape, Cockpit Country, Jamaica, Maroon War, guerrilla tactics

1. TROPICAL KARST LANDSCAPES

Karst landscapes in the humid tropics are among the most spectacular anywhere in the world because high temperatures favor chemical erosion, high humidities and rainfall provide abundant water for dissolution, and the environmental conditions encourage high levels of biological activity (Day, 2000). Extensive areas of humid tropical karst occur in southern Mexico and Central America, in the Caribbean, in Southeast Asia and southern China,

D.R. Caldwell et al. (eds.), Studies in Military Geography and Geology, 79–88.
© 2004 *Kluwer Academic Publishers.*

with other significant expanses in South America, Madagascar, New Guinea and northern Australia.

The most diagnostic elements of tropical karst terrain are large more-or-less enclosed and polygonal depressions (sinkholes) and systems of dry valleys. Bordering these negative topographic features are sinuous ridges and interconnected or isolated residual hills, generally known as cones or towers. Relative relief may exceed 100 m, slopes may be near vertical, surface collapses may occur, and the ground surface is often unstable or treacherously pitted, with jagged indentations and protrusions (karren). Surface drainage is uncommon, with peripheral sinking streams and springs but, perversely, flash floods may occur during intense rainstorms. Vegetation is variable in composition and density, but is generally moist to dry tropical forest, with its attendant floral and faunal characteristics. Large and extensive cave systems are integral to many tropical karst areas, which are generally highly cavernous.

2. MILITARY CAMPAIGNS IN TROPICAL KARST LANDSCAPES

Tropical karstlands pose unusual military problems (see Ehlen and Abrahart, 2004, this volume for military operations in another type of karst terrain). The irregular topography, restricted surface water supply, and the suitability of caves for refuge and ambush, afford strategic offensive and defensive advantages to native combatants familiar with the surface and underground terrain, and pose tactical problems for unfamiliar foreign forces (Day and Kueny, 2004). Caves and karstlands are especially well suited to guerilla warfare by small, mobile local units, and partisans have hidden and lived in caves since time immemorial (Kempe, 1988).

Tropical karstlands, particularly those in the Caribbean, Central America, and Southeast Asia, have hosted several notable military campaigns in which outnumbered native forces have harassed, repelled, and otherwise resisted better-equipped and numerically superior invading troops. During the Philippine-American War, for example, karstlands were strongholds of native partisans, and they continue to harbor pockets of internal resistance. Tropical karst terrain has also featured in international and internal conflicts in Vietnam, Indonesia, China, Cuba, and Guatemala.

Terrain plays a significant role in military engagements (Winters et al., 1998; Doyle and Bennett, 2002) but:

> Despite the widespread recognition of the importance of terrain within military action, it has rarely been used as an historical tool to help deconstruct events, actions and outcomes of military engagements, yet

clearly its potential to impact on our understanding of such actions is considerable (Doyle and Bennett, 2002, 1).

3. THE COCKPIT COUNTRY

Jamaica's Cockpit Country is the spectacular humid tropical "type example" of what Ford and Williams (1989) describe as the "egg-box" style of polygonal karst terrain. Centered on Trelawny Parish in northwestern Jamaica (Fig. 1), the Cockpit Country covers about 600 km^2, and is developed largely in Eocene carbonates of the White Limestone Fm., although some of the southern area is developed on older rocks of the Yellow Limestone group (Sweeting, 1958; Versey, 1972; Day, 2004).

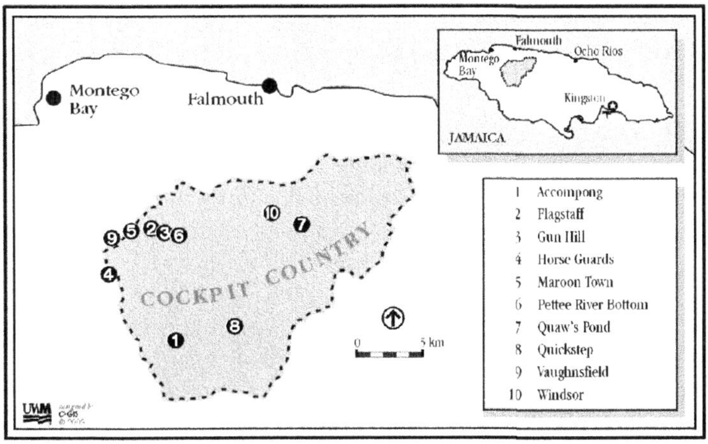

Figure 1. Location of the Cockpit Country karst in Jamaica.

The cockpits, which dominate the landscape, are steep-sided, more-or-less enclosed lobate depressions, some over 100 m deep and 1 km in diameter (Fig. 2). Formed by dissolution of the bedrock and surrounded by residual hills or ridges, they are so named because they resemble the arenas formerly used for cock fighting. The residual hills and ridges are notched by elevated saddles, and many cockpits are connected to one or more of their neighbors by a lower corridor. Some cockpits are elongated, reflecting structural influences or inheritance from abandoned surface drainage courses. Cockpit slopes and the surrounding hilltops and ridges are highly irregular, and slopes consist of combinations of vertical cliffs, inclined bedrock surfaces, and "staircases" or talus accumulations (Figs. 3 and 4). By contrast, the depression bases often have a deep regolith cover, and some contain relict, debris-choked vertical shafts.

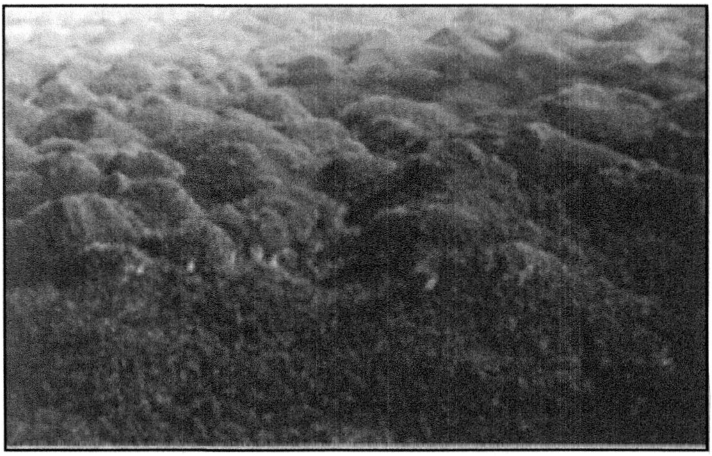

Figure 2. Cockpit Country terrain from the air.

Cockpit drainage is centripetal, although dominantly vertical (Day, 1979). Regional drainage is dominantly internal, underground, and generally northward, although there are some external inputs on the southern boundary. Near the edge of the Cockpit Country there is ephemeral surface drainage, associated with temporary elevation of the water table, but the interior has no perennial surface water sources, with the exception of Quaw's Pond, east of Windsor and Pettee River Bottom, east of Maroon Town (Fig. 1). On the northern periphery, underground drainage emerges at a series of springs, which feed rivers draining to the north coast (Fig. 5).

Cave systems associated with the Cockpit Country include some active river caves flanking the karst to the south and some large abandoned conduit systems around the periphery (Fig. 6). Large subterranean river passages, flowing approximately down dip presumably exist at depth beneath the surface, but most known caves within the Cockpit Country proper are either debris-choked pits within cockpits or fragmentary dry passages within the residual hills and ridges (Fincham, 1997).

Cockpit Country vegetation is tropical wet-dry limestone forest, with considerable species diversity (Proctor, 1986). Moister, denser forest occurs in cockpit and valley bottoms, where soil cover is greatest, with increasingly sparse and xeric vegetation on talus slopes and hill summits. Human impacts on the forest have been limited by inaccessibility and the lack of surface water.

Figure 3. Cliffed cockpit slope.

Figure 4. Residual hill summit.

Figure 5. A peripheral karst spring.

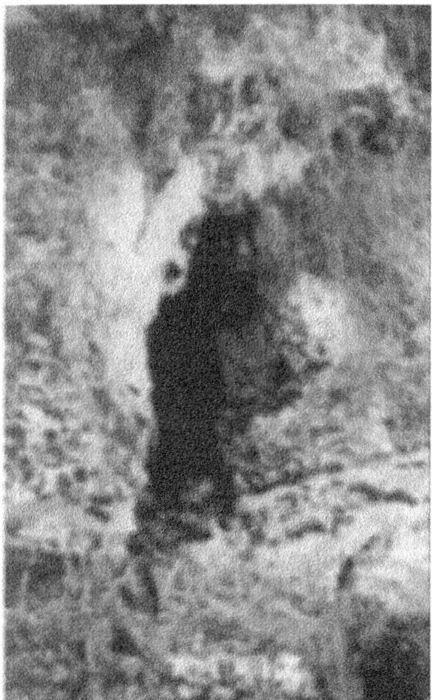

Figure 6. Cave entrance at Windsor.

4. THE MAROON WARS

The Maroon Wars in Jamaica encompassed a protracted series of variously energetic military engagements that lasted for over a century between 1690 and 1796. The protagonists were, on the one side, British troops, Colonial militia, and auxiliaries, and, on the other, free Africans and escaped slaves. Hostilities were not continuous, nor were most confrontations prolonged. The Maroons adopted hit-and-run guerrilla tactics, and the British, following the abject failure of "search and destroy" missions, fell back to secure peripheral bases (Dallas, 1803; Bridges, 1828). Although skirmishes occurred throughout the country, the principal conflicts were centered in and around the Cockpit Country, which became a principal Maroon stronghold.

The early Maroons were a diverse people whose name is of obscure origin, but whose territorial presence beyond the limits of Colonial control pre-dated the British occupation of Jamaica in 1655 (Eyre, 1980; Campbell, 1988). They encompassed free Africans and escaped slaves, with some native Caribbean component, and their descendants.

As the plantation system expanded in the 1700s, Maroon territory contracted into the most inaccessible interior forests. Skirmishes occurred prior to 1690, but a slave rebellion in that year swelled the Maroon ranks and marked the effective start of serious hostilities, which escalated into a period of sustained military action between 1730 and 1739 - the First Maroon War (Dallas, 1803; Robinson, 1969; Eyre, 1980).

The Maroon Wars were the only significant British colonial conflict conducted in humid tropical forests before World War II, and the inexperienced troops' adversaries were the terrain and the climate as much as the Maroons themselves (Eyre, 1980). Local temperatures hovered around 30°C, with humidities around 76% and annual rainfall in excess of 2000 mm; insects and disease were antagonistic and debilitating. Moreover, they found the unaccustomed terrain confusing and difficult to traverse, becoming disoriented and injured, without adequate water supplies. By contrast, the Maroons made tactical advantage of their intimacy with the terrain, utilizing refuges in the least accessible reaches, but maintaining access to the few water sources. They selected ambush locations in which the British were confined to single file, particularly within narrow rocky corridors, and from which the Maroons themselves could make speedy egress into the maze of cockpits, hills, and caves.

During the First Maroon War, Cudjoe and the main body of the Maroons established their Cockpit Country base in Pettee River Bottom, a particularly defensible although rather atypical karst depression near the western margin of the Cockpit Country and not more than 4 km distant, as the crow flies, from the British military base at Vaughansfield. The term "bottom" in the Cockpit Country generally denotes a glade - an elongated although enclosed karst depression with at least a partially alluviated floor and with seasonal, if not perennial, surface water - and such is Pettee River Bottom (Fig. 7). About 1.5 km in length, but for much of that less than 100 m in width, the Bottom has a perennial water source, is flanked by rugged ridges and hills, which provide convenient lookout points, and is accessible only via narrow corridors at the southern and northwestern ends. The Maroons used these narrow corridors to ambush and halt advancing British troops, inflicting casualties as much by the toppling of loose rocks (talus) as by use of firearms or machetes (Robinson, 1969; Eyre, 1980).

Figure 7. Part of Pettee River Bottom

The British, under the command of Colonel John Guthrie, established barracks around Flagstaff, little more than 2 km from the northwestern entrance to the Bottom, near what was then known as Cudjoe's Town, a scattered settlement which the Maroons abandoned for their redoubt during British incursions. Infantry efforts to penetrate the Maroons' stronghold were repeatedly repelled, with mounting losses (Eyre, 1980). Thus thwarted, the British negotiated an uneasy truce, and the parties signed a treaty on March 1, 1739.

The cessation of hostilities lasted over 50 years, with intermittent interruptions, until 1795, when the year-long Second Maroon War broke out. This time the British deployed both infantry and cavalry forces, numbering about 3000, against perhaps 400 Maroons, again ensconced in their Cockpit Country bastion. The initial results were the same - Pettee River Bottom proved impregnable, and the British incurred mounting losses, including both the infantry and cavalry commanders, Colonels Fitch and Sandford.

Finally admitting that conventional confrontation and pursuit were ineffectual in the Cockpit Country forest, the new British commander, Major-General Walpole, embarked upon a new strategy. Slaves gradually cleared the forests surrounding the approaches to the Maroon fastness, but infantry incursions continued to fail, and an almost incomprehensible cavalry assault ended in a debacle. Nonetheless, the defoliation was not without impact, in that Walpole, seeing the terrain exposed, realized that it would be possible to launch an artillery bombardment upon the Pettee River Bottom from a hilltop not far from the British base at Flagstaff. Accordingly, what Dallas (1803) termed a "howitzer" was dragged to the summit of what is still known as Gun Hill, whence the Maroons were steadily shelled until they reluctantly withdrew from their redoubt deeper into the interior of the Cockpit Country.

Having thus been displaced from their primary water supply, the Maroons became increasingly dependent on the intermittent rainfall and upon the few ephemeral water sources in the Cockpit Country interior with which they were less familiar. The British further weakened the Maroon situation by encircling the Cockpit Country with a road network, stationing troops at strategic points and restricting access to the peripheral springs. A campaign of attrition set in, with the British entrenched around the periphery and the Maroons, although compromised by limited water and food supplies and further decimated by measles contracted from the British, continuing to harass them.

Having achieved this stalemate, and apparently fearing that the remaining Maroons might take advantage of the impending dry season to set ruinous forest fires, the British negotiated the duplicitous Pond River Treaty of 1796. Failing to honor the treaty terms, the British seized Maroon lands and transported Maroon leaders and their adherents to Nova Scotia, thence to Sierra Leone (Black, 1965; Robinson, 1969). Thus, having achieved "no victory of importance" (Dallas, 1803, 130) the British attained "...by guile what they could not achieve by force of arms" (Eyre, 1980, 87).

5. POSTSCRIPT

One lasting geographic legacy of the Maroon Wars is that of Cockpit Country place names. On the British side, the names *Horse Guards*, *Flagstaff*, and *Gun Hill* remain, as do vestiges of the barracks and much of the encircling road network, although parts of it are in disrepair. The atmosphere of apprehension is reflected in the place names *Quick Step, Me No Sen You No Come, Don't Come Back,* and the District of *Look Behind.* By comparison, *Maroon Town* (once *Trelawny Town*) is a significant settlement (population 3500) on the western edge of the Cockpit Country, and the settlement of *Accompong*, still very much a Maroon community, is named after Cudjoe's brother, also a Maroon military leader.

ACKNOWLEDGEMENTS

This paper draws extensively upon L. Alan Eyre's 1980 account in the Jamaica Historical Journal, and I acknowledge my considerable debt to that source, which deserves to be more widely recognized in the geographic literature. I am also grateful to Sean Chenoweth and the University of Wisconsin-Milwaukee Cartography and GIS Center for their assistance with Fig. 1, which is based on an original created by and used with the permission of the Windsor Research Station, Sherwood Content, Jamaica. Figs. 3 and 4 are reproduced by courtesy of Conrad Aub-Robinson.

REFERENCES

Black, C.V. 1965. *The Story of Jamaica from Prehistory to the Present.* London: Collins Publishers, Ltd.

Bridges, G.W. 1828. *The Annals of Jamaica.* London: Murray Publishers, Ltd. (reprinted 1968, London: Cass, Ltd.).

Campbell, M.C. 1988. *The Maroons of Jamaica, 1655-1796: A History of Resistance, Collaboration and Betrayal.* Granby, MA: Bergin and Garvey.

Dallas, R.C. 1803. *The History of the Maroons.* London: Longman and Rees (reprinted 1968, London: Cass, Ltd.).

Day, M.J. 1979. The hydrology of polygonal karst depressions in northern Jamaica. Zeitschrift für Geomorphologie 32:25-43.

Day, M.J. 2000. Tropical Karst. In *The Oxford Companion to the Earth*, P.L. Hancock and B.J. Skinner, eds., Oxford: Oxford University Press, 1057-1058.

Day, M.J. 2004. Cockpit Country cone karst, Jamaica. In *The Encyclopedia of Caves and Karst Science*, J. Gunn, ed., New York: Taylor and Francis, 233-235.

Day, M.J. and Kueny, J.A. 2004. Military uses of caves. In *The Encyclopedia of Caves and Karst Science*, J. Gunn, ed., New York: Taylor and Francis, 513-515.

Doyle, P. and Bennett, M.R. 2002. Terrain in military history: An introduction. In *Fields of Battle - Terrain in Military History*, P. Doyle and M.R. Bennett, eds., Dordrecht, The Netherlands: Kluwer Academic Publishers, 1-7.

Ehlen, J. and Abrahart, R.J. 2004. Terrain and its affect on the use of artillery in the American Civil War - The Battle of Perryville 8 October 1862. In *Studies in Military Geography and Geology*, D.R. Caldwell, J. Ehlen, and R.S. Harmon, eds., Dordrecht, The Netherlands: Kluwer Academic Publishers, 155-172.

Eyre, L.A. 1980. The Maroon Wars in Jamaica: A geographical appraisal. Jamaica Historical Review 12:80-102.

Fincham, A.G. 1997. *Jamaica Underground: The Caves, Sinkholes and Underground Rivers of the Island.* Kingston, Jamaica: The Press, University of the West Indies.

Ford, D.C. and Williams, P.W. 1989. *Karst Geomorphology and Hydrology*. London: Unwin Hyman, Ltd.

Kempe, D.R.C. 1988. *Living Underground: A History of Cave and Cliff Dwelling*. London: Herbert Press.

Proctor, G. 1986. Cockpit Country and its vegetation. In *The Forests of Jamaica.* D.A. Thompson, P.K. Bretting, and M. Humphreys, eds., Kingston, Jamaica: Jamaican Society of Scientists and Technologists, 43-48.

Robinson, C. 1969. *The Fighting Maroons of Jamaica.* London: Collins Publishers, Ltd.

Sweeting, M.M. 1958. The karstlands of Jamaica. Geographic Journal 124:184-199.

Versey, H.R. 1972. Karst in Jamaica. In *Karst: Important Karst Regions of the Northern Hemisphere,* M. Herak and V.T. Stringfield, eds., Amsterdam, The Netherlands: Elsevier, 445-466.

Winters, H.A., Galloway, G.E., Jr., Reynolds, W.J., and Rhyne, D.W. 1998. *Battling the Elements: Weather and Terrain in the Conduct of War.* Baltimore, MD: Johns Hopkins University Press.

Chapter 8

A MILITARY GEOGRAPHY OF THE HUDSON HIGHLANDS

Focal Point in the American War of Independence

Eugene J. Palka
US Military Academy

Abstract: During the American War of Independence (1775-83), the Hudson Highlands were a focus of military activity as both the Continental and British forces struggled to gain control of the Hudson River. Continental and British commanders alike recognized the strategic importance of the waterway as a major thoroughfare into the interior of the Colonies, as a vital link between New England and the Middle-Atlantic, and as the major connector between New York harbor and Canada. West Point was regarded as the most decisive locale throughout the entire Hudson Valley because of its commanding position along the banks of the river. This paper is an example of historical military geography that focuses on understanding how the physical geography of the region influenced military decisions and activities.

Key words: Hudson Highlands, American War of Independence, historical military geography

1. INTRODUCTION

The Hudson River served as a vital artery during the growth and expansion of Colonial America. The military significance of the river was apparent throughout the era, as it provided the only non-oceanic avenue of approach for the French into the Colonies, or the British into Canada during the Colonial Wars (Rutledge, 1956; Kagan, 1966; Miller et al, 1988; Crackel, 2002). Consequently, the Hudson Valley had been at the center of hostilities throughout a century of Colonial conflicts, known by their American names as King William's War (1689-97); Queen Anne's War (1702-13); King George's War (1744-48); and the French and Indian War (1754-63) (Crackel, 2002). The river and the valley through which it flows

D.R. Caldwell et al. (eds.), Studies in Military Geography and Geology, 89–103.
© 2004 *Kluwer Academic Publishers.*

were also decisive terrain features during the American War of Independence. At the outset of the war, control of the river was almost a foregone conclusion, if not an integral part of both the Colonial and British military strategies.

This paper examines the Highlands section of the Hudson Valley from a military geographic perspective within the context of the American War of Independence. I begin with a brief historical geography of the Colonies prior to the outbreak of hostilities. Then, I describe the strategic setting from both American and British viewpoints. I subsequently explain aspects of the military geographic analysis that underscored the nature and location of the Continental Army's defensive positions throughout the Hudson Highlands.

2. HISTORICAL CONTEXT

The Treaty of Paris, ending the Seven Years' War, was signed on 10 February 1763. Ironically, even while British peoples on both sides of the Atlantic celebrated the triumph over France, their relationship began to deteriorate rapidly as interests diverged (Meinig, 1986). In the aftermath of the war, the Colonists became agitated by the presence of British troops and a succession of acts (such as the Revenue Act of 1764, the Stamp Act of 1765, the Tea Act of 1773, and the Quartering Act of 1774) passed by Parliament. Rebellious activities ensued throughout the Colonies, but the hotbed of dissention appeared to be concentrated in New England.

Prior to the American War of Independence, the Thirteen Colonies stretched nearly 1200 mi along the Atlantic coast, with approximately 2.5 million people (Lemon, 2001) settled in a long sweeping arc from Boston in the north to Savannah in the south (Palka and Galgano, 2001; Fig. 1). One source estimated the population (excluding Native Americans) to be somewhere between 2,325,000 and 2,600,000 in 1775, with the Mason-Dixon Line (the Pennsylvania-Maryland border) dividing the population almost equally between North and South (Meinig, 1986). Residents were clustered in New England towns, or lived on dispersed farmsteads, plantations, or in small villages throughout the Middle and Southern Colonies. All of the prominent cities were on the coast or along navigable waterways (Brown, 1948); but only four cities (Boston, New York, Philadelphia, and Charleston) had populations of over 10,000 people (Palmer and Tripp, 1977). The Hudson River and Chesapeake Bay, with its main tributary, the Susquehanna River, formed natural territorial divisions, separating the inhabited portions of the Colonies into three regions. The regions differed in terms of climate, soil, and natural resources.

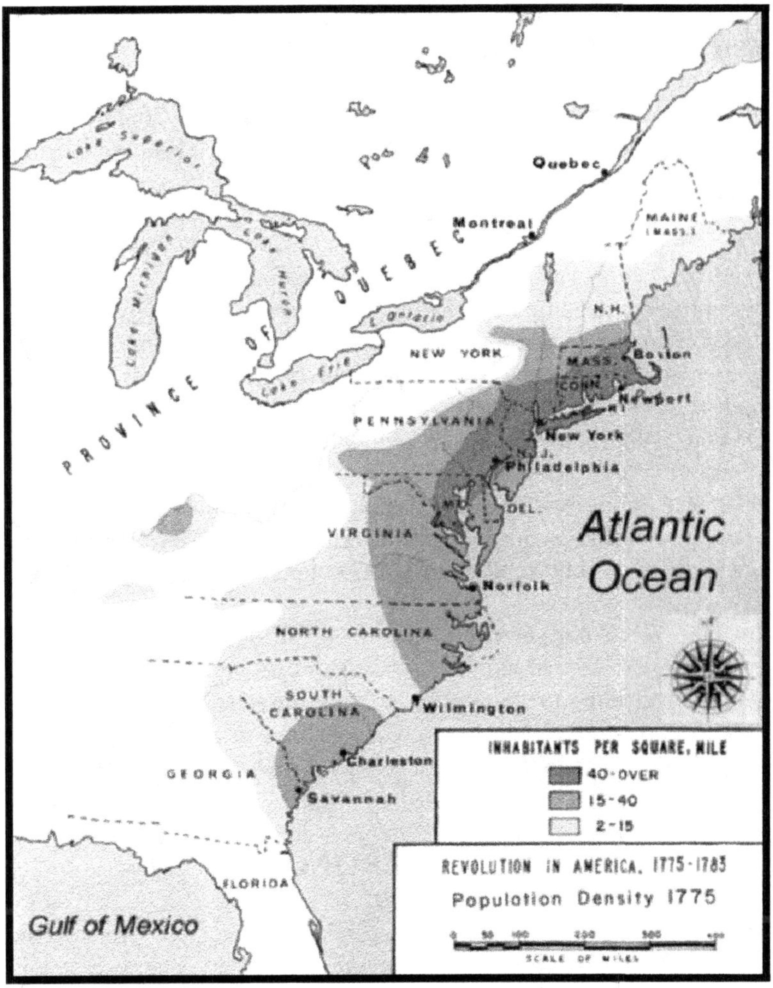

Figure 1. Population density in Colonial America. Adapted from USMA (1998).

As populations increased, regional settlement patterns, and unique economic and political systems provided early evidence of distinct cultural geographic regions. Eventually three culture hearths emerged - New England, Middle Atlantic, and Tidewater-Virginia. The principal water routes (the Connecticut, Hudson, Delaware, Potomac, and James Rivers) that penetrated deeply into the interior shaped these regions of Colonial America by influencing settlement patterns, facilitating commercial activities, and providing the mediums for spatial interaction.

Regional exchange patterns emerged by the mid-18th century, although subsistence farming was widespread throughout the Colonies. The Middle and Southern Colonies were the leading producers of corn, wheat, rice, and tobacco. Small mills and factories in New England provided the Continental Army with clothing and military equipment and through its ports passed munitions and armaments from France, destined for the Continental Army west of the Hudson. New England received substantial food supplies from the more productive agricultural lands in the Middle Colonies, yet exported lumber and fish (Palmer, 1969; McIlwraith and Muller, 2001).

The Hudson River was one of the critical links to inter-regional commerce, since it was navigable for large sailing vessels from New York harbor upriver to Albany. Long before the American War of Independence it had been a frequent invasion route between Canada and the Colonies for both French and British expeditions. Use of the waterway had opened settlement north of the Hudson Highlands, and by 1775, New York's population within the Hudson River Valley was estimated at 185,000 (Thompson, 1966). The leading edge of settlement included a narrow corridor penetrating west into the fertile Mohawk and Cherry Valleys. The powerful and hostile Iroquois Nation had been steadily forced westward by advancing settlers, but conflict between colonial frontiersmen and Native Americans was continual.

On the eve of the American War of Independence, the Hudson Valley was also home to a large number of British Loyalists. In absolute numbers, New York had three to four times as many Loyalists as the next ranking Colony (Meinig, 1986). Most were concentrated in and around New York City and extended north through the Hudson Valley to Albany and Troy.

3. THE STRATEGIC SETTING

Prior to the American War of Independence, navigable rivers served as the principal transportation networks on the North American continent (Brown, 1948). A few bridges spanned only the narrowest streams near the larger settlements. Ferries or fords were used exclusively to cross major rivers. Spatial interaction entailed slow overland movement, sailing from port to port along the eastern seaboard, or tedious movement inland via major river systems. In the latter case, break-in-bulk points, smaller boats, and occasional portages were involved.

The Hudson River was the main inland thoroughfare within the Colonies (Fig. 2). The river linked New York Harbor and Canada either through Lake Champlain, or via the Mohawk River, Lake Oneida, Oswego River, and Lake Ontario route. Ferry-crossing sites along the Hudson also linked New

Figure 2. The strategic setting. Adapted from Dunwell (1991).

England with the Middle Atlantic colonies. The main crossing points between New York City and Albany were near Newburgh and at Verplanck's Point. King's Ferry (between Verplanck's Point and Stony Point) linked a trunk road from Massachusetts and Connecticut to one extending southwest into New Jersey and Pennsylvania (Stowe, 1955; Adams, 1996a). The Albany Post Road paralleled the Hudson River on the east, passing through Peekskill, Continental Village, and Fishkill. As such, the river was vital to both north-south and east-west travel and communications (Fig. 3).

At the outset the Revolution, American and British commanders alike recognized the strategic importance of the Hudson River as a major thoroughfare into the interior of the Colonies and as a vital link between New England and the Middle Atlantic. The Hudson River-Lake Champlain line was unquestionably the most important strategic objective of the war from the British perspective, and Sir William Howe proposed to make it the main objective of his campaign in 1776 (Coakley and Conn, 1992). Meanwhile, a concerted effort was made by the Continental Army to construct and fortify positions along the lower Hudson in order to protect crossing sites, ensure the continued flow of logistics and commerce, and prevent the British from using the river as a major thoroughfare to transport troops and supplies (Palka, 2001).

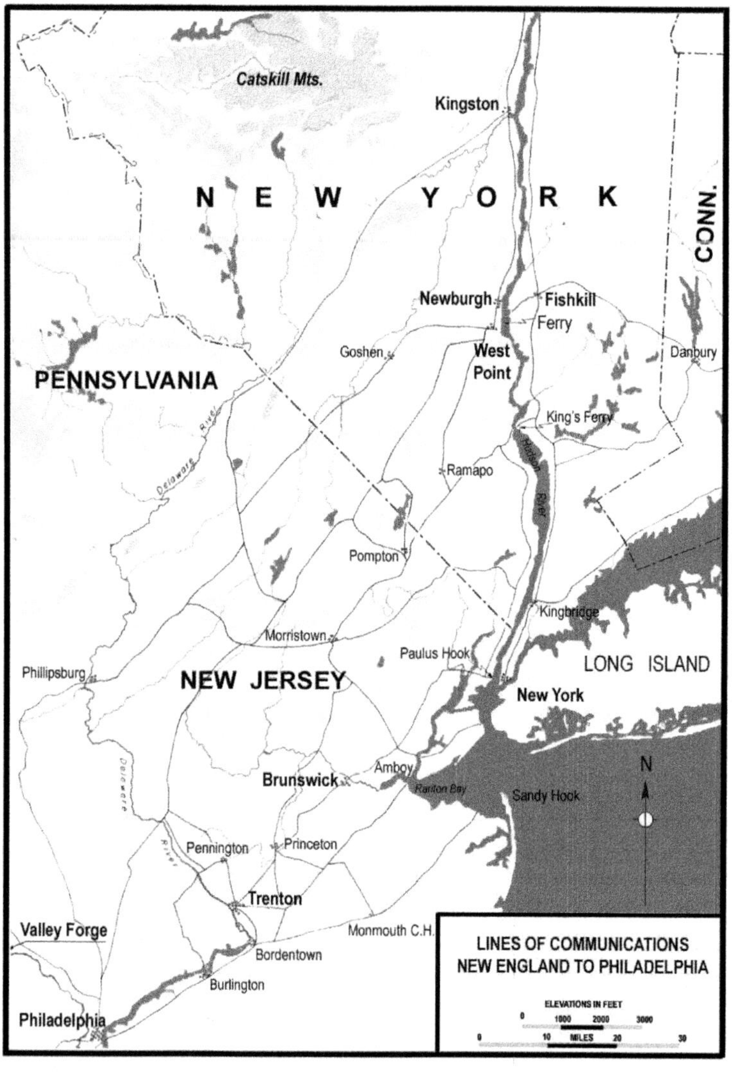

Figure 3. Lines of communication. Adapted from Palmer (1991).

3.1 The British Strategy

The British sought to control the Hudson River for both strategic and practical reasons (Coakley and Conn, 1992; Palka, 2001; Crackel, 2002). First, by seizing control of the Hudson, the British could literally divide the Colonies in half and isolate the rebellion in New England. Second, the Hudson was the most efficient link necessary to reinforce their Indian allies in upstate New York. The Saint Lawrence Seaway provided an alternative route; however, the latter required several portages and was rendered unusable by ice during winter. Third, by controlling the Hudson, the British could deny agricultural supplies from the interior of New York, commercial trading between the New England and Middle Atlantic Colonies, and logistics reinforcements to the Continental Army (Fig. 2).

3.2 The Colonial Strategy

In one respect, the Colonial effort to control the Hudson was intended to deny British use of the river to accomplish tactical and strategic objectives. The river was a vital transportation corridor that constituted "decisive terrain" for both the British and the Continental forces. George Washington and the Continental Congress were specifically concerned with the following. First, the Hudson River was the crucial link for providing logistical support to the Continental Army in the field. Second, it was necessary to maintain the flow of commerce throughout the Colonies, where patterns of regional complementarity were established. Third, the Continental Army wanted to cut the supply lines between the British and their Indian allies in the interior. Fourth, it was important for the Continental Army to maintain the flexibility to maneuver. Maneuverability was essential for either massing or economizing forces in response to British attacks throughout New England or the Middle Atlantic, while avoiding the prospect of being cut-off or divided from other American units. Finally, Colonial control of the Hudson required the British to garrison troops in Canada to deter a potential Continental invasion (Miller et al., 1988).

4. A MILITARY GEOGRAPHIC ANALYSIS

If the Hudson River was strategically important, so was the dominant high ground that facilitated control of the river at key locations along its course. From this perspective, the Hudson Highlands constituted decisive terrain relative to the river. The Highlands are part of a larger geologic province that extends from southeast Pennsylvania northeastward into New

England. The region has been forged by a variety of geologic events including: Volcanism, metamorphism and granitic intrusion deep within the earth's crust, folding and faulting, erosion by the Hudson River, and glaciation (LaMoe and Mills, 1988). The Highlands are comprised of granitic and gneissic complexes that were significantly modified during the Pleistocene epoch. Indicators of Pleistocene glaciation include: Smoothed, striated, and quarried uplands; asymmetrically eroded hills; and over-steepened fiord-like walls along the Hudson River valley (LaMoe and Mills, 1988). Chatter marks, rock polish, kettle ponds, and kame terraces provide further evidence of both glacial erosion and deposition.

The Hudson River bisects the Highlands and extends more than 300 mi from its source in the Adirondacks into the Atlantic Ocean at New York City, dropping only 6 ft in elevation over the course of its last 150 mi (Wilstach, 1969). The navigable portion of the river is actually a fiord with bedrock along its entire length and more than 765 ft of sediment accumulated above the bedrock in the river channel (LaMoe and Mills, 1988). The Hudson attains a maximum depth of 216 ft in the vicinity of West Point, only about 50 mi from its mouth (Wilstach, 1969). The river is not especially wide for a tidal river of its length, but it is extremely deep, facilitating the passage of ocean vessels from New York City to Albany, a distance of approximately 150 mi.

Given the strategic significance of the Hudson to both the British and the Colonies, it was necessary for the latter to assess the physical characteristics of the river, as well as the adjacent terrain, in order to develop an effective defensive scheme. The Hudson is tidal from its mouth into the Atlantic Ocean south of New York City, upriver to Albany. Consequently, tides, current, and winds affected sailing ships. From a defender's perspective, ideal locations included those locales where the river was narrow, ebb tide was at its strongest, wind was unpredictable and treacherous, and where adjacent terrain had a commanding view of the river and could be easily fortified (Miller et al., 1988). This military geographic analysis directed the Continental Army initially towards three specific locations: Dunderberg, Anthony's Nose, and Martelaer's Rock (which was later renamed Fort Constitution) (Fig. 4).

In addition to securing defensive positions to attack British vessels sailing up the Hudson, it was also necessary for the Colonists to protect east-west lines of communications. As such, river-crossing sites constituted key terrain and required protection. Most important were the ferry crossings between Stony Point and Verplank's Point in the south (known as Kings Ferry), and between Fishkill Landing and Plum Point in the north. The above analysis explains the eventual construction of fortifications at Stony Point,

Peekskill, Fort Clinton, Fort Montgomery, West Point, Fort Constitution, and Plum Point (Palka and Galgano, 2001).

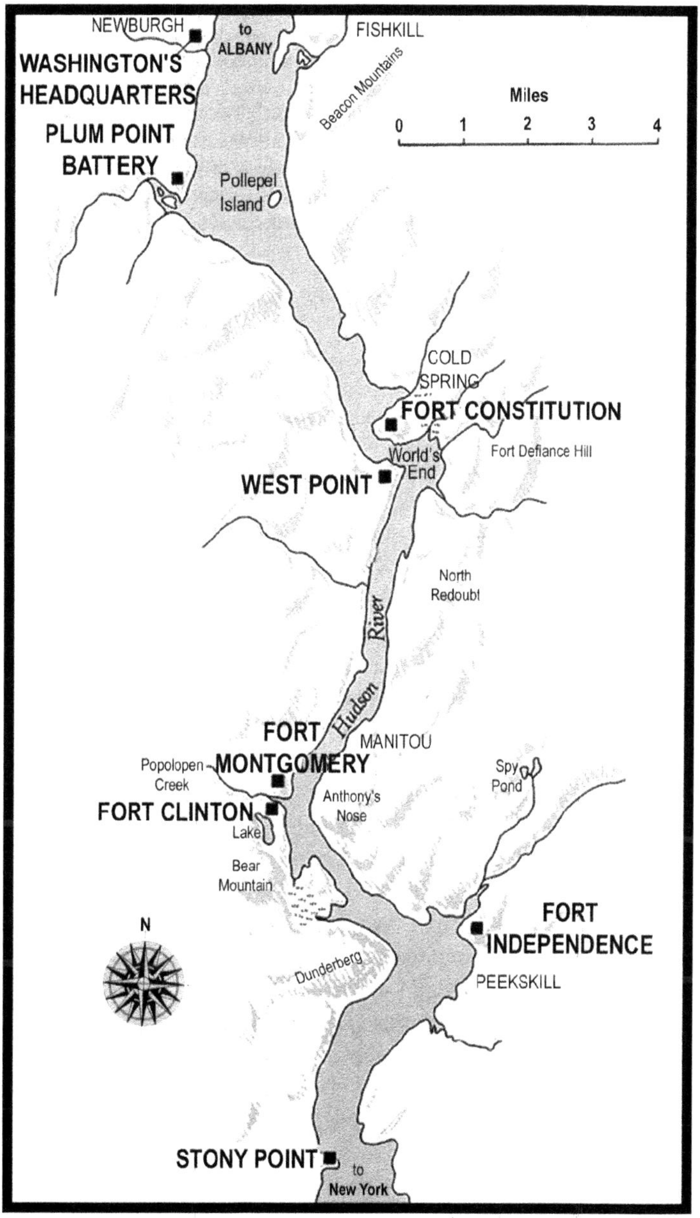

Figure 4. Key terrain and fortifications along the Hudson River. Adapted from Dunwell (1991).

The intent of the Colonial defensive effort was to simply render the Hudson River unusable to the British, while trying to maintain control of the river to suit their own needs. Control of the Hudson, however, did not necessarily entail employing naval vessels to operate on the river. The Continental Navy was inferior to the British fleet in virtually every respect and was incapable of accomplishing such a feat. As such, by understanding the characteristics of the river and then seizing and reinforcing the key terrain throughout the Hudson Highlands, it was perceived that the Continentals could successfully engage the powerful British naval vessels from fortified positions on land. Thus, dominant high ground, natural obstacles, and restrictive terrain provided the means to execute the Colonists' strategy, while simultaneously negating the superior advantage enjoyed by the British Navy.

5. SUPERIMPOSING A MILITARY GEOGRAPHIC IMPRINT

The Hudson Highlands are comprised of a rugged, dissected, and glaciated upland underlain by a complex sequence of Precambrian metamorphic and igneous rocks (LaMoe and Mills, 1988). From a military perspective, the irregular nature of the landscape and variable relief provided cover, concealment, and long-range observation for defenders, but served to severely canalize and/or impede the movement of an attacker.

With the idea of focusing the initial effort on the treacherous part of the river known as "World's End," work to fortify Martelaer's Rock (later named Fort Constitution) began in August 1775, and the island was formally garrisoned on 21 September 1775 (Miller, 1972; Adams, 1996b; Crackel, 2002; Galgano, 2004, this volume). In June 1776, Continental soldiers began construction of Forts Montgomery and Clinton (Dunwell, 1991; Coakley and Conn, 1992; Galgano, 2004, this volume). In August of that same summer, efforts to erect Fort Independence were also undertaken (Dunwell, 1991). Additionally, between 1775 and 1777, 27 redoubts were established and occupied throughout the Highlands (Dunwell, 1991). Redoubts were small fortifications that were located on principal mountaintops throughout the region. These outposts functioned as observation points, each manned by a group of five to seven Continental soldiers, and in some cases, the redoubts protected key terrain features such as a mountain pass (USMA, 1998). In an effort to protect the Kings Ferry, Stony Point was fortified on the west bank of the Hudson while Fort Lafayette was constructed on the east bank at Verplank's Point.

The defensive positions and fortifications along the Hudson were concentrated between Newburgh in the north and Stony Point in the south. Redoubts were situated throughout the Hudson Highlands in an effort to effectively link the forts into an overarching defensive scheme. All of the forts were focused on the river in anticipation of a naval attack from the south. As history would show, however, most were extremely vulnerable to ground attacks.

On 6 October 1777, British forces under the command of Sir Henry Clinton seized Forts Montgomery and Clinton, and Constitution Island (USMA, 1983). The British employed a highly successful naval diversion along the river between Verplanck and Peekskill during a period of limited visibility. Meanwhile, they landed more than 2000 troops ashore at Stony Point. The latter marched overland to attack the twin forts from the rear. The Colonials suffered more than 300 casualties and were forced to set ablaze their ships, frigates, and an armed sloop rather than risk capture (Kick et al., 1994). After brief occupations, the forts were destroyed and the British forces withdrew (Fig. 5).

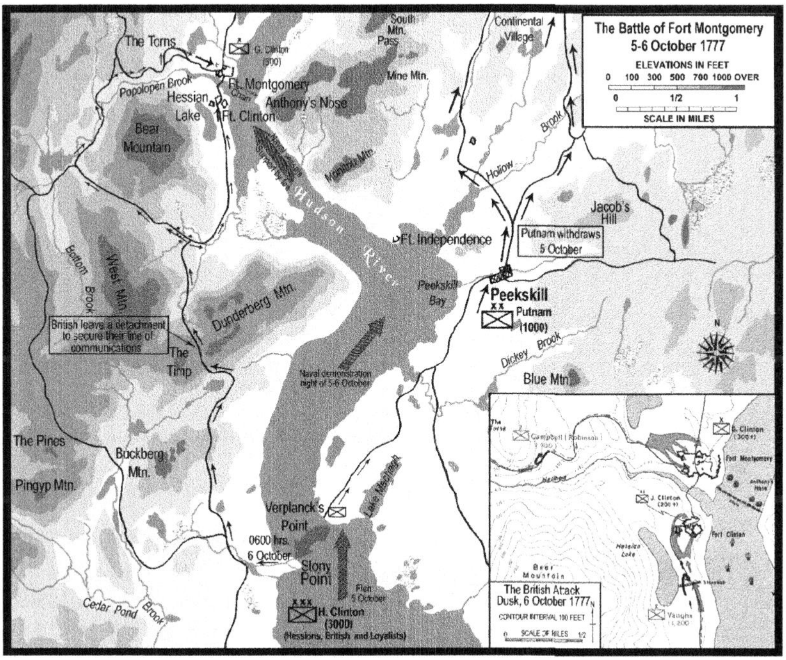

Figure 5. British attacks on Forts Montgomery and Clinton. Adapted from USMA (1998).

In the aftermath, Washington recommended the fortification and defense of West Point, and so the latter was occupied in January 1778 (Galgano, 2004, this volume). Preparations commenced three months later on the construction of Fort Putnam on the dominant piece of terrain overlooking West Point, Constitution Island, and the Hudson River. Additionally, several mutually supporting redoubts were emplaced and in the spring of 1778, Colonel Thaddeus Kosciusko directed the construction of Fort Arnold on the level of the present-day parade ground (Adams, 1996b; Fig. 6).

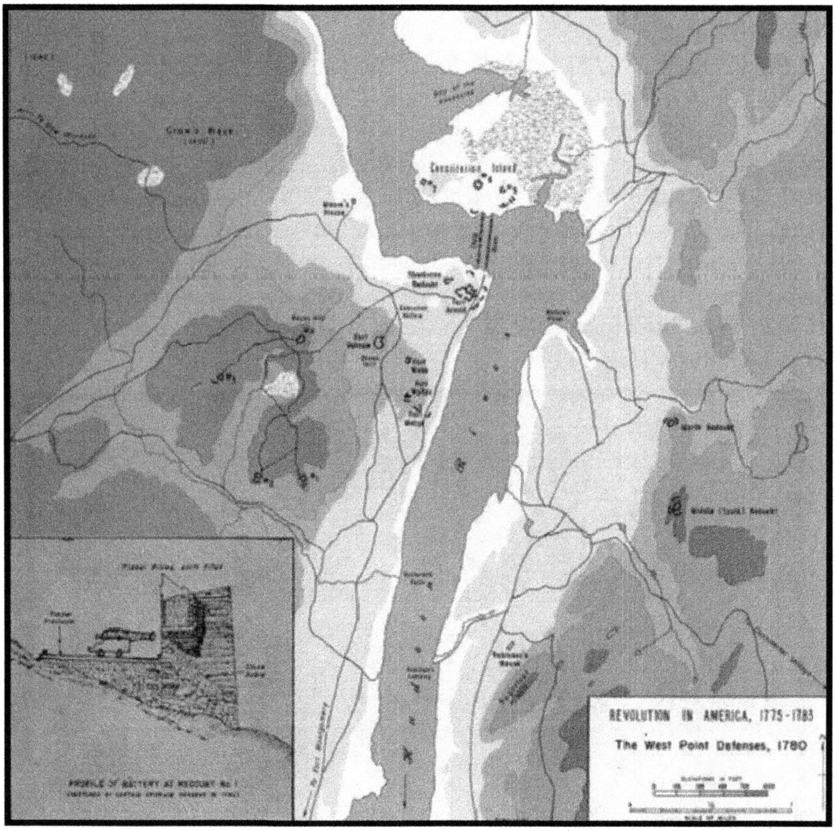

Figure 6. The defensive plan for the West Point. Adapted from USMA (1998).

"Fortress West Point," however, was never attacked. The Continentals did not commit the same mistakes made earlier when fortifying Constitution Island along the east bank of the Hudson. By comparison, the fortifications on the "west point" benefited from dominant high ground that afforded defenders excellent cover, concealment, and line of sight (Galgano, 2004, this volume). Moreover, it was possible to construct mutually supporting, inaccessible positions providing defense against overland attacks from multiple directions, as well as naval assaults from the river.

Additionally, to thwart the anticipated naval attack, the Continentals reinforced the restrictive nature of the terrain by emplacing a major obstacle so as to render British vessels vulnerable to fire as the latter attempted to negotiate the river's treacherous S-curve in a slow and cumbersome fashion. The "Great Chain," emplaced on 30 April 1778, was 500 yd long, comprised of 2.25 in square iron bars, and weighed approximately 65 tons (Diamant, 1989). The Great Chain, like Fortress West Point, was never tested in battle.

The last major action within the Lower Hudson Valley began on 1 June 1779, when the British captured Forts Lafayette and Stony Point (Adams, 1996). The latter was retaken by Brigadier General "Mad Anthony" Wayne and about 2000 Continental soldiers during a daring night attack on 16 July 1779. After winning the battle decisively, Wayne evacuated the fort three days later (Fig. 7).

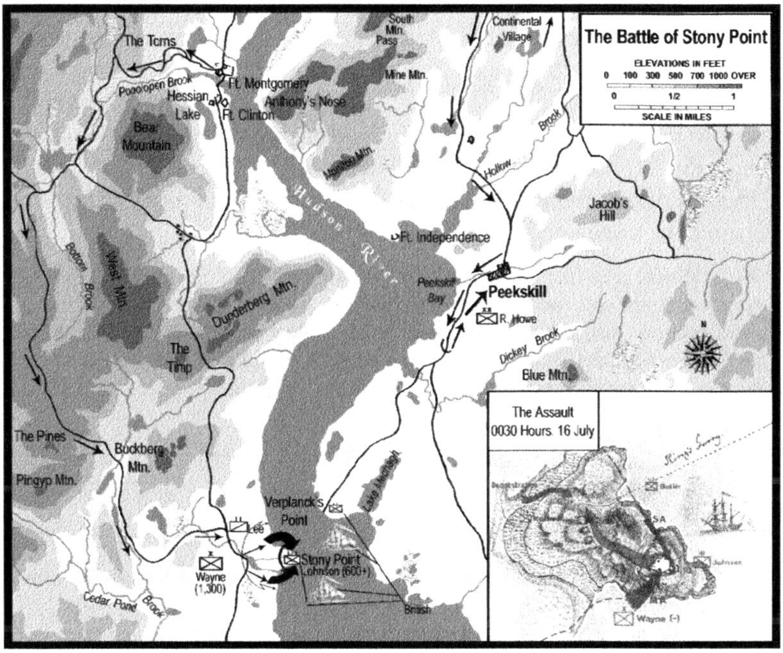

Figure 7. The battle for Stony Point. Adapted from USMA (1998).

Meanwhile at West Point, where Wayne had trained his forces only a month earlier, fortifications continued to be improved and expanded under the guidance of Kosciusko throughout the remainder of 1779 and the first half of 1780 (Galgano, 2004, this volume). In August, command of West Point was transferred to Benedict Arnold, who had already distinguished himself throughout the Hudson Valley. Unbeknownst to Washington and the Continentals, however, Arnold had already corresponded with Major John

André in June, in an effort to provide the information necessary for British forces to seize Fortress West Point, considered at that time as "America's Gibraltar." On the eve of 21 September, Arnold met with André near Verplank's Point to make final coordination and hand over the plans to West Point (Crackel, 2002). André was later seized as he tried to return to British lines after the meeting, and was subsequently tried and hanged. Arnold successfully escaped aboard the H.M.S. Vulture, but perhaps more important to Washington and the Continental Army, he did not succeed in compromising West Point's plans and fortifications

6. CONCLUSION

The Hudson Highlands provided the context for a pivotal era in American history. During the course of the American War of Independence, the pristine, scenic valley experienced a concerted effort by the Colonial government and Continental Army to reinforce the natural terrain in an attempt to deny the British use of the Hudson River, while maintaining the strategic thoroughfare to suit their own needs.

The historian, Sidney Fisher, in his study of the American War of Independence, concluded that:

> West Point and the Highland Passes constituted the most important American strategic positions. Fortifications, redoubts, battlefields, and an assortment of built structures endure as tangible evidence that others shared the same perspective. If Benedict Arnold's treachery had succeeded in delivering West Point to the British, the war might have ended sooner and otherwise. (Fisher, in Steele, 1951, 14)

REFERENCES

Adams, A.G. 1996a. *The Hudson throughout the Years*. New York: Fordham University Press.

Adams, A.G. 1996b. *The Hudson River Guidebook*. New York: Fordham University Press.

Brown, R. H. 1948. *Historical Geography of the United States*. New York: Harcourt Brace and World, Inc.

Brown, W. 1969. *The Good Americans: The Loyalists in the American Revolution*. New York: William Morrow.

Coakley, R.W. and Conn, S. 1992. *The War of the American Revolution*. Washington, DC: Center of Military History.

Crackel, T.J. 2002. *West Point: A Bicentennial History*. Lawrence, KS: University of Kansas Press.

Diamant, L. 1994. *Chaining the Hudson*. New York: Carol Publishing Group.

Dunwell, F.F. 1991. *The Hudson River Highlands*. New York: Columbia University Press.

Galgano, F.A., 2004, Decisive terrain - A military geography of Fortress West Point, 1775-1797. In *Studies in Military Geography and Geology*, D.R. Caldwell, J. Ehlen, and R.S. Harmon, eds., Dordrecht, The Netherlands: Kluwer Academic Publishers, 105-120.

Kagan, H.H., ed. 1966. The *American Heritage Pictorial Atlas of United States History*. New York: American Heritage Publishing Company.

Kick, P., McMartin, B., and Long, J.M. 1994. *50 Hikes in the Hudson Valley*, 2nd edition. Woodstock, VT: Backcountry Publications.

LaMoe, J.P. and Mills, R.W. 1988. *Field Guide to the Geology of West Point*. West Point, NY: US Military Academy, Department of Geography and Computer Science.

Lemon, J.T. 2001. Colonial America in the 18th century. In *North America: The Historical Geography of a Changing continent*, 2nd edition, T.F. McIlwraith and E.K. Muller, eds., Lanham, MD: Rowman and Littlefield Publishers, Inc.

McIlwraith, T.F. and Muller, E.K. 2001. *North America: the Historical Geography of a Changing continent*, 2nd edition. Lanham, MD: Rowman and Littlefield Publishers, Inc.

Meinig, D.W. 1986. *The Shaping of America: Volume 1 (Atlantic America, 1492-1800)*. New Haven, CT: Yale University Press.

Miller, LTC C.E., Jr. 1972. *The Fortification of Constitution Island 1775-1783*. West Point, NY: US Military Academy, Department of History.

Miller, C.E., Jr., Lockey, D.V., and Visconti, J., Jr. 1988. *Highland Fortress: The Fortification of West Point during the American Revolution, 1775-1783*. West Point, NY: US Military Academy, Department of History.

Palka, E.J. 2001. Decoding the cultural landscape of the Lower Hudson Valley: Military geographic remnants of the Revolutionary War. In *From the Hudson to the Hamptons: Snapshots of the New York Metropolitan Area*, I.M. Miyares, M. Pavlovskaya, and G.A. Pope, eds., Washington, DC: Association of the American Geographers.

Palka, E.J. and Galgano, F.A. 2001. *The Military Geography of Fortress West Point*. West Point, NY: US Military Academy.

Palmer, D.R. 1969. *The River and the Rock: The History of Fortress West Point, 1775-1783*. New York: Greenwood Publishing Corporation.

Palmer, D.R. 1991. *The River and the Rock*, 2nd edition. West Point, NY: Association of Graduates with Hippocrene Books.

Palmer, D. R. and Tripp, R.L. 1977. *Early American Wars and Institutions*. West Point, NY: US Military Academy, Department of History.

Rutledge, J.L. 1956. *Century of Conflict: The Struggle between the French and British in Colonial America*. Garden City, NY: Doubleday Books.

Stowe, G.C. and Weller, J. 1955. Revolutionary West Point: The key to the continent. Military Affairs 19: 81-92.

Thompson, J.H., ed. 1966. *Geography of New York State*. Syracuse, NY: Syracuse University Press.

US Military Academy, Department of Geography and Computer Science (USMA). 1983. *Guidebook to the Historical Geography of Fortress West Point*. West Point, NY: US Military Academy Press.

US Military Academy, Department of History (USMA). 1998. *West Point Fortifications Staff Ride*. West Point, NY: US Military Academy.

Wilstach, P. 1969. *Hudson River Landings*. Port Washington, NY: Ira J. Friedman, Inc.

Chapter 9

DECISIVE TERRAIN

A Military Geography of Fortress West Point, 1775-1797

Francis A. Galgano
US Military Academy

Abstract: The Hudson Highlands were the key terrain of the American War of Independence. The Highlands form a 15 mi wide barrier bisecting the lower river valley; West Point is the most dominating position. Initial Colonial efforts at fortification were unsound and all but ignored West Point; the Colonists fortified Constitution Island, overlooking West Point's commanding position. This historical military geography uses maps, historical records, and a digital elevation model to examine the fortifications at West Point and evaluate their effectiveness. The results indicate that from the river perspective, it is easy to incorrectly infer that Constitution Island was key terrain; however, positions there are dominated by terrain on either bank, by West Point in particular. Furthermore, batteries positioned on Constitution Island are masked by West Point and lack range to control effectively the river bend. Conversely, fortifications on West Point easily dominate Constitution Island and the all-important bend in the river.

Key words: Hudson Highlands, Hudson River, military geography, fortifications, West Point, digital elevation model

1. INTRODUCTION

Perhaps the most significant operation of the American War of Independence was the fortification of the Hudson Highlands, which led to the creation of Fortress West Point at the site of the modern US Military Academy. These fortifications were important because the Hudson River was the center of gravity of the war, and retention of this region was a decisive factor in the attainment of American victory (Palmer, 1969; Palka, 2004, this volume).

D.R. Caldwell et al. (eds.), Studies in Military Geography and Geology, 105–120.
© 2004 *Kluwer Academic Publishers.*

Not long after the war began, British and Colonial forces labored to control the strategic Hudson River-Lake Champlain-St. Lawrence River waterway (Miller et al., 1979). This corridor was essential to the strategic geometry of the American War of Independence for two reasons. First, the Hudson River was the natural dividing line between New England and the mid-Atlantic Colonies. By controlling the Hudson, the British could drive a wedge between the manufacturing and agricultural centers of the Colonies, thus rupturing the Colonial war effort. Second, this waterway physically connected the British military centers in New York City and Montreal (Fig. 1). Colonial control of the waterway was necessary to prevent concentration of British military power and thus, to fragment British ability to act in unison (Diamant, 1994).

The Hudson Highlands, which extend from Tappan Zee to Newburgh (Fig. 2), was the natural place for the Colonists to exert control. The terrain here forms a 15 mi barrier with rugged mountains and three sharp river bends. West Point is unquestionably the strongest position because here the river forms its sharpest bend - narrowed further by Constitution Island - and is dominated by imposing terrain on each bank. Although many recognized West Point's dominating position, early attempts to fortify the area focused instead on Constitution Island. This was the major flaw in the Colonial effort to secure control of the Hudson River.

1.1 A Strategic Place on the Hudson River

The strategic importance of the Hudson River came into focus for the American Colonists in the spring of 1775. In May, the Continental Congress received information that Colonial militia under Ethan Allen and Benedict Arnold had seized Fort Ticonderoga, a wilderness fortress 225 mi north of New York City. Upon hearing this news, Colonial leaders realized that the colonies faced the imminent prospect of war. Hence, they immediately appointed a committee to consider the defense of New York. The committee was chaired by George Washington, who believed that the Colonists would have to secure the Hudson Highlands.

From their experience, Colonial leaders understood that the natural invasion route between the Colonies and Canada was along the Hudson Valley. Furthermore, they appreciated that the Highlands controlled important communication links that crossed the Hudson from Fishkill to Newburgh in the north and from Verplanck's to Stony Point in the south (Fig. 2; Palka, 2004, this volume). Should the British cut these links, they might prevent the Colonists from moving supplies between New England and the mid-Atlantic Colonies and also block the movement of reinforcements. Ensuring freedom of movement along this corridor was particularly important because New England was to provide most of the troops during the war (Boynton, 1863).

Washington recommended that the Continental Congress take steps to defend New York. Consequently, a resolution was sent to the New York Provincial Convention suggesting that:

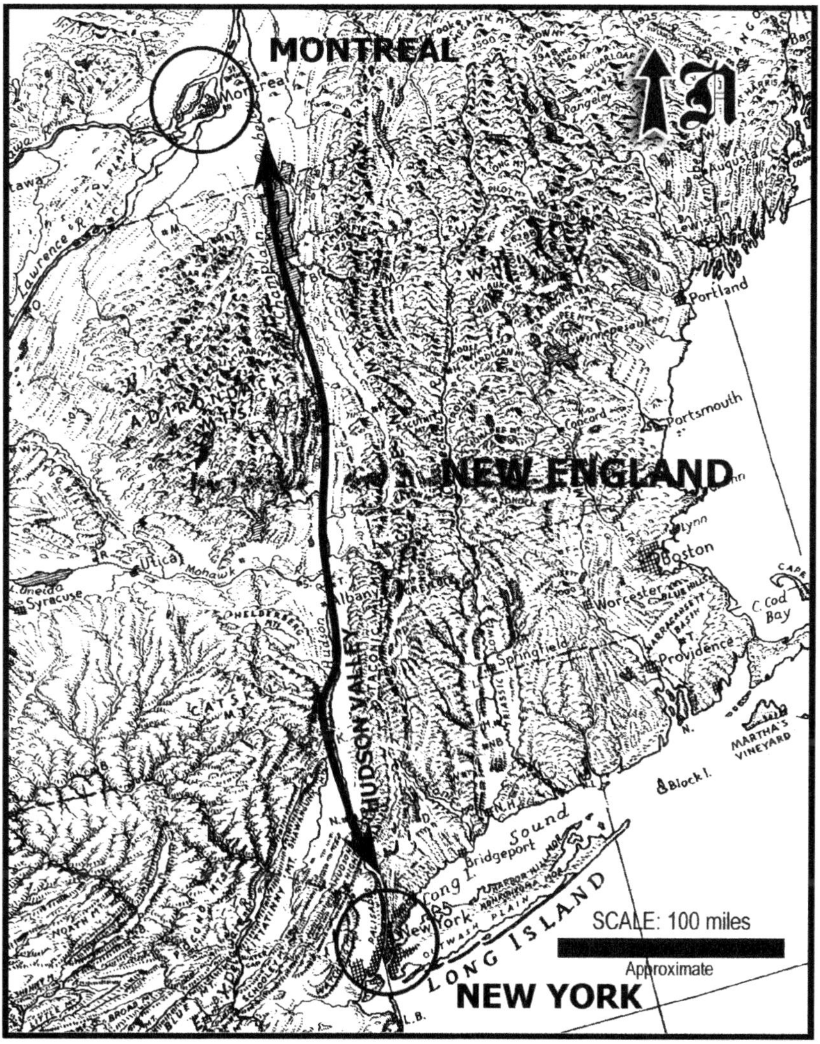

Figure 1. Terrain map of the Hudson Valley. The Hudson River was the geographic link between major British bases in New York City and Canada, and separated the New England from the mid-Atlantic Colonies. Modified from Raisz (1957).

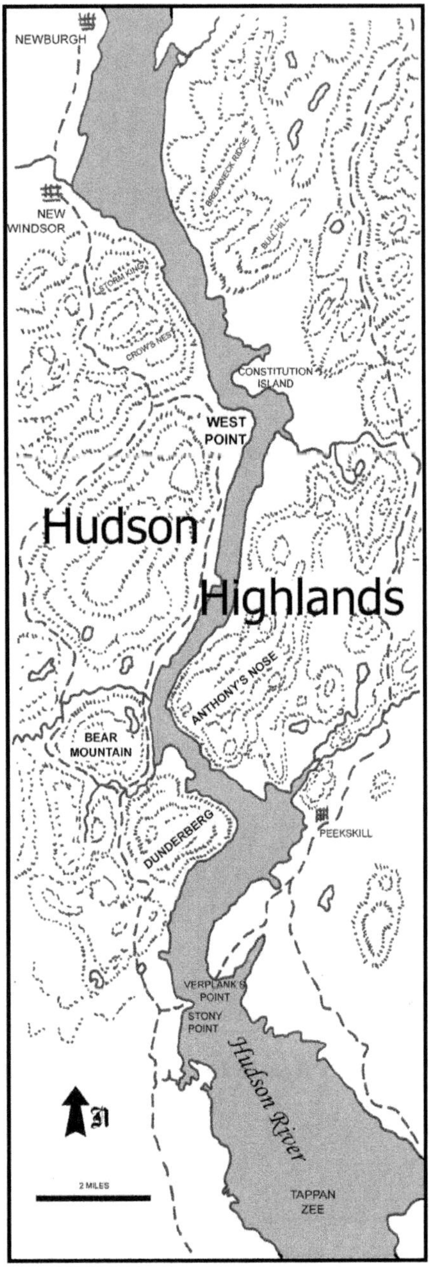

Figure 2. The Hudson Highlands are a 15 mi wide area of rugged granite and gneissic hills that bisect the Hudson River between Tappan Zee and Newburgh. Adapted from Palmer (1969).

> . . . a post be taken in the Highlands, on each side of the Hudson River, and batteries be erected; and that experienced persons be immediately sent to examine said river, in order to discover where it will be most advisable and proper to obstruct the navigation. (Force, 1837, 231)

The New York Convention responded and on 2 June 1775, Colonel James Clinton and Christopher Tappen sailed north to reconnoiter positions in the Highlands (Boynton, 1863). Clinton and Tappen observed several potential sites - Stony and Verplanck's Point, Anthony's Nose, the mouth of Popolopen Creek, Martelaer's Rock (i.e., Constitution Island), and West Point (G&CS, 1983). What the two men were really looking for was a location where both banks of the river could be fortified at a single point, thus closing it to the British. The best choice appeared to be "the West Point," where the river narrows considerably into a sharp S-shaped bend between Martelaer's Rock and the west bank. Here the river is buffeted by unpredictable winds and subject to difficult tidal currents. In addition to recommending the emplacement of batteries on "the West Point" and on Martelaer's Rock, Clinton and Tappen urged that a boom be constructed to block the river (Bradley, 1976). The Clinton-Tappen plan was sound and demonstrated that they had a keen eye for terrain. However, in their official report they inexplicably recommended that the largest garrison be established on Martelaer's Rock (Miller et al., 1979).

1.2 The "West Point" on the Hudson River

Before the American War of Independence, West Point was not a significant place and not greatly populated. Farming was limited because of the near absence of level, arable land, and habitation was established largely to satisfy the terms of land grants. The first known use of the term "West Point" was recorded by Goldsbrow Banyar - deputy secretary of the Province of New York - in his diary on 6 August 1757. He noted that "At 7 this Evening came to an Anchor at the W. Point of Marbling's Rock [Martelaer's Rock]" (Forman, 1950, 180). The place name "the West Point" was used exclusively in deeds and military records with the definite article retained. After many years, common usage established the name as West Point (Berard, 1886).

The physical geography of the Hudson Highlands lent logic to their fortification. The region is part of a belt of granite and gneiss mountains stretching northeast from Pennsylvania, across northern New Jersey and southeastern New York, into western New England (Berkey and Rice, 1919; Miller, 1924). In intersecting the Hudson Valley, the Highlands form a 15 mi barrier. Distinguishing features of the river's effect on the landscape are the deep gorge, narrow mountain passes, and small rocky islands. Additionally, the Highlands include characteristic fluvial and glacial terraces about 50-100 ft above the modern river, one of which is the West Point Plain, and three remarkably sharp turns in the river - West Point, Anthony's Nose, and Dunderberg - where crystalline rock has withstood the erosive powers of water and glacial ice (Fig. 2).

2. DEFENDING THE HIGHLANDS

West Point is the most dominating position in the Highlands because of the structure of the terrain and the configuration of the river channel. Here the Hudson, which normally trends north and south, turns abruptly west and then back again to the north (Fig. 3). This is important because in the days of sail, ships were vulnerable to shore batteries when forced to slow and navigate this turn (Nickerson, 1928). Furthermore, sailing around the bend is problematic because of complicated winds and tidal currents.

Washington recognized these important defensive characteristics, and urged fortification of West Point. He made this recommendation as head of the committee appointed by the Continental Congress to examine the defense of the Highlands. Washington introduced his proposal to erect batteries and prevent British vessels from using the river on 25 May 1775 (JCC, 1904). However, the British were thinking along similar lines and were also planning an operation to secure the Highlands (Force, 1837).

2.1 Early Blunders: Constitution Island

The Clinton-Tappen committee examined the terrain in June 1775 and developed the first plan for the river batteries on Martelaer's Rock (i.e., Constitution Island) and North and South Redoubts on the east bank (Fig. 3). They also recommended construction of defensive works at Forts Montgomery and Clinton - north and south of the mouth of Popolopen Creek - 6 mi south of West Point. The Clinton-Tappen committee also suggested emplacement of a physical barrier across the river to impede boat traffic and urged, "that . . . by means of four or five booms, chained together on one side of the river, ready to be drawn across, the passage can be closed up to prevent any vessels passing or repassing" (JPCNY, 1775, 64). This survey is important because, for some reason, Clinton and Tappan overlooked the commanding position at West Point, focusing instead on Constitution Island and other, more vulnerable positions (i.e., Popolopen Creek). The New York Congress accepted their proposal and ordered the necessary work to be completed under the direction of a Dutch "engineer," Bernard Romans.

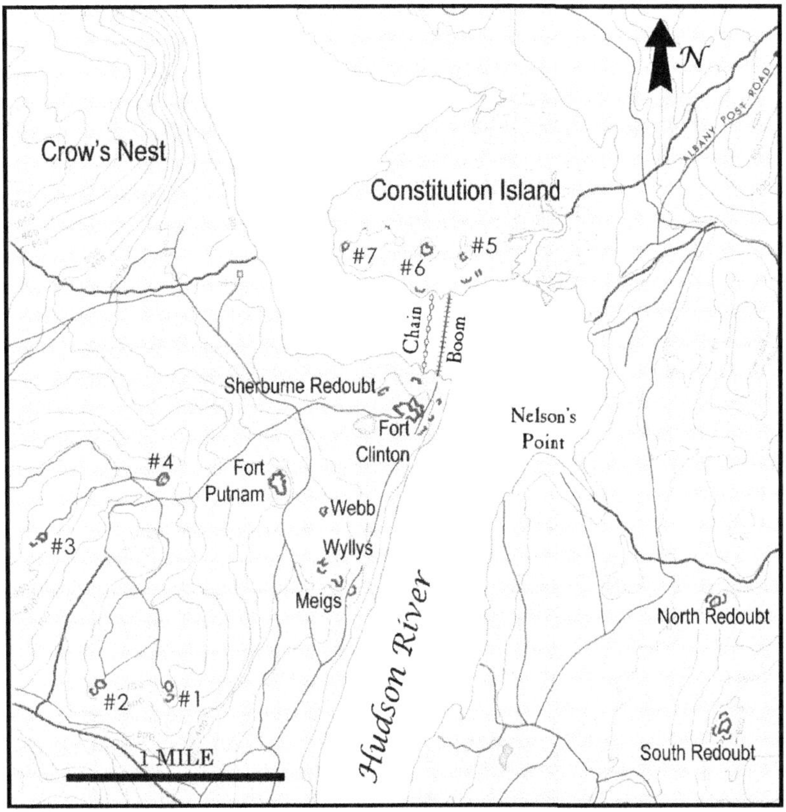

Figure 3. The fortifications at West Point and Constitution Island. Modified from Palmer (1969).

The establishment of fortifications on Constitution Island and at the mouth of Popolopen Creek - leaving West Point essentially undefended - was the seminal defect in Colonial strategy to defend the Hudson Highlands (Miller et al., 1979). This flaw would prove nearly fatal to the Colonial cause. The 1775 Clinton-Tappen plan focused on Constitution Island because they, and those that followed them, mistakenly believed that batteries at that position would dominate the bend in the river. Their mistake is understandable: From the river perspective, it is easy to see how one could infer that Constitution Island was key terrain. At first glance, it appears that batteries there would dominate naval craft sailing north. In reality, however, West Point masks the approach of ships from two-thirds of the positions on the island, and river batteries cannot engage a ship until it is too late (Fig. 4). More importantly however, Constitution Island is dominated by commanding terrain on both banks of the river, and especially by West Point (Fig. 5).

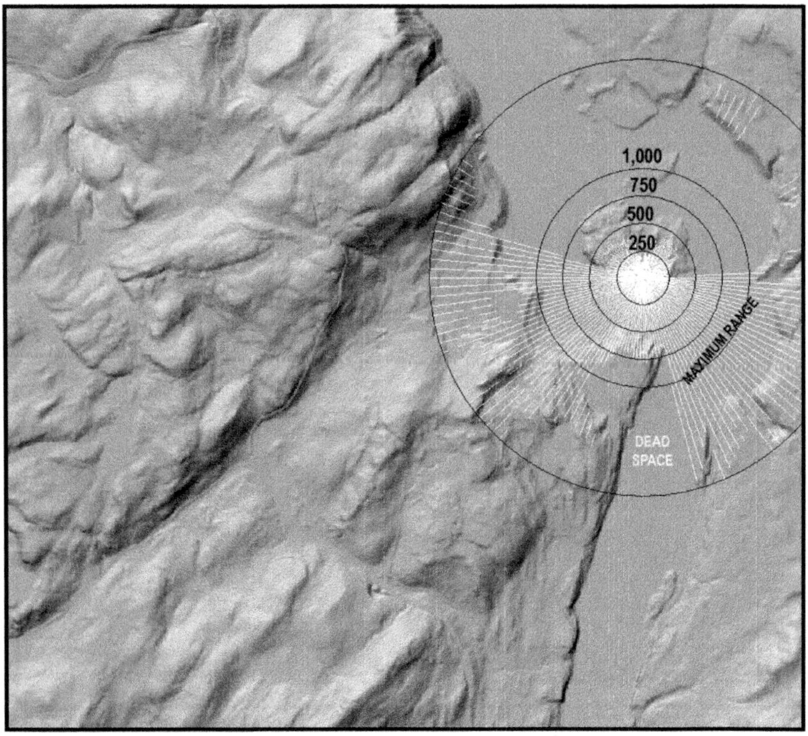

Figure 4. Range fan diagram illustrating the artillery engagement fan from Fort Constitution. The maximum effective range of the largest guns was only 1000 m. The range fan lines have been extended to 2000 m to highlight the masking effect of West Point.

Thus, a military force that occupies nearly any location on either bank of the river can dominate Constitution Island. This tactical disadvantage is illustrated in Fig. 6. Positions on Constitution Island can be controlled by artillery emplaced on either bank, but especially from West Point (Point B, Fig. 6). Furthermore, Fig. 6 reveals conclusively how commanding positions on West Point dominate the river bend. Thus, notwithstanding the construction of North and South Redoubts on the east bank (Fig. 3), positions on Constitution Island could be destroyed or compelled to surrender by the British should they occupy West Point.

Despite these apparent shortcomings, Colonial military leaders were drawn to Constitution Island. This attraction was fostered by Romans, the Dutch cartographer and naturalist who planned and directed construction of the first fortifications. His role as "engineer" was problematic because he

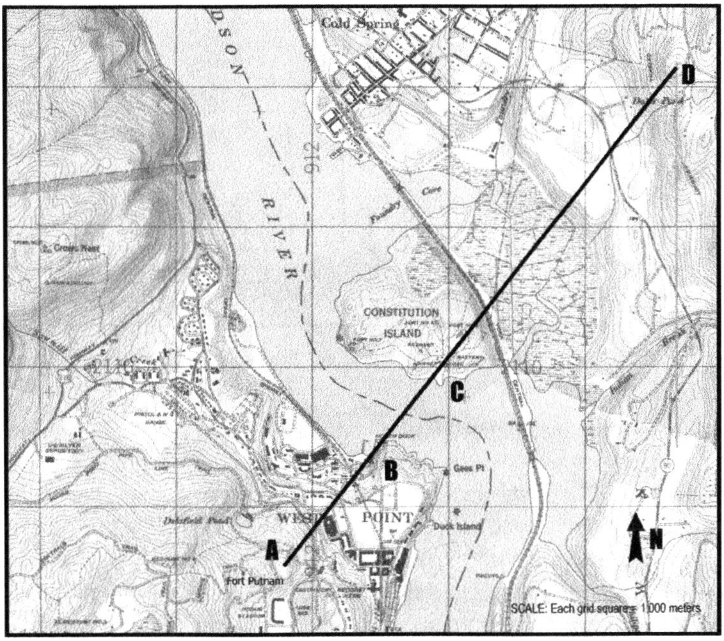

Figure 5. Topographic map of the West Point-Constitution Island area demonstrating Constitution Island's terrain disadvantage. A digital elevation model line-of-sight diagram of Line A-D is given in Fig. 6.

was not trained to establish such important works. However, he was previously employed by the British in various technical capacities and there were few trained engineers available: At the time, he was the only person available with any qualifications (G&CS, 1983). Romans and his work party arrived at Martelaer's Rock and preliminary construction began on 29 August 1775. Official reports from the island after that date were headed "Fort Constitution" (JPCNY, 1775).

2.2 Late 1775: Doubts about Fort Constitution

Fort Constitution was intended to be Romans' grand bastion, but in reality, it was a poorly sited, badly constructed fort that failed to take advantage of the range of its guns and the constricted Hudson River channel (Fig. 4). Moreover, Romans inexplicably ignored recommendations made by Washington, Clinton-Tappen, and others to fortify the dominating heights at West Point (Bradley, 1976). Consequently, if the Colonists had been required to rely on Romans' bastion to stop the Royal Navy in 1775, an aggressive British commander probably would have been able to run the river gauntlet with ease and the Highlands would have fallen (Palmer, 1969).

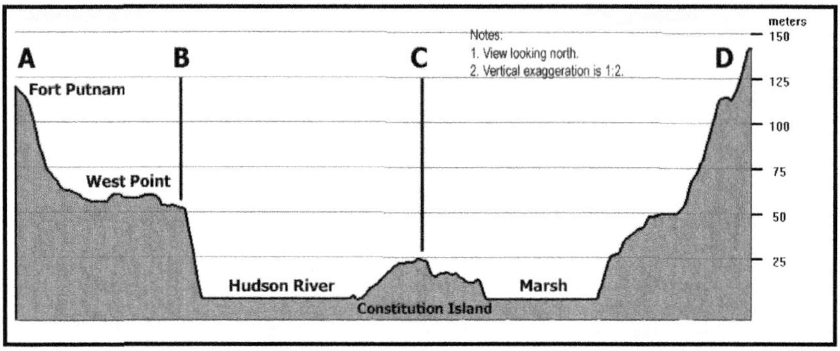

Figure 6. Line-of-sight diagram drawn along the line between Points A and D (Fig. 5). Constitution Island is clearly at a tactical disadvantage given its location and elevation.

Even as Romans began implementing his plan, some military leaders began to doubt its wisdom, because they perceived the tactical shortcomings of the position. An acute difference of opinion arose between Romans, local commanders, and the Congressional Commission over the division of authority and prudence of the plan (Force, 1837). This dispute contributed to doubts about the usefulness of the works, but more importantly, instead of focusing on solving the problem, it delayed construction of all fortifications and further endangered the security of the Highlands (Forman, 1950).

The debate compelled the Continental Congress to scrutinize fortification of the region again, and the commissioners did not like what they found. The works were incomplete, badly constructed, and so poorly sited that in reality, Fort Constitution was not a barrier to the Royal Navy. Furthermore, the garrison had not secured the landward approach from the east. Like Clinton and Tappen before them, the commissioners saw that the ground at West Point dominated Romans' position. Before they left the island, the inspectors concluded that Romans could not handle the job of fortifying the Highlands (Bradley, 1976). These findings were reported on 23 November 1775 and include the first official suggestion recommending occupation of West Point:

> The fortress [on Constitution Island] is unfortunately commanded by all the grounds about it; but the most obvious defect is that the grounds on the West Point are higher than the Fortress, behind which an enemy may land without the least danger. In order to render the position impassible, it seems necessary that this place should be occupied, and batteries thrown up on the shore opposite [at West Point]. (Force, 1837, 175)

However, recognizing a problem and implementing practical alternatives are distinctly different. The disagreement over the efficacy of the fortifications on the island exposed a fundamental problem which plagued the Colonists throughout the course of the war - the difficulty of obtaining qualified engineers to oversee construction of important fortifications

(G&CS, 1983). Thus, when Romans refused to deviate from his plan, Captain William Smith replaced him in January 1776. Smith too was quickly replaced, and there followed a succession of officers. Thus, the defense of the Highlands remained at risk throughout 1776 because numerous changes in leadership and lack of expertise resulted in inaction, and worse; inertia meant that fortification of Constitution Island would continue in spite of its obvious defects (Palmer, 1969).

2.3 1776-1777: Dilution of Effort and Near Disaster

The most important outcome of the debate over Constitution Island was a dangerous dilution of effort and the nearly disastrous concentration on other positions in the Highlands. The Congressional report of 23 November troubled the New York Provincial Assembly. They thus appointed their own committee (the third) to evaluate defenses in the Highlands. This committee was unable to agree on a solution, but did recommend fortification of Popolopen Creek instead of Constitution Island. The idea of fortifying the area near Popolopen Creek was a good one (certainly more effective than Constitution Island), but the inspectors still ignored West Point. To encumber further the logical fortification of the Highlands, the Continental Congress also approved building new fortifications near Popolopen Creek, a project that eventually brought work on Constitution Island to a standstill (Bradley, 1976).

Shortly thereafter, Washington heard about the disjointed efforts in the Highlands, and assigned a new officer to command the area (Fitzpatrick, 1934). He sent an inspection team under Brigadier General William Alexander Lord Stirling to examine the situation. Lord Stirling and his assistants conducted a systematic evaluation (the fourth) of the terrain and ongoing efforts in the Highlands. They recommended that works be built at Stony and Verplanck's Point and that a new position be built on the south side of Popolopen Creek, on high ground that overlooked Fort Montgomery. At Constitution Island, Stirling's party, like previous ones, saw that West Point dominated the island and recommended that a redoubt be placed there. Unfortunately, in his report to Washington, Stirling did not mention West Point and the Colonists again ignored this critical area. Consequently, during the spring of 1776, George Clinton, Governor of New York, in cooperation with his brother, Colonel James Clinton, erected Forts Clinton and Montgomery at the mouth of Popolopen Creek, 6 mi south of West Point (Bradley, 1976).

In 1775 and 1776, the British could have easily taken control of the Hudson River. Fortunately for the Colonists, the British too were overcome by a curious inability to act decisively. Since holding the Highlands was so important, it is hard to believe that the Colonists accomplished so little in 1775 and 1776. Even historic problems such as raising a new army and lack of money do not explain the pervasive lack of focus and just plain inefficiency demonstrated by the Colonists. Yet, they could not count on British inaction forever, and as 1776 drew to a close, they were galvanized into action because preparations in New York indicated that the British were about to make a concerted effort to seize the Highlands.

The British invasion of the Highlands began in fall 1777 when General Sir Henry Clinton, the British commander in New York, departed to capture the Highlands (Dalton, 2004, this volume). Early in October, the British landed at Verplanck's Point (Fig. 2) and drove out the garrison. They quickly crossed to Stony Point, moved north, and took Forts Montgomery and Clinton from the landward side. On 8 October, two thousand men under General Tryon proceeded up the river to Constitution Island to complete demolition of the Colonial fortifications (Heath, 1798).

The British victory was short-lived, however, because Burgoyne's defeat at Saratoga unhinged Sir Henry's plan (Dalton, 2004, this volume). Consequently, he had to abandon the Highlands after a 20 day occupation. Nonetheless, Colonial control of the Highlands, and perhaps the entire war effort, hung in the balance for those 20 days. This temporary British success thrust "the West Point" into a position of prominence, and crystallized Colonial opinion on its strategic importance. Thus on 2 December 1777, Washington directed construction of new works on the Hudson River, making special mention of the west bank, and recommending that a ". . . strong fortress should be erected at the West Point, opposite to Fort Constitution" (Fitzpatrick, 1931, 236).

2.4 1778: The Decision to Build Fortress West Point

The Colonists should have acted immediately to build new and better fortifications in the Highlands after Sir Henry Clinton retired to New York, but they did little during the remainder of 1777. Notwithstanding Washington's orders, a strong debate emerged over where to build major forts. Even though most local leaders wanted to abandon Forts Montgomery and Clinton and build a new fortification at West Point, expert opinion held otherwise. Washington's chief engineer in the Highlands, French Lieutenant Colonel de la Radiere, wanted to fortify the Popolopen Creek area again (Palmer, 1969). After much discussion, local military leaders prevailed, and it was decided that West Point would be fortified after all. Finally, in mid-January 1778, Radiere outlined the trace of a new fort on the plateau at West Point. Later that month, Colonial soldiers marched across the frozen river to

West Point for the first time, and established a post that has been occupied continuously ever since (Bradley, 1976).

The first unit to occupy West Point was a Massachusetts brigade under the command of General Samuel Parsons. They crossed the frozen river on 20 January 1778, and occupied the tip of West Point (Palka and Galgano, 2001). West Point's fortifications began to take shape throughout the winter and spring, yet there were still problems to overcome. Radiere continued to discount the logic of fortifying West Point. Arguments between Radiere and Colonial leaders, reminiscent of earlier disputes about Constitution Island, slowed progress. However, in March 1778, things began to improve. Colonel Thaddeus Kosciuszko, a French-trained Polish engineer who distinguished himself at Ticonderoga and Saratoga, arrived to assume the duties of Chief Engineer at West Point (Bradley, 1976). Kosciuszko too clashed with Radiere, and Washington tried to persuade the two to work together. In late April, Radiere was finally removed, and West Point now had an expert engineer who could organize the important task ahead (Bradley, 1976). Under Kosciuczko's supervision, an integrated system of fortifications began to take shape. As construction began in April 1778, Thomas Machin emplaced the Great Chain between West Point and Constitution Island (Palmer, 1969). The links were forged during the winter at the Sterling Iron Works, in the mountains about 25 mi from West Point. The chain (Fig. 3), which weighed 140-150 tons, was mounted on logs, and each spring, until the end of the war, it was stretched across the Hudson and taken up again before the river froze (Bradley, 1976).

2.5 Kosciuszko's Integrated Fortification System

Kosciuszko's innovative defensive scheme linked an integrated system of forts, each sited on commanding terrain. The lynchpin of this system was a bastion located on the tip of West Point. This bastion was sufficiently complete by July 1778 to receive the name Fort Arnold to honor Benedict Arnold, the hero of Quebec. Renamed Fort Clinton following Arnold's defection, it consisted of tree trunks piled on a rock ledge and hand-hewn stones (Palka and Galgano, 2001). Fort Clinton was the most important river fortification and it was supported by four river-line batteries (Fig. 3). Thus, Fort Clinton, the river-line batteries, and the soon-to-be rebuilt works on Constitution Island, when combined with Machin's floating boom-and-chain apparatus, offered for the first time a "system" to control the Hudson River (Diamant, 1994). However, this network of forts needed protection from landside attack - a lesson dearly learned the year before at Popolopen Creek.

The immediate rear of Fort Clinton was protected by Colonel Henry Sherburne's redoubt also on the level of the Plain. On higher ground to the west (Fig. 3), Colonel Rufus Putnam's 5th Massachusetts Regiment built a large stone fort that would eventually bear his name. Fort Putnam was a substantial, well-sited structure, whose ramparts enclosed a powder magazine, cistern, and garrison quarters. The batteries commanded the open plain behind Fort Clinton as well as the two major land routes from the south. Battery Sherburne and Fort Putnam were completed during summer 1779 (Palmer, 1969). Along the ridgeline south of Fort Clinton, covered by Fort Putnam, three Connecticut regiments built Forts Webb, Wyllys, and Meigs, naming them after their colonels (Fig. 3). These redoubts covered the southern approaches to West Point along the river terrace and the dead space south and east of Fort Putnam (Bradley, 1976).

West Point's fortifications attained their highest level of development by the end of 1779, after redoubts (Nos. 1, 2, 3 and 4) were completed on the hills south and west of Fort Putnam (Fig. 3). Kosciuszko originally intended to build Redoubt 4 in 1778, but it was not completed until 1779. This redoubt was especially critical to the defense of Fort Putnam because it stood on higher ground where enemy cannon could be placed to fire into the fort below. That same year (1779) the garrison also built the southern and western redoubts - Nos. 1, 2, and 3 - with their batteries and outlying works (Fig. 3). These new fortifications added depth to the West Point position and further protected the approaches to Fort Putnam, which was the key to the defense of Fort Clinton (Bradley, 1976). Thus, after nearly three years and numerous missteps, a credible bastion was finally built on the river (Steele, 1951).

The final element of the defensive network was the troublesome position on Constitution Island. Nearly three years after their unpromising beginning, these fortifications were finished. The island was reoccupied in 1778, and soldiers partially rebuilt the Marine and Gravel Hill Batteries to cover the river line and chain. They also constructed Redoubts 5, 6, and 7 along the crest of the island to protect the vital river batteries from a landside attack (Fig. 3). At the same time, soldiers completed North and South Redoubts on the high ridge above the eastern shore, further solidifying Colonial control of this most critical point in the Hudson Highlands (Bradley, 1976).

3. CONCLUSION

In the final analysis, the completed fortifications on West Point and its surrounding terrain, along with the Great Chain, discouraged a repetition of Sir Henry Clinton's 1777 campaign. In fact, after 1778, the security of the Highlands was never again in doubt. One thing is clear; as a fortress, West Point was far ahead of its time, because as 19th and 20th century soldiers discovered, a fortified position consisting of mutually supporting strong points is the basis of a modern defensive system and is considerably stronger than a single position built in the 18th century tradition. In this regard, Kosciuszko's fortification concept was at the same time revolutionary, complex, and elegant. The West Point fortification system - unlike earlier attempts on Constitution Island and at Popolopen Creek - was truly integrated, mutually supporting, and took fullest advantage of the commanding terrain and configuration of the river.

REFERENCES

Berard, A.B. 1886. *Reminiscences of West Point in the Olden Time*. East Saginaw, MI: Putnam.

Berkey, C.B. and Rice, M. 1919. Geology of the West Point quadrangle, New York. New York State Museum Bulletin 225: 226.

Boynton, E.C. 1863. *History of West Point and its Military Importance during the American Revolution*. New York: D. Van Nostrand.

Bradley, J.H. 1976. *West Point and the Hudson Highlands in the American Revolution*. West Point, NY: US Military Academy.

Dalton, J.B., Jr. 2004. Saratoga - A military geographic analysis. In *Studies in Military Geography and Geology*, D.R. Caldwell, J. Ehlen, and R.S. Harmon, eds., Dordrecht, The Netherlands: Kluwer Academic Publishers, 121-131.

Department of Geography and Computer Science (G&CS). 1983. *Guidebook to the Historical Geography of Fortress West Point*. West Point, NY: US Military Academy.

Diamant, L. 1994. *Chaining the Hudson*. New York: Carol Publishing Group.

Fitzpatrick, J.C., ed. 1931-1944. *The Writings of George Washington*. Washington, DC: US Government Printing Office, vols. 10, 16, 21.

Force, P., ed. 1837-1853. *American Archives*. Washington, DC: US Government Printing Office, vols. 4, 5.

Forman, S. 1950. *West Point: A History of the United States Military Academy*. New York: Columbia University Press.

Heath, W. 1798. *Memoirs of Major General Heath*. Boston, MA: Unpublished Manuscript.

Journals of the Continental Congress (JCC), 1774-1789, 1904-1922, 2:60.

Journal of the Provincial Congress of New York (JPCNY), May 25, 1775 to June 13, 1776.

Miller, W.J. 1924. *The Geological History of New York State*. New York State Museum Bulletin 255.

Miller, C.E., Lockey, D.V., and Visconti, J. 1979. *Highland Fortress: The Fortification of West Point during the American Revolution, 1775-1783*. West Point, NY: US Government Printing Office.

Nickerson, H. 1928. *The Turning Point of the Revolution*. Boston, MA: Rand Publishers.

Palka, E.J. 2004. A military geography of the Hudson Highlands - Focal point in the American War of Independence. In *Studies in Military Geography and Geology*, D.R. Caldwell, J. Ehlen, and R.S. Harmon, eds., Dordrecht, The Netherlands: Kluwer Academic Publishers, 89-103.

Palka, E.J. and Galgano, F.A. 2001. *The Military Geography of Fortress West Point*. West Point, NY: US Military Academy.

Palmer, D.W. 1969. *The River and the Rock: The History of Fortress West Point, 1775-1783.* New York: Greenwood Publishing Corporation.

Raisz, E. 1957. *Landform Map of the United States*. Danvers, MA: GEOPLUS.

Richards, S. 1903. Personal narrative of an officer in the Revolutionary War. United States Magazine 4:721-33.

Steele, M.F. 1951. *American Campaigns.* Washington, DC: Combat Forces Press.

Chapter 10

SARATOGA
A Military Geographic Analysis

James B. Dalton, Jr.
US Military Academy

Abstract: The Campaign and Battle of Saratoga (19 September-7 October 1777) serves as a valuable case study from which one can gain a better understanding of the critical geographic factors affecting Continental Army commanders during operations against Lieutenant General John Burgoyne's British Army at the micro and macro geographic levels. The Battle of Saratoga illustrates the impact that the physical and cultural landscapes have on the outcome of battle at each of these scales of analysis. The analysis used historical data, digital elevation data, vertical photos, and battlefield observations to illustrate these aspects of the battle.

Key words: military geography, Saratoga, American War of Independence, terrain, weather, cultural features

1. INTRODUCTION

The Battle of Saratoga serves as a valuable case study for gaining a better understanding of critical geographic factors that impacted Continental Army commanders during the American War of Independence. Colonial operations against British forces provide lessons at the tactical (micro-geographic factors), operational, and strategic levels (macro-geographic factors). The Battle of Saratoga (19 September to 7 October 1777) in the vicinity of Bemis Heights, NY illuminates the impact geographic factors both physical and cultural can have on the outcome of battle. The battle occurred at the location chosen by the Continental Army commander and this choice resulted in a major change in the course of the American War of Independence. Many historians have correctly identified this battle as the decisive turning point of the war (Ketchum, 1997; Morrissey, 2000;

D.R. Caldwell et al. (eds.), Studies in Military Geography and Geology, 121–131.
© 2004 *Kluwer Academic Publishers.*

Schnitzer, 2002). A geographic analysis supports the historian's contention of its importance to US history.

2. STRATEGIES AND BATTLE OVERVIEW

A review of the battle provides context for examination of the geographic factors affecting this conflict. Saratoga was part of a larger British strategy to isolate New England from the rest of the Colonies (see Palka, 2004, this volume). The British viewed the New England Colonies as the source of revolution and, believed that if isolated, the rest of the colonies would capitulate. To accomplish this, British high command purposed a three-pronged offensive (Fig. 1) to drive a wedge between the rebellious New England Colonies and the remaining Colonies.

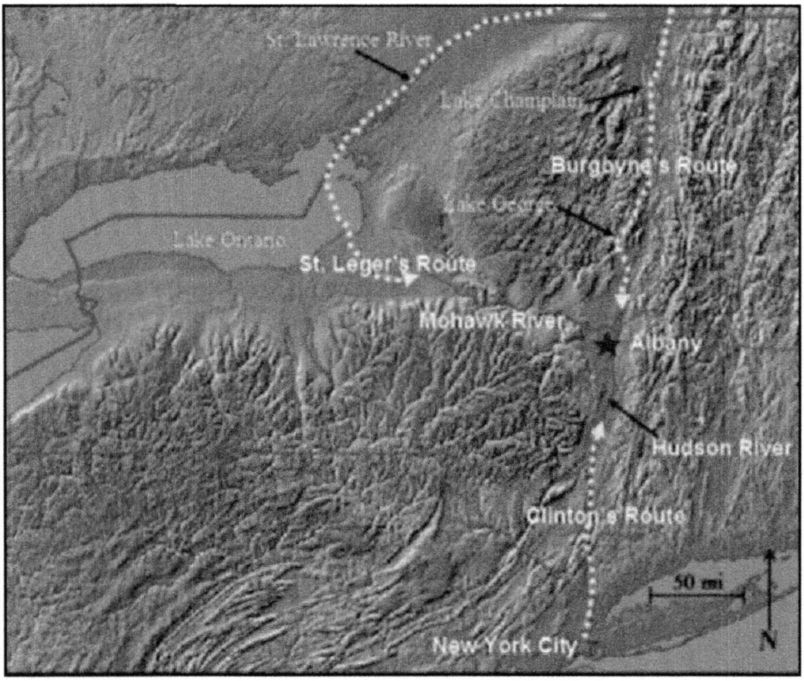

Figure 1. British strategy to divide the Colonies. Modified from Shaded Relief Map copyright © 1995 by Ray Sterner, Johns Hopkins University Applied Physics Laboratory.

The three-pronged attack was planned to occur simultaneously so as to converge on Albany and take control of the entire Hudson River Valley. With British control of the Hudson, the lines of communications between New England and the remaining Colonies would effectively be cut. British

Lieutenant General John Burgoyne was to lead the first prong of the offensive. His army would attack from the north across Lake Champlain and Lake George, then down Wood Creek to the Hudson River and then onto Albany (Fig. 1). Brigadier General Barry St. Leger was to lead the second prong of the attack. His mission was to move west using the St. Lawrence River (Fig. 1) into Lake Ontario, then turn east and march on Albany via the Mohawk River valley. General William Howe was to lead the final prong north from New York City up the Hudson River to attack Albany from the south (Morrissey, 2000). In fact, Howe took the majority of his army to Pennsylvania, leaving this task to Lieutenant General Sir Henry Clinton and the remaining British forces. The force commanded by St. Leger, arguably traveling the furthest, made it as far as Ft. Stanwix (Fig. 2) in New York in August 1777. St. Leger was turned back after Continental Army Major General Benedict Arnold lifted the British siege on the fort (Schnitzer, 2002). Clinton made a half-hearted attempt to move north up the Hudson from New York City taking Colonial positions at Forts Montgomery and Clinton at Popolopen Creek (Fig. 2; see Palka, 2004, this volume), but did not proceed far enough north to threaten Albany. The reason for Clinton's

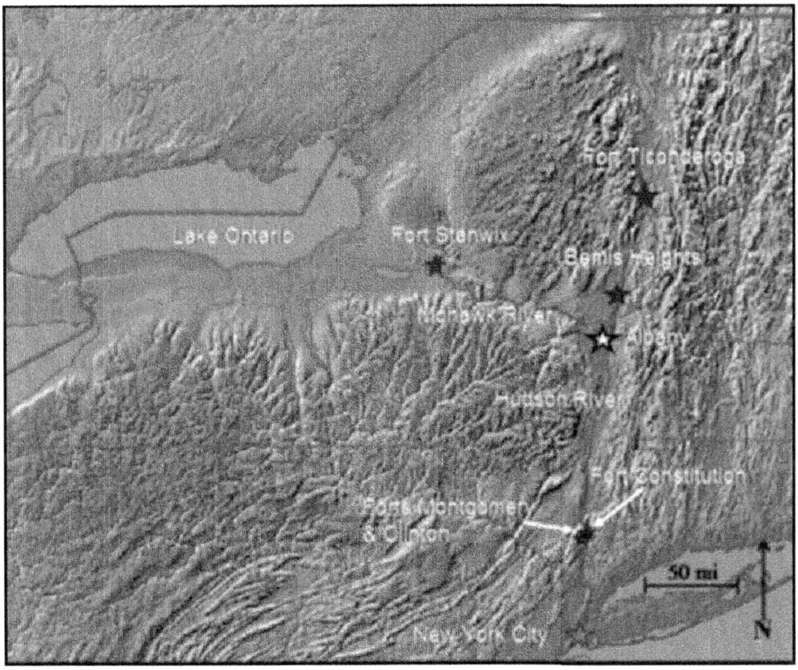

Figure 2. Colonial strategy to defend Albany. Modified from Shaded Relief Map copyright © 1995 by Ray Sterner, Johns Hopkins University Applied Physics Laboratory.

inaction is outside the scope of this paper but, nevertheless, Burgoyne based several key decisions on the assumption that Clinton was on his way north from New York City (Morrissey, 2000; Schnitzer, 2002).

The Colonial strategy was simple - defend all approaches to Albany (Fig. 2). Albany was a strategic city controlling access to the Mohawk River Valley and the upper reaches of the Hudson River. Albany controlled the commerce and the key crossings at the confluence of the Hudson River and Mohawk River. To protect the southern approaches numerous fortifications were established in the lower Hudson River Valley with the strongest located at Forts Montgomery and Clinton on Popolopen Creek south of Fort Constitution (Fig. 2) across from the west point of the Hudson River (Galgano, 2004, this volume). Ft. Stanwix was built at what is today Schuyler, NY to provide protection from attacks along the western approach through the Mohawk River Valley. Fort Ticonderoga anchored the northern approach. As previously noted, the Continental Army successfully defended the western approach in August 1777 and Clinton's advance from the south, even with victories at Forts Montgomery and Clinton, ended with his return to New York City. As a result, the decisive battle took place at Bemis Heights in the vicinity of Saratoga, NY (Morrissey, 2000).

Continental Army Major General Horatio Gates commanded the Northern Army that opposed Burgoyne's advance. His principal subordinate commanders were Major General Benedict Arnold, Colonel Daniel Morgan, and Lieutenant Colonel Henry Dearborn. They commanded forces roughly equal in size to Burgoyne's army (Morrissey, 2000). Gates moved north from Stillwater, NY upon learning of the British advance from Canada. With the help of Polish engineer Colonel Thaddeus Kosciuszko, Gates chose Bemis Heights to defend against Burgoyne's army. The Colonial position was superior and located on the best defensive ground between Lake Champlain and Albany. After only a week, Gates effectively fortified this key terrain to create a choke point along the only road Burgoyne could take to Albany (Morrissey, 2000; Schnitzer, 2002).

The British had neither accurate intelligence of Colonial dispositions on the route to, and in the vicinity of, Bemis Heights, nor knowledge of the exact nature of the terrain. They knew the terrain was covered by broadleaf virgin forest that would make movement difficult except along established roads or via the river. There was only one direct road that paralleled the Hudson River. The British lacked sufficient water craft to use the river exclusively, so had to rely on both road and river for movement (Morrissey, 2000; Schnitzer, 2002).

Once the British left Lake Champlain, resupply was difficult as supply lines became long and portages were required to bypass portions of the route not connected by water. The further south the British moved, the more their support from the local population dwindled and the greater the likelihood of a Colonial interdiction. Burgoyne's greatest concern was the security of his stores and protection of his lines of communication (Schnitzer, 2002). To dispel his concern, Burgoyne depleted his combat-capable forces to garrison strong points along his route of march and designated a permanent force to guard the army's supplies.

Faced with the superior Colonial defensive position at Bemis Heights, Burgoyne had two choices. He could stay on the road and continue to move south along the river and then attack directly into the Colonists' strongest defensive point (Fig. 3). Or, he could leave the road and attack to the west of the strongly defended bluffs. He chose the latter. Subordinate commanders were Brigadier General Simon Frasier attacking on the west flank, Brigadier General Hamilton attacking in the center, and Major General Baron Fredrick von Rediesel moving along the road on the east flank.

Figure 3. British move west off the road to Albany and engage the Continental Army at Freeman's Farm.

The British began the attack on 19 September 1777 by moving west off the Albany road and then south through the heavily wooded terrain (Fig. 3). The conflict looked as if it was going in favor of the Colonists, but the arrival of the Germans under von Rediesel as darkness set in gave a slight advantage to the British. Since the British retained the terrain at the end of the day, they felt they had won the initial battle. However, the British suffered twice as many casualties as the Colonists and the Continental Army still held the best defensive terrain blocking the road to Albany. From 20 September until 7 October no major action occurred and only minor skirmishes took place. The Colonial forces seemed content to wait while continuing to improve their superior position. The British took the opportunity to build defensive positions of their own on the recently gained terrain (Ketchum, 1997; Morrissey, 2000; Schnitzer, 2002).

After much debate, Burgoyne decided to send out a reconnaissance-in-force on 7 October to determine the strength and disposition of the Colonial forces. His intent was to follow-up the next day with a full scale attack. In addition, this 1500 man reconnaissance force was to forage for food. The rations for the British force were quickly being depleted and winter was approaching. The Colonists quickly recognized the British intent and rushed forward to meet them (Fig. 4).

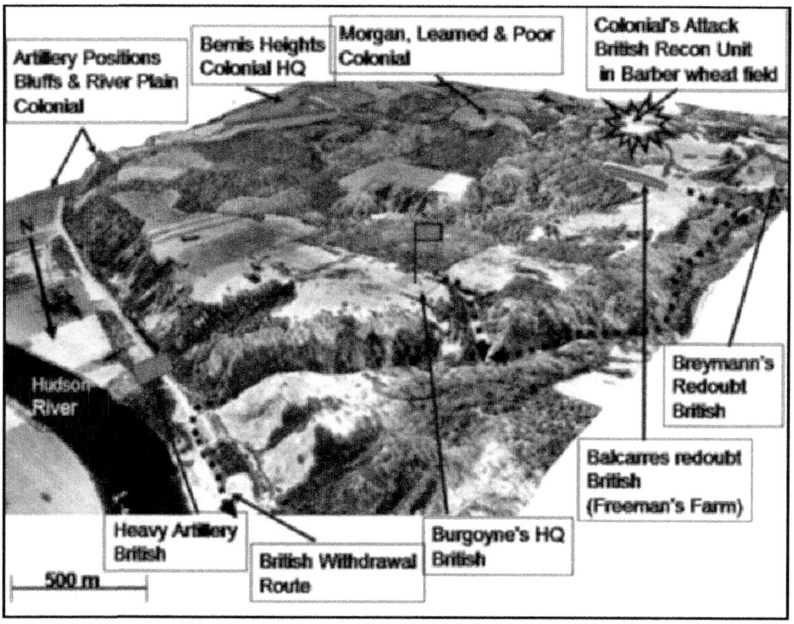

Figure 4. View of the battlefield from the north with positions on 7 October 1777.

In the subsequent battle, the Colonial forces seized the initiative and gained the upper hand. As night fell and rain began to fall, the British withdrew. The Colonists kept up the pressure and pursued the British. Burgoyne saw no hope of help from the south and with winter quickly approaching, he sued for surrender. After lengthy negotiations, terms were reached and Gates accepted Burgoyne's surrender on 16 October 1777 ending the Battle of Saratoga (Ketchum, 1997; Morrissey, 2000; Schnitzer, 2002).

3. MICRO-GEOGRAPHIC FACTORS

3.1 Cultural Factors

The battle site provided the optimum area for employment of Colonial weapons and tactics against the British. Specifically, Bemis Heights and Freeman's Farm had been modified by human activities creating a battlefield giving the Continental Army a defensive, tactical advantage. Colonial tactics used in the battle were heavily influenced by three specific aspects of the cultural landscape. First, farms had been created out of the forest by cutting trees to clear fields, and fences were erected to protect the fields. Second, the choice of crops that were planted determined the vegetation over much of the battle area. Finally, the buildings that were constructed and their placement on the farms influenced the battle.

The presence of fenced farm fields created an area in which British and Colonial forces could be formed in the traditional European style of multiple lines to mass musket fire. The Colonists seized on this factor; namely, the fences provided some cover and concealment early in the battle. On the other hand, the fences were a hindrance to quick movement of the British regulars.

The second factor was used most effectively by the Colonists under Morgan. His Virginia riflemen were able to see and engage the British as they walked into the wheat fields even though the wheat had not been harvested and while they were beyond the range of the British Brown Bess Rifle. Had corn been growing in the fields as is more common today, the British would have been afforded some degree of concealment. From the cover and concealment provided by the heavy wooded areas and fences surrounding the fields and with the support of Dearborn's light infantrymen armed with muskets, Morgan's riflemen were able to inflict heavy casualties upon the British.

The final cultural factor, the farm buildings, was used by the British for intelligence gathering and as positions from which to view and engage the Colonists. Other buildings away from the immediate battle area were used by both forces throughout operations in the area (Ketchum, 1997).

3.2 Physical Geographic Factors

The British commanders underestimated the physical landscape; they misjudged the many natural obstacles (micro-scale factors) and the distances between locations, discussed later in this paper (macro-scale factors). On the micro scale, the physical landscape provided many challenges to the attacking British forces. First, the off-road routes of advance were dissected with deep ravines. Second, the ravines were heavily wooded with thick undergrowth. Third, the Albany road crossed numerous streams (Fig. 5) and fourth, the bluffs provided the Americans the high ground. The physical landscape was influenced by the meteorological conditions throughout the engagement. Each of these had specific consequences and considerations.

Figure 5. One of the many streams bisecting the Albany Road.

3.2.1 Ravines

The ravines that dissect the British avenues of approach to the west of the Albany Road were formed by erosion after the last Pleistocene glacier receded. They run generally west to east and the depths vary with the deepest to the east (200 m) and shallowest to the west (1-2 m). The impact of the ravines on the British was to (1) slow their advance, (2) hinder command and control between and within the advancing columns, and (3) reduce their ability to see and to detect Colonial forces. Colonial operations were not hindered by the ravines; in contrast, they used the ravines to conceal their movement. In addition, the Colonists used the ravines in the period between the two major engagements to conduct probing and harassing attacks. They gained the most by fighting in the heavily dissected terrain.

3.2.2 Heavily Wooded Terrain

The forest was essentially virgin, never having been cleared by European settlers (Braun, 1950; Russell, 2001). As such, not only were the forests thick, but the forest floor had thick undergrowth combined with abundant vegetation debris that further hindered cross-country movement. The British experience in European forests did not prepare them for the North American forest. Just as the ravines hindered movement, command and control, and visibility, the dense forest did the same. The ravines and forest combined to tire the British soldiers mentally and physically, resulting in reduced effectiveness. The Colonists were more accustomed to the conditions and used them to their advantage with few negative consequences (MacDonald, 1984; Morrissey, 2000; Schnitzer, 2002). As an example, Morgan's forces took full advantage of the heavily wooded terrain to maneuver and engage the British while regular Continental Army forces fought the British in the open farm fields.

3.2.3 Streams

The numerous streams along the Albany Road (Fig. 5) had to be bridged to get artillery and supply wagons forward. Between Sword's Farm, approximately 2 km north of Bemis Heights, and Bemis Heights several major streams had to be crossed. Existing bridges had been destroyed by the Colonists as part of their defensive strategy. The impact on the British was a delayed advance and fewer fighting men: Soldiers had to be committed to building and guarding bridges. Burgoyne was conscious of the negative consequences of losing his supplies and heavy artillery, making the bridges essential to his lines of communication. For their part, the Colonists did not need the road north of Bemis Heights for their defense, and when they needed to move forces north of the British position, they could use the east bank of the Hudson River.

3.2.4 Bluffs

Although never tested by the British, the Colonial defenses along the high ground created by the bluffs (named Bemis Heights, after the original European owner), were critical at the micro level of analysis. These bluffs were significant not due to their great height, but due to their location at a key choke point along the Hudson River. Colonel Kosciuszko recognized the ability to control the northern approach to Albany from this terrain feature. The bluffs, averaging 250 m above the Hudson River, provide a clear view of the narrow, 500 m wide river flood plain. This narrow strip of level land was the route for the road to Albany. If you controlled the bluffs, you controlled the road (Schnitzer, 2002).

3.2.5 Weather

Finally, the weather immediately following the 9 October engagement gave an edge to the Continental Army and in the end provided a final knockout blow to Burgoyne's army. Heavy rain turned the trails and roads into thick mud. British soldiers already tired from combat and low rations, were further demoralized by these conditions, which slowed their retreat. The British army's energy, resolve, and morale were quickly sapped by this meteorological factor that led to their surrender.

4. MACRO-GEOGRAPHIC DISCUSSION

At the macro geographic level, the three independent British forces planned to conduct operations against Colonial forces in a unified, coordinated manner using the three approaches to Albany discussed earlier (Fig. 1). The planned outcome was to divide Continental forces and then to defeat the states separately in turn. The length of each avenue of approach and physical terrain helped create the situation for the Battle of Saratoga. Burgoyne's army encountered terrain that hindered and slowed their movement, and reduced their forces as mentioned earlier. The British command did not have a unified effort and lacked communications between the three armies, making operational control of their effort non-existent.

A map analysis of the routes taken by British forces shows they underestimated the physical landscape. The approximate distance between Burgoyne's forces and Clinton in New York was 290 km through Colonist-controlled territory: This made communication between the two leaders impractical. Burgoyne had two choices for communications: He could send a dispatch back through Canada and then by sea to New York, or he could send couriers overland. The Canada-sea route was long but secure, whereas the overland route was dangerous because couriers had to pass through Colonist-controlled territory and were thus subject to interception. Communications for the Colonists, on the other hand, were much easier since they had interior lines of communication and greater support from the local population.

At the strategic level, the outcome of this one battle dramatically changed the course and context of the war. The major geostrategic result of the battle at Saratoga and British surrender was French recognition of the United States. This seemly simple act was essentially a French declaration of war against England (Morrissey, 2000; Schnitzer, 2002). The French commitment to the Colonial cause was solidified, and help came in many forms - diplomatic recognition, military aid, and a morale boost for the Colonists.

5. CONCLUSION

The Battle of Saratoga study illuminates the effect micro- and macro-geographic factors (physical and cultural) have on the outcome of battle. They clearly support the historians' claim that this was a decisive battle and that the outcome of the American War of Independence was profoundly altered as a result. This geographic analysis highlights the factors that affected the battle and the people who fought it. The physical and cultural landscapes were recognized as important then and serve as a valuable learning tool today.

REFERENCES

Auburey, T. 1963. *With Burgoyne from Quebec: An Account of the Life at Quebec and the Famous Battle of Saratoga.* Toronto: Macmillan.

Braun, E.L. 1974. *Deciduous Forests of Eastern North America.* New York: Hafner Press.

Galgano, F.A. 2004. Decisive terrain - A military geography of Fortress West Point, 1775-1797. In *Studies in Military Geography and Geology*, D.R. Caldwell, J. Ehlen, and R.S. Harmon, eds., Dordrecht, The Netherlands: Kluwer Academic Publishers, 105-120.

Ketchum, R.M. 1997. *Saratoga, Turning Point of America's Revolutionary War.* New York: Henry Holt and Company.

MacDonald, J. 1985. *Great Battlefields of the World.* London: Marshall Editions, Ltd.

Morrissey, B. 2000. *Saratoga 1777, Turning Point of a Revolution.* Oxford: Osprey Publishing.

Palka, E.J. 2004. A military geography of the Hudson Highlands - Focal point in the War of American Independence. In *Studies in Military Geography and Geology*, D.R. Caldwell, J. Ehlen, and R.S. Harmon, eds., Dordrecht, The Netherlands: Kluwer Academic Publishers, 89-103.

Russell, E.W.B. 2001. Applications of historical ecology to the land-management decisions in the northeastern United States. In *Applying Ecological Principles to Land Management*, U.H. Daleand and R.A. Haeuber, eds., New York: Springer-Verlag, 119-135.

Schnitzer, E. 2002. *Battling for the Saratoga Landscape 1777* (Unpublished manuscript). Saratoga National Historic Park, NY: National Park Service.

Chapter 11

THE IMPACT OF GEOLOGY ON THE MARCH TO THE BATTLE OF EUTAW SPRINGS

Irene B. Boland[1] and Charles A Boland[2]
[1] *Winthrop University*
[2] *Rock Hill, SC*

Abstract: Atlantic Coastal Plain geology impacted Continental and British strategy in many ways prior to the Battle of Eutaw Springs during summer 1781. The elevation of the Upper Coastal Plain High Hills of Santee influenced General Nathanael Greene to encamp his Continental army there to gain relief from the heat, humidity, and diseases associated with the adjacent Middle Coastal Plain swamps. The strategic advantage offered by high ground inside a large Middle Coastal Plain meander lobe of the Congaree River influenced Lieutenant Colonel Alexander Stewart to encamp his British force there in anticipation that Greene would attack from across the river and swamps. However, Greene marched around the swamps because the sandy roads facilitated advancement. To avoid being trapped, Stewart moved to high ground at Eutaw Springs, where a stream emerging from a cave in the Santee Limestone provided abundant fresh water and nearby Coastal Plain terraces offered fresh food.

Key words: Plain, High Hills of Santee, karst, Santee Limestone, Eutaw Springs

1. INTRODUCTION

The geology of three Atlantic Coastal Plain landforms - the High Hills of Santee, the Wateree, Congaree, and Santee swamps, and karst topography developed on Santee Limestone at Eutaw Springs - influenced Continental and British strategy from July to 8 September 1781, prior to the American War of Independence Battle of Eutaw Springs. This paper describes the geology of those landforms and discusses its strategic influence.

D.R. Caldwell et al. (eds.), Studies in Military Geography and Geology, 133–146.
© 2004 *Kluwer Academic Publishers.*

2. GEOLOGIC OVERVIEW

Extending southeast from the Fall Zone to the Atlantic coast, the stratigraphic units of the Atlantic Coastal Plain (Fig. 1) compose a sequence of Upper Cretaceous to Holocene sedimentary rocks and sediments deposited on the eastern flank of the Upper Proterozoic to Middle Cambrian crystalline rocks of the Piedmont province. The Coastal Plain dips gently to the southeast; hence, the formations become progressively younger and increase in thickness toward the coast. Two major marine transgressions and numerous minor episodes have sculpted the Coastal Plain landscape into a stair-step series of terraces and scarps (Colquhoun, 1974; Soller and Mills, 1991; Hockensmith, 2001).

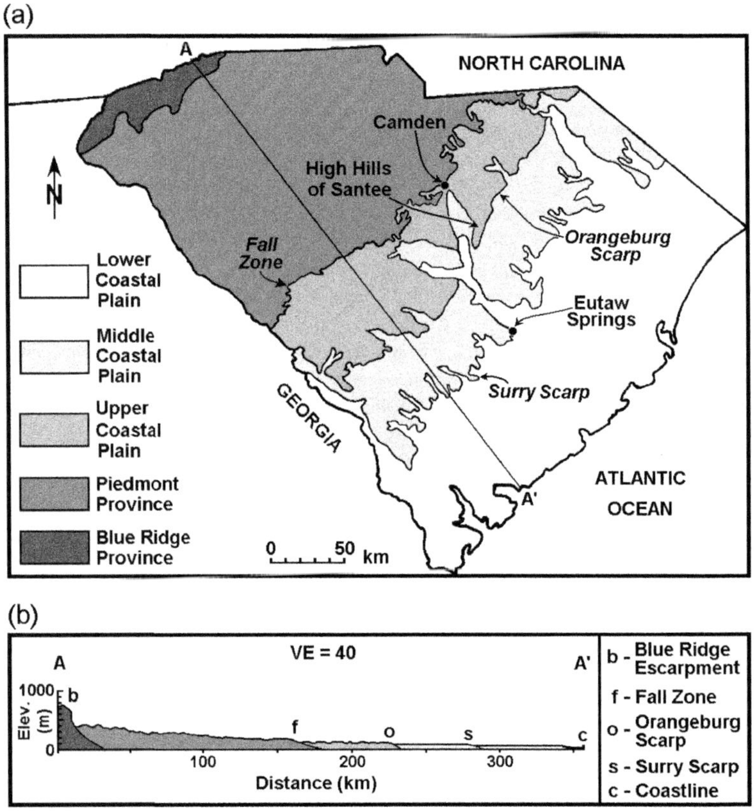

Figure 1. (a) Geomorphic provinces of South Carolina. Modified after DuBar et al. (1974) and Colquhoun (1974). (b) Generalized cross-section of the geomorphic provinces.

The Fall Zone, bordering the northwest margin of the Upper Coastal Plain, is a sinuous erosionally shaped boundary. Streams typically have moderately steep gradients and rapids where crossing this boundary. As shown in Fig. 1, the Coastal Plain comprises three distinct geomorphic regions: The Upper, Middle, and Lower Coastal Plains (Colquhoun, 1969).

The Upper Coastal Plain, located between the Fall Zone and Orangeburg Scarp, is characterized by an erosional topography consisting of rolling hills with several hundred feet of relief and sandy soils that differ markedly from the red clay soils typical of the Piedmont. In stark contrast to the hilly topography of the Piedmont and Upper Coastal Plain, the relatively flat topography of the Middle and Lower Coastal Plains (Fig. 2) is characterized by swamps, such as those associated with the Wateree, Congaree, and Santee Rivers, and vast flat terraces with relief of 5 m or less. The terraces were shaped by wave scour and deposition under shallow coastal conditions; the sedimentary sequence comprises terrigenous sands and clays, beach sands, marl, and limestone deposited under shallow coastal to deep off-shore marine conditions (Colquhoun and Muthig, 1991).

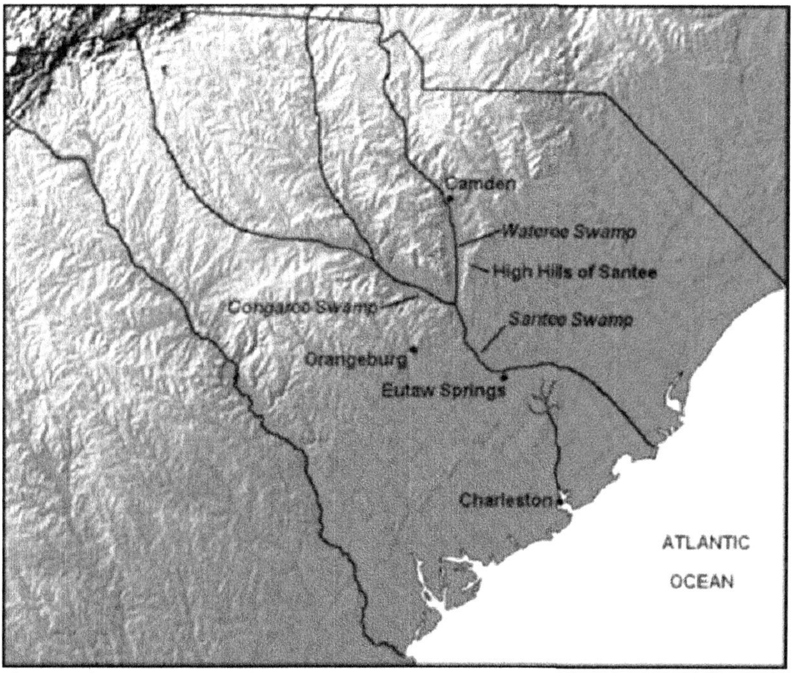

Figure 2. Relief map of South Carolina, c. 1781. Modified after Sterner (1995).

2.1 The High Hills of Santee

The High Hills of Santee are a prominent landform of the Upper Coastal Plain. They are remnants of an Upper Cretaceous to Upper Miocene sequence (Nystrom and Willoughby, 1992) bounded on the east by the Orangeburg scarp and on the west by the Middle Coastal Plain floodplain of the Wateree River (Figs. 1 and 2). They form a ridge about 32 km long, less than a kilometer wide to the south and almost 6.4 km wide to the north. The High Hills are up to 90 m higher in elevation than the adjacent Middle Coastal Plain and swamp. Relief of 30 m or more over short distances is common, and there is a paucity of large flat land areas. The High Hills exists because younger erosion resistant clayey-sandy Tertiary sediments protect the underlying sandy Cretaceous sediments from erosion (Nystrom and Willoughby, 1992).

2.2 The Congaree, Wateree, and Santee Swamps

The average gradient of the Wateree and Congaree Rivers within the Fall Zone is 2 m/km. Below the Fall Zone, in the Congaree, Wateree, and Santee swamps on the Middle Coastal Plain (Fig. 2), the average gradient is 0.2 m/km. Where the Wateree and Congaree meet to form the Santee River the swamp is 10 km wide. Elsewhere the swamps are about 5 km wide.

2.3 The Santee Limestone and Eutaw Springs

The Santee Limestone is a major sedimentary unit of the Middle and Lower Coastal Plains (Fig. 3). Based on the fossil assemblage, Charles Lyell assigned a Middle Eocene age to the Santee Limestone in 1845 (Siple, 1957). The Santee Limestone is the northern member of the Floridan aquifer (Hockensmith, 2001). The high permeability and porosity of the aquifer are due to dissolution of the limestone bedrock by acidic groundwater (Aucott et al., 1987). Such dissolution leads to the formation of karst topography, typified by caves and sinkholes, prominent landforms in areas shallowly underlain by Santee Limestone. Subterranean streams flow through caves or cave groups in the Santee Limestone from the aquifer head (the interfluves between the Santee and Edisto Rivers) to tributaries in the Santee drainage system and to the south shore of Lake Marion (Siple, 1975).

As shown in Figure 3a, the top and bottom of the Santee Limestone sequence are macrofossiliferous and the middle of the sequence is microfossiliferous (Edwards et al., 1997). The macrofossiliferous nature of the top and bottom of the sequence suggests deposition under shallow high-energy conditions. The microfossiliferous nature of the middle of the sequence suggests deposition in deep water. Thus, the fossiliferous nature of the Santee Limestone emphatically records sea level changes during the Eocene Tejas transgressive-regressive cycle (Sloss, 1963).

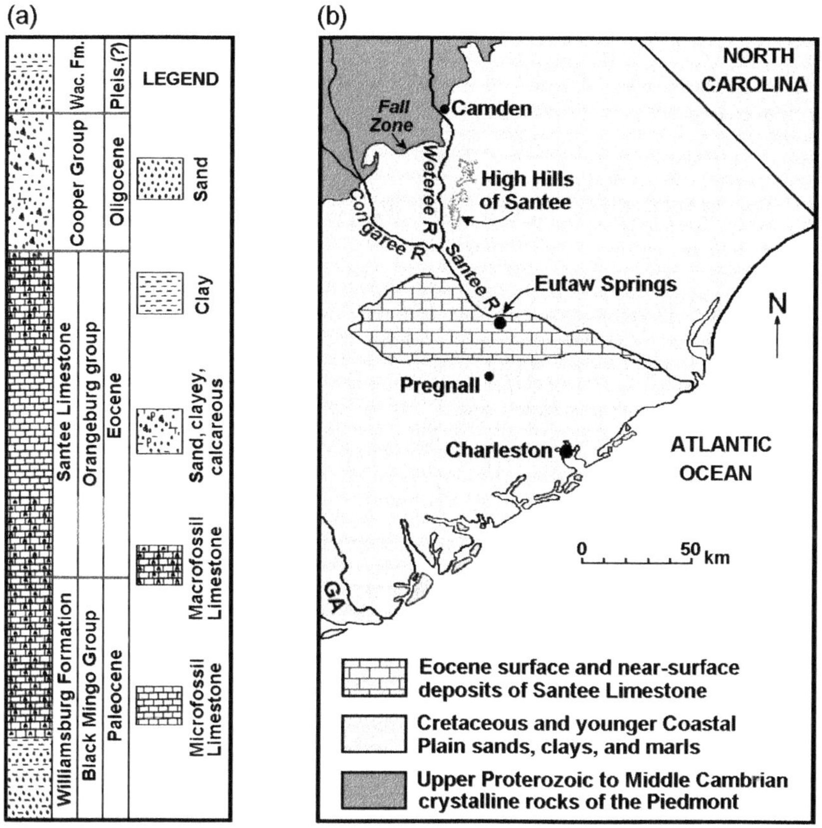

Figure 3. (a) Stratigraphic column from the Pregnall No. 1 Corehole. Modified after Edwards et al. (1997). (b) Surface and near-surface extent of the Santee Limestone in South Carolina. Modified after Heron (1962).

Eutaw Springs is located on the Lower Coastal Plain in a zone of karst topography near the base of the Surry Scarp (Figs. 1, 2, 3, and 6). The springs for which Eutaw Springs is named were inundated in 1941 when the Santee River was dammed to create Lake Marion. Prior to that, a flow great enough to turn a large mill wheel was discharged from a subterranean stream into a 9 m deep ravine at the head of Eutaw Creek (Lossing, 1859). From there the creek flowed 3 km to the Santee River across the Wicomico terrace (Fig. 6). Lossing (1859) observed that water issuing from a small spring flowed approximately 50 m, disappeared into a cavern and then reappeared at a large spring on the other side of a ridge after traversing 150 m underground. According to legend, Eutaw Indian braves made sport of diving into the cavern, swimming under the ridge with the strong current and reappearing at the large spring (Glenn, 1970). Dense woods grew along the creek and open woods grew within 5 km of the springs in a region pockmarked by sinkholes (Lossing, 1859). Vast flat farm fields lay 5 km to the west and 2 km to the south on the marine-scoured Okefenokee and Wicomico Terraces (Colquhoun and Duncan, 1966).

3. OVERVIEW OF THE WAR IN THE CAROLINAS

The British Southern Campaign to subdue the southern Colonies and advance northward began in 1780 (Fig. 4). The Southern Campaign began in March 1780 at Charleston, SC where Continental troops and Militia ultimately surrendered in May. Following the Continental defeat at Camden in August, General Nathanael Greene, from Rhode Island, assumed command of the Southern Continental Army in December 1780 near Charlotte, NC. The British, led by Lord Cornwallis, pursued Greene into North Carolina and Virginia. In March 1781, both armies met at Guilford

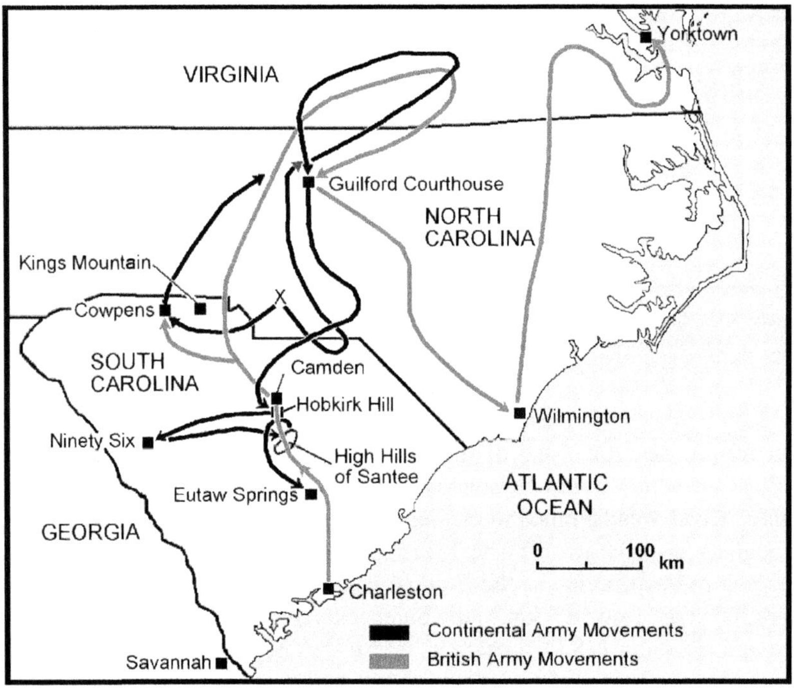

Figure 4. The Southern Campaign. Modified after Divine (2000).

Courthouse, NC, where the British won, but at cost. The weakened main British army retreated to Wilmington, NC and then eventually to Yorktown, VA. With the main British army out of the way, Greene returned to South

Carolina and, aided by Militia and State troops, began attacking the remaining small British outposts, with the objective of forcing the British to retreat to Charleston. After the Battle of Hobkirk Hill and the Siege at Ninety Six, Greene encamped in the High Hills of Santee on 16 July 1781 (Figs. 4 and 5) to await reinforcements, rest and recuperate his troops, and formulate his next move.

4. THE MARCH TO EUTAW SPRINGS

By late August 1781, Greene had assembled 1500 Continental regulars and local militia; he knew that a British force of 2000-2300 troops, under the command of Lieutenant Colonel Alexander Stewart, was camped at Thompson's plantation in a large meander lobe on the south side of the Congaree, near its confluence with the Wateree to form the Santee (Conrad, 1997; Russell, 2000). The forces were within 24 km of each other but separated by the 10 km wide Congaree-Wateree-Santee swamp at the confluence (Fig. 5).

Feeling strong enough to go on the offensive again, Greene set out on 23 August to confront Stewart (Fig. 5). However, his direction of approach was influenced by the fact that a hurricane had drenched South Carolina on 10 August and the swamps were still full of water, ". . . up to a horse's belly for miles in the lowlands" (Conrad, 1997, 242). Lacking sufficient boats to cross the swamps, but knowing that the sandy roads surrounding them would allow for easy travel, Greene opted to march around the swamps - traversing first north for 48 km to cross the Wateree at Camden, then southwest for 56 km to cross the Congaree at Howell's Ferry, and finally southeast for 24 km to Stewart's camp on the south side of the Congaree. When Greene finally reached the British camp on 2 September, 10 days after leaving the High Hills, he learned that Stewart's force had moved south 58 km to Eutaw Springs. Greene followed Stewart and bivouacked at Burrell's Tavern, 11 km north of Eutaw Springs on 7 September, 15 days and 203 km after leaving the High Hills (Conrad, 1997). Along the march from the High Hills, Greene's forces were joined by 900 militia and state troops, increasing his army to 2400 prior to attacking Stewart at Eutaw Springs (Russell, 2000).

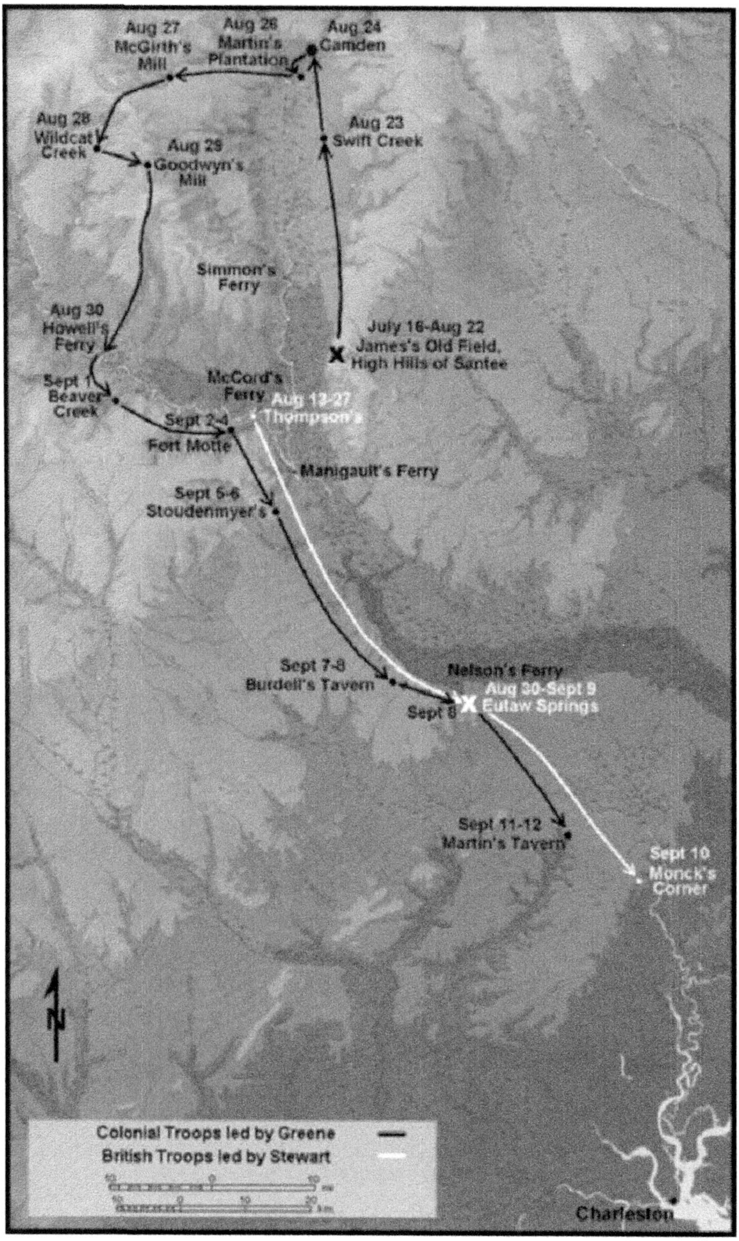

Figure 5. Colonial and British troop movements in South Carolina during the summer of 1781. Based on accounts in Conrad (1997).

5. THE BATTLE OF EUTAW SPRINGS

At Eutaw Springs, a garden and two-story brick house capped the interfluve between the springs and a large flat clearing straddled the road, the main road to Charleston (Figs. 6 and 7). Stewart's troops had been camped in the clearing since 30 August. Stewart sent troops out each morning to gather sweet potatoes. However, in unfriendly territory and lacking reliable intelligence, he was unaware of Greene's exact location and progress toward his camp. Thus, on the morning of 8 September, Greene's force met Stewart's rooting party 6 km west of the springs, gathering potatoes from fields near the main road (Fig. 6). Greene's troops quickly overran the rooting party and skirmishers sent 4 km out by Stewart to meet him after being alerted to his presence by the rooting party cavalry escort (Ripley, 1983; Morrill, 1999).

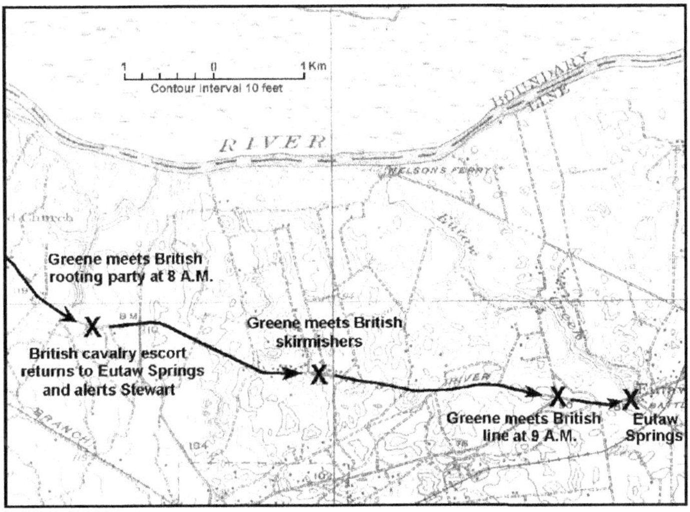

Figure 6. Topographic map of Eutaw Springs and vicinity, pre Lake Marion, showing the karst features and time-distance relationships at the onset of the Battle of Eutaw Springs (USGS, 1921). Based on accounts by Ripley (1983) and Morrill (1999).

Both sides formed battle lines while exchanging artillery fire. Stewart set his line of troops astride the road about 0.8 km west of his camp. His right flank on the north was anchored by the steep, thickly wooded bank of Eutaw Creek; his left flank on the south was in open woods; Greene formed two lines astride the road, with the militia units in front and the Continental regulars behind (Fig. 7). The militia units advanced but were halted and slowly pushed back by the British, fighting with bayonets. Continental units, also fighting with bayonets, moved up to support the militia, halted the

British advance and forced them back through their camp. The Continentals, then thinking they had won the battle, stopped to rummage in the British tents. A detachment of British troops occupied the house and directed fire on the rummagers, giving Stewart an opportunity to re-form and counterattack. Stewart's troops subsequently pushed Greene's troops back until Continental reserves came up and stabilized the line. By this time, Greene had lost numerous officers; sensing disorganization, he left a cavalry unit at the springs to provide a presence and retreated back to Burrell's tavern for much needed water and rest (Ripley, 1983; Lumpkin, 1987; Morrill, 1999). Stewart stayed at the springs that night. The next morning, he left his wounded and retreated toward Charleston. Greene followed, almost to Moncks Corner, but neither side was in a position to continue battle. Greene then returned to his camp in the High Hills. Stewart followed, but did not cross the Santee, and eventually returned to Charleston (Conrad, 1997).

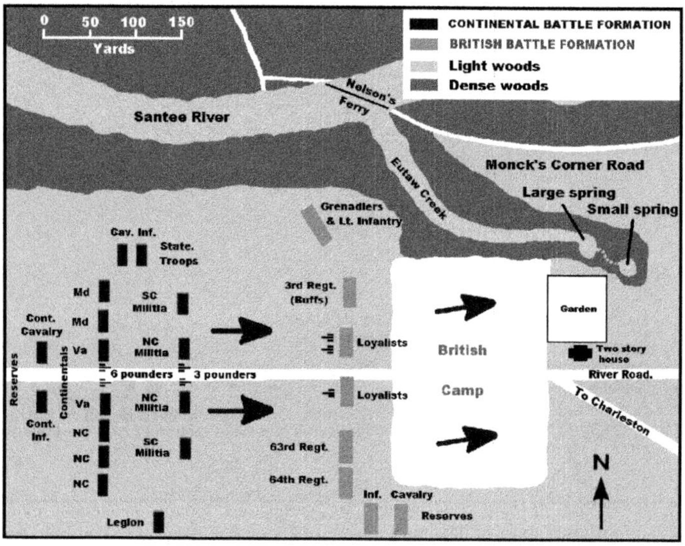

Figure 7. Battle map of the Battle of Eutaw Springs. Modified after Ward (1952).

6. DISCUSSION

The geology of the Atlantic Coastal Plain had a major impact on Continental and British strategies during the summer of 1781 prior to the Battle of Eutaw Springs. For example, the healthful attributes of the High Hills of Santee influenced Greene to use them as a retreat to bide the summer, await reinforcements, and rest his troops. Heat, humidity, and sickness were major concerns during the Southern Campaign (Ward, 1952).

The temperature during the day is as much as 5° C cooler in the High Hills than in the adjacent swamps. The word "santee" is a derivation of the French word for "health"; during summer, Low Country residents used the High Hills as a retreat from the heat, mosquitoes, yellow fever, and malaria associated with the swamps and marshes (Wright, 1976). Greene knew of the healthful attributes of the High Hills through his association with General Thomas Sumter, who owned a summer home in the High Hills at Stateburg.

The day before Greene broke camp on 22 August, he notified Colonel Henry Lee that he was going to march by way of Camden to attack Stewart, noting that the swamps were still flooded, the road to Camden was good, the river crossing at Camden was easy, and that he would meet reinforcements along the way (Conrad, 1997). Stewart's camp was on the south side of the Congaree, 24 km southwest of Greene's position on the east side of the Wateree (Fig. 5). Following the hurricane of 10 August, the dense vegetation, flat topography, and low gradient (0.2 m/km) of the 10 km wide swampy Wateree and Congaree floodplains separating the two camps exacerbated and prolonged the flood. In contrast, at Camden, where the Wateree crosses the Fall Zone, the gradient is steeper (0.5 m/km), the floodplain is much narrower (0.6 km), and the surface elevation of the Wateree is 15 m higher than where it meets the Congaree. As Greene expected, the flood had abated at Camden by 24 August; and his army crossed the Wateree there in two days. Also as he anticipated, the permeable sandy soils on the roads along the way did indeed enable easy advancement, allowing his army to march up to 40 km in a day.

Stewart had strategically camped on high ground in a meander lobe on the south side of the Congaree, expecting Greene to attack from across the swamp and river. When Stewart learned that Greene was marching around the swamp and about to cross the Congaree at Howell's Ferry to his west, to avoid being trapped in the lobe, he moved his force to Eutaw Springs (Ripley, 1983). He likely chose that location for a number of reasons. Eutaw Springs was a prominent location marked on both British and Colonial maps. Moreover, it was on the main road and 58 km closer to Charleston, and he was expecting reinforcements to arrive from Charleston. At Eutaw Springs, Stewart camped in a large, flat clearing astride the road and adjacent to the springs. The road provided an easy escape route to Charleston. The site was defendable, protected on the north side by the steep bank and thick vegetation along the creek and river. The other sides, though pockmarked with sinkholes, were surrounded by open woods. Furthermore, the springs provided abundant fresh groundwater, and the flat farm fields on the marine-scoured terraces located nearby provided fresh food for easy gathering (Ripley, 1983; Morrill, 1999).

The Battle of Eutaw Springs, the last major battle in South Carolina between Continental and British regulars, lasted about five hours. Casualties (dead, wounded, and missing in action) were heavy. Greene lost 554, or 23% of his force; Stewart lost 693, or 30% of his force (Ripley, 1983). Although whether there was a tactical winner is debatable, strategically the battle was another in which the British suffered irreplaceable losses that ultimately convinced them to abandon occupation of South Carolina.

7. CONCLUSIONS

The geology of the Atlantic Coastal Plain impacted Colonial and British strategy prior to the Battle of Eutaw Springs in the following ways:

(1) Greene used the Upper Coastal Plain High Hills of Santee as a retreat to rest his army because elevations above those of the adjacent Middle Coastal Plain swamps provided relief from the oppressive heat and diseases associated with the swamps.

(2) Although six times farther, Greene marched his army around, rather than across, the Wateree and Congaree swamps to attack Stewart because the swamps were still flooded 13 days following a hurricane - due to their flat topography, low gradient, and vegetation - and the permeable sandy Coastal Plain soils surrounding the swamps provided an excellent roadbed for easy troop movement.

(3) Aware that Greene was camped east of the swamps in the High Hills, and expecting him to attack from the north across the swamps and rivers, Stewart strategically chose to encamp on the south side of the Congaree River on high ground in a large meander lobe.

(4) When he learned that Greene instead was about to cross the Congaree to his west, to avoid being trapped in the meander lobe, Stewart moved his troops south to Eutaw Springs.

(5) At Eutaw Springs, Stewart most likely chose to encamp on high ground in a large flat clearing astride the road to Charleston because the site was defendable (protected on the north by steep banks and dense vegetation along Eutaw creek), the springs provided abundant fresh groundwater, and the flat farm fields on the nearby marine-scoured sandy plains of the Coastal Plain terraces provided fresh food.

ACKNOWLEDGEMENTS

The authors thank Robert Whisonant, Judy Ehlen, and Russell Harmon for insightful reviews that greatly improved the quality of this paper.

REFERENCES

Aucott, W.R., Davis, M.E., and Sperian, G.K. 1987. Geohydrologic framework of the Coastal Plain aquifers of South Carolina. US Geological Survey Water-Resources Investigations, Report 85-4271.

Colquhoun, D.J. 1969. Geomorphology of the Lower Coastal Plain of South Carolina. Columbia, SC: South Carolina State Board of Development, Division of Geology, MS-15.

Colquhoun, D.J. 1974. Cyclic surficial stratigraphic units of the Middle and Lower Coastal Plains, central South Carolina. In *Post-Miocene Stratigraphy, Central and Southern Atlantic Coastal Plain,* R.Q. Oaks, Jr. and J.R. Dubar, eds., Logan, UT: Utah State University Press, 179-190.

Colquhoun, D.J. and Duncan, D.A. 1966. Geology of the Eutawville Quadrangle, South Carolina. Columbia, SC: South Carolina State Board of Development, Division of Geology, MS-12.

Colquhoun, D.J. and Muthig, M.G. 1991. Stratigraphy and structure of the Paleocene and Lower Eocene Black Mingo Group, South Carolina. In *The Geology of the Carolinas,* J.W. Horton, Jr. and V.A. Zullo, eds., Knoxville, TN: University of Tennessee Press, 241-250.

Conrad, D.M., ed. 1997. *The Papers of Nathanael Greene, Vol. IX, July-December 1781.* Chapel Hill, NC: University of North Carolina Press.

Divine, R.A., Breen, T.H., Fredrickson, G.M., and Williams, R.H. 2000. *America Past and Present.* New York: Addison Wesley Longman, Inc.

Dubar, J.R., Johnson, H.S., Jr., Thom, B.G., and Hatchell, W.O. 1974. Neogene stratigraphy and morphology, south flank of the Cape Fear Arch, North and South Carolina. In *Post-Miocene Stratigraphy, Central and Southern Atlantic Coastal Plain.* R.Q. Oaks, Jr. and J.R. Dubar, eds., Logan, UT: Utah State University Press, 139-173.

Edwards, L.E., Bybell, L.M., Gohn, G.S., and Frederiksen, N.O. 1997. Paleontology and physical stratigraphy of the USGS-Pregnall No. 1 Core (DOR-208), Dorchester County, South Carolina. US Geological Survey Open-File Report 97-145.

Glenn, E.S. 1970. *A Legend of Eutaw Springs.* Orangeburg, SC: Orangeburg County Tri-centennial Programs.

Herron, S.D. 1962. Limestone resources of the Coastal Plain of South Carolina. Columbia, SC: South Carolina State Board of Development, Division of Geology, Bulletin 28.

Hockensmith, B.L. 2001. Potentiometric map of the Floridan Aquifer and Tertiary Sand Aquifer in South Carolina - 1998. Columbia, SC: South Carolina Department of Natural Resources, Water Research Report 23.

Lossing, B.J. 1859. *The Pictorial Field-Book of the Revolution, Volume II.* New York: Harper and Brothers Publishers.

Lumpkin, H. 1987. *From Savannah to Yorktown.* New York: Excel Press.

Morrill, D.L. 1999. *Southern Campaigns of the American Revolution.* Mount Pleasant, NC: Nautical and Aviation Publishing Company of America.

Nystrom, P.C. and Willoughby, R.H. 1992. Cretaceous and Tertiary stratigraphy of the High Hills of Santee, Western Sumter and Lee Counties, South Carolina. Columbia, SC: South Carolina Geological Survey Field Trip Guidebook 24.

Ripley, W. 1983. *Battleground - South Carolina in the Revolution.* Charleston, SC: Evening Post Publishing Company.

Russell, D.L. 2000. *The American Revolution in the Southern Colonies.* Jefferson, NC: McFarland Company Publishers.

Siple, G.E. 1957. Carolina Geological Society guidebook for the South Carolina Coastal Plain field trip, November 16-17, 1957. Columbia, SC: South Carolina State Development Board Division of Geology Bulletin 24.

Siple, G.E. 1975. Ground-water resources of Orangeburg County, South Carolina. Columbia, SC: South Carolina State Board of Development, Division of Geology Bulletin 36.

Sloss, L.L. 1977. Sequences in the cratonic interior of North America. Geological Society of America Bulletin 74: 93-113.

Soller, D.R. and Mills, H.M. 1991. Surficial geology and geomorphology. In *The Geology of the Carolinas*, J.W Horton, Jr. and V.A. Zullo, eds., Knoxville, TN: University Tennessee Press, 290-308.

Sterner, Ray. 1995. Black and white landform map of South Carolina. (Online Version: http://fermi.jhuapl.edu/states)

Ward, C. 1952. *The War of the Revolution*, Vol. II. New York: MacMillan Company.

Wright, L.B. 1976. *South Carolina -A Bicentennial History.* Nashville, TN: W.W. Norton and Company.

Chapter 12

THE 1815 BATTLE OF NEW ORLEANS

A Physical Geographical Analysis

Richard W. Dixon
Texas State University

Abstract: The closing battle of the War of 1812 occurred on the floodplain of the Mississippi River 7 mi south of New Orleans. Strategically, control of New Orleans determined control of the Mississippi River and, by extension, most of the western territory. Tactically there were six approaches to the city. All had some geographic disadvantage, mostly relating to the disparity between the draft of the vessels available to the British and the water depths in the various lakes, channels, and bayous. The eventual choice of approach via Lake Borgne and Bayou Bienvenue placed the invading army in logistical peril as all supplies had to be ferried over long distances in small, open boats. The battleground itself was beset with a number of obstacles of a physical geographic nature such as a shallow depth to water table, inadequate cover, poor soils, and unseasonable weather.

Key words: New Orleans, Mississippi River, weather, floodplain geomorphology, bayous

1. BACKGROUND AND INTRODUCTION

The American Congress passed a declaration of war against Great Britain on 18 June 1812 in response to perceived British interference in American free trade. At the time the British were embroiled in a protracted war with France that had started in 1793. The Americans had attempted to remain neutral as that war spread to involve most of the major European powers. In the end, the Americans found themselves caught between the competing strategies of England and France. Combined with an increasing American nationalism and a British disdain for its ex-Colony, the Americans were led into a war they were not prepared to fight (Caffrey, 1977; Hickey, 1989). For two years, the British and Americans clashed in a series of inconclusive battles, but British attention was clearly focused on Europe and the French.

D.R. Caldwell et al. (eds.), Studies in Military Geography and Geology, 147–154.
© 2004 *Kluwer Academic Publishers.*

Following the defeat and exile of Napoleon to Elba in April 1814, the British turned full attention to the Americans, and planned a three-pronged attack to fracture the country and lead to American concessions as a condition of peace. The British were not successful in prosecuting this strategy as two disastrous defeats befell them in September 1814. An attempt to isolate the New England region by an attack down the Hudson River valley to seize New York City was thwarted by the destruction of the British naval force at the Battle of Lake Champlain. In the mid-Atlantic, Washington, DC was seized and burned by the British sending the American government into disarray. But British plans in this theater were dealt a mortal blow with their defeat at Baltimore and the death of Major General Robert Ross. Ross was eventually replaced by Lieutenant General Sir Edward Pakenham for the third prong of the attack, an assault against New Orleans. New Orleans was the key to the American western frontier. Control of New Orleans assured control of the Mississippi River and large sections of the interior. Tactically, there were six approaches to the city. All had some geographic disadvantage mostly relating to the disparity between the draft of the vessels available to the British and the water depths in the various lakes, channels, and bayous needing to be traversed to attack the city.

The Battle of New Orleans began on 23 December 1814 with the arrival of a British invasion force at the Villeré Plantation located on the east bank of the Mississippi River at Chalmette about 7 mi downriver from New Orleans. The Americans, under the command of Major General Andrew Jackson, established a defensive position upriver along the Rodriguez Canal between the Chalmette and Macarty Plantations. There followed, over the next 16 days, a series of engagements culminating with the final unsuccessful assault by the British on 8 January 1815. Many aspects of this engagement, from the selection of the site to the conduct of the actual battle, were influenced by the physical geography of the area. Geographic factors such as relative location, landforms, soils, vegetation, weather, and hydrography impacted the decisions made by the commanders in the field. This paper highlights three physical geographic factors that played crucial roles in the outcome of the battle - hydrography, landforms, and weather.

2. HYDROGRAPHY

Fig. 1 details the six possible approaches to the city (Brooks, 1961); the Mississippi River, Lake Borgne, Lake Pontchartrain, or one of three winding routes through bayous and swamps from the Bay of Barataria, Bayou La

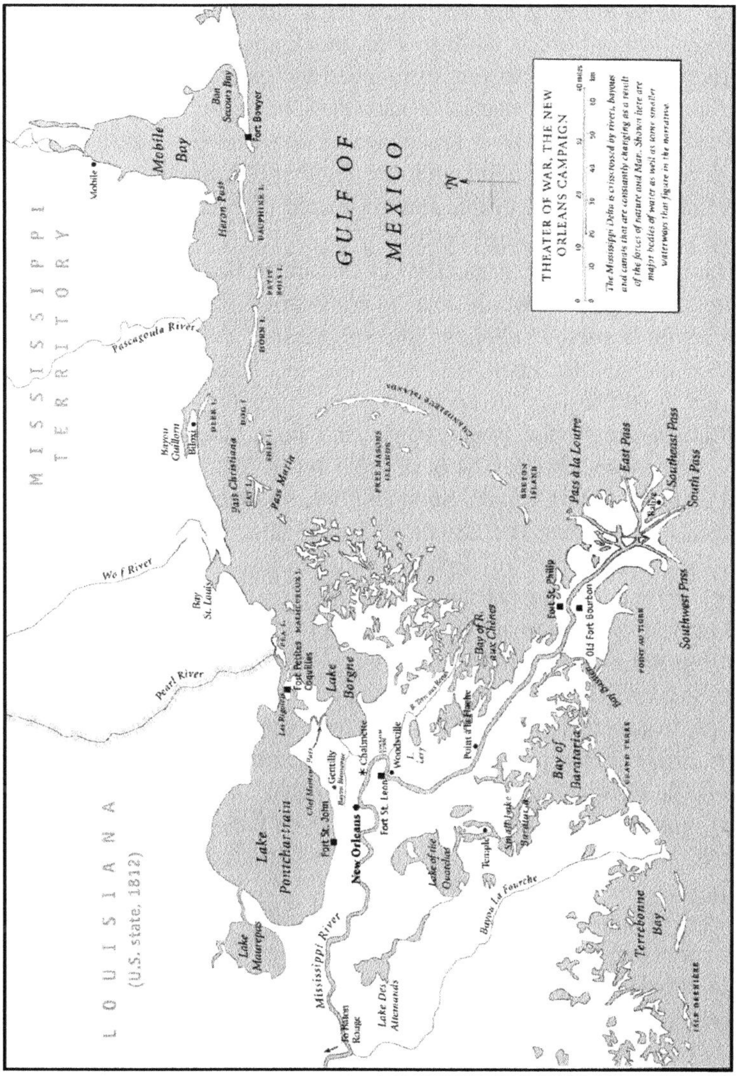

Figure 1. Map of approaches to New Orleans from Reilly (2002). Used with permission of Robin Brass Studio, Inc., Toronto.

Fourche, or Bayou Terre Aux Boeufs. New Orleans lies 90 mi from the mouth of the Mississippi and three major obstacles are present in this reach of the river. First, the bar at the mouth of the river had a depth of 12 ft (Brown, 1969), preventing the largest of British warships from entering. Second, any ships moving upriver would have to attack and destroy the American garrison at Fort St. Phillip. This attack had to be a bombardment from the river because the fort was surrounded by impenetrable swamp. Any fleet overcoming these obstacles would then have to navigate through English Turn, a bend in the river requiring sailing vessels to wait for a wind shift for successful passage. In addition, another American garrison was located at English Turn. All of this activity would take place on a river with variable water levels and strong currents.

The swamp approach from the Bay of Barataria, Bayou La Fourche, or Bayou Terre Aux Boeufs would lead to trafficability problems. A typical Louisiana bayou is narrow, shallow, and often filled with silt deposits. The banks of the bayous are lined with tall, reed-like vegetation that inhibits visibility and would require the construction of temporary cordoroy roads to support the passage of the many troops and their supplies. The narrow, twisting waterway and limited lines of sight would result in disorientation for all but experienced watermen. Such an invasion route also required more shallow-draft vessels than the British had available. Prior to sailing from England, the commander of British naval forces, Vice Admiral Sir Alexander Cochrane, requested that his invasion fleet be outfitted with flat-bottomed landing craft upon arrival in Jamaica (the jumping-off point for the invasion), but none were provided (Brown, 1969). In addition to shallow-draft craft, the British would also require the assistance of local guides to navigate the twisting passages of the swamps and bayous. The notorious pirate Jean Lafitte was approached by the British and offered $30,000 for his cooperation (de Grummond, 1961). Lafitte refused and later provided men and critical material to Jackson's defense. A third factor against the choice of approach through the swamps was the almost assured loss of surprise by the invading forces.

An approach via Lake Pontchartrain would also require shallow draft vessels to successfully navigate the narrow passage (Les Rigolets) between Lakes Pontchartrain and Borgne. This area was also known for its treacherous tidal currents and strong winds (Reilly, 1974). However, this was widely believed to be the most logical approach and Jackson had major troop concentrations in the area at Fort Petites Coquilles and Fort St. John (Walker, 1856).

After dismissing the other five options, the British settled on the approach via Lake Borgne and connecting bayous. Their lack of landing craft required the use of ship's barges and other small ship's boats for transport. Admiral Cochrane organized a "flotilla" of these shallow-draft craft and sailors to row the almost 60 mi from the fleet anchorage at Pea Island (near the mouth of the Pearl River) via Lake Borgne and Bayou Bienvenue to a landing site a short distance from the Villeré Plantation on the east bank of the Mississippi River, 7 mi downriver from New Orleans.

Due to the limited number of shallow-draft craft available, the invasion force was split into three brigades with about 2000 men in each. The brigades were transported one at a time by Cochrane's sailors. The first brigade arrived at Villeré Plantation on the morning of 23 December 1815. Their arrival was a complete surprise, but they were too fatigued from the long water journey and too few in number to effect an attack on the city. It would take most of the day to land the remaining two brigades. By then the element of surprise was lost and Jackson was preparing to attack the invaders. This inability to land the entire invasion force and strike a decisive early blow gave Jackson time to organize his strong defensive position along the Rodriguez Canal between the Chalmette and Macarty Plantations. This position was directly upriver from the adjacent Villeré Plantation Thus the British would be forced to attack and carry Jackson's position before marching on New Orleans.

3. LANDFORMS

The armies faced off on a narrow strip of floodplain bordered on the British left by the 4 ft high river levee and on their right by thick cypress woods and swamp. The width of the floodplain varied from 0.5 mi at Jackson's line to near 1 mi at the British position at the Villeré Plantation. The plantation fields were covered in cane stubble and crisscrossed by drainage ditches and low fences (Brooks, 1961). There was little shelter for cover and concealment and the water table depth of approximately 1 ft prevented the construction of any trenches. Jackson had constructed an earthen berm (Fig. 2) using slaves from nearby plantations to haul soil from surrounding fields. He set his defense behind this berm, anchoring it to the levee on his right and the swamp on his left. The berm was fronted by the 10 ft wide Rodriguez Canal and measured 8 ft from the bottom of the canal to the top of the berm (de Grummond, 1961). The British supply lines were impossibly long, yet they managed to supply their troops and even bring forward heavy naval artillery to attempt to breach Jackson's earthworks. The artillery duel of 1 January 1815 was a disaster for the British as their heavy guns became mired in the soft alluvial soils rendering them useless. Unable to breach Jackson's defenses, the only recourse for the British was a frontal attack. Their attack on 8 January 1815 was a resounding defeat and resulted in 2037 casualties (291 killed, 1262 wounded, 484 taken prisoner). Among the dead were most of the senior British officers including the commanding general Sir Edward Pakenham (Remini, 1999).

Figure 2. View of Jackson's line looking toward the river along the constructed berm.

4. WEATHER

Tree ring chronologies indicate the winter of 1814-15 was an El Nino winter (Logh, 1992). Such winters in the south tend to be cooler and wetter than normal (Ropelewski and Halpert, 1986) with frontal storm passages occurring at twice the frequency of storms in non-El Nino winters (Muller and Rohli, 2002). All of the previous referenced sources contain numerous examples of the wet and cold weather endured by the troops. Table 1 is an extract of daily weather conditions from those sources. This weather timeline constructed from the various reports indicates a series of cold front passages marked by rain or sleet followed by cold, blustery, north winds. Frequent mention is made of frost overnight and in the early morning hours. For the British, on an exposed field with little shelter and at the end of a very long supply chain, these conditions surely hampered morale and fighting effectiveness. Hardest hit by this weather were the two West Indian regiments that joined the British expedition at Jamaica and were not equipped for cold, wet weather (Brooks, 1961). The frequent rain also contributed to mobility problems on the battlefield and trafficability issues for the British supply lines, which stretched over 60 mi by boat and foot to their supply base and fleet anchorage at Pea Island.

Table 1. Weather conditions during the New Orleans Campaign, December 1814-January 1815. Compiled from Walker (1856), Brooks (1961), de Grummond (1961), Rankin (1961), Brown (1969), and Reilly (1974).

Date	Temperature	Precipitation	Wind	Visibility	Remarks
1814					
22 Dec	Cold	PM rain	North		Night freeze
23 Dec	PM warm			Clear	Night fog
24 Dec	Cold		Northerly		Night freeze
25 Dec	Cold	PM rain	Northwest		
26 Dec	Cold	Rain	Northerly		Night freeze
27 Dec			Northerly		
28 Dec	Cool			Clear	AM frost
29 Dec	Cool				Fair
30 Dec	Cool				Fair
31 Dec	Cool		Calm	Fog in PM	
1815					
01 Jan		PM rain		Fog in AM	
02 Jan	Cold	Rain			
03 Jan					
04 Jan					
05 Jan					
06 Jan					
07 Jan					
08 Jan		Rain		Fog	
09 Jan					
10 Jan		Heavy rain			Uncovers graves

5. CONCLUSIONS

The Battle of New Orleans was strongly impacted by the physical geography of the landscape. British General John Keane in his Journal of Operations report to the Duke of Wellington cited six factors in the defeat of the British forces: 1) length of supply lines through the swamp, 2) lack of intelligence, 3) soil and weather conditions, 4) loss of surprise, 5) lack of landing craft, and finally, 6) a short enemy line, strongly flanked and impossible to turn (Brooks, 1961). Five of the reasons cited have their roots in the distinctive physical geography of the New Orleans battlefield site. Indeed, Major C. R. Forrest, Assistant Quartermaster General for the British, wrote in his diary: "It cannot therefore with justice be said that it was owing to any other than the natural obstacles of the Country that the object of the Expedition was unsuccessful" (Rankin, 1961, 44).

REFERENCES

Brooks, C.B. 1961. *The Siege of New Orleans.* Seattle, WA: University of Washington Press.

Brown, W.S. 1969. *The Amphibious Campaign for West Florida and Louisiana, 1814–1815.* Tuscaloosa, AL: University of Alabama Press.

Caffrey, K. 1977. *The Twilight's Last Gleaming: Britain vs. America 1812–1815.* New York: Stein and Day.

de Grummond, J.L. 1961. *The Baratarians and the Battle of New Orleans.* Baton Rouge, LA: Louisiana State University Press.

Hickey, D.R. 1989. *The War of 1812: A Forgotten Conflict.* Urbana, IL: University of Illinois Press.

Lough, J.M. 1992. An index of the Southern Oscillation reconstructed from western North American tree-ring chronologies. In *El Nino: Historical and Paleoclimatic Aspects of the Southern Oscillation*, H.F. Diaz and V. Markgraf, eds. New York: Cambridge University Press.

Muller, R.A. and Rohli, R.V. 2002. Synoptic associations of regional flooding rainstorms in Louisiana, 1911-2000. *Southwestern Geographer* 6:1-32.

Rankin, H.F. 1961. *The Battle of New Orleans: A British View.* New Orleans, LA: The Hauser Press.

Reilly, R. 1974. *The British at the Gates.* New York: G.P. Putnam's Sons.

Reilly, R. 2002. *The British at the Gates.* Toronto: Robin Brass Studio, Inc.

Remini, R.V. 1999. *The Battle of New Orleans: Andrew Jackson and America's First Military Victory.* New York: Penguin Books.

Ropelewski, C.F and Halpert, M.S. 1986. North American precipitation and temperature patterns associated with the El Nino/Southern Oscillation (ENSO). Monthly Weather Review 114:2353–62.

Walker, A. 1856. *Jackson and New Orleans.* New York: J.C. Derby.

Chapter 13

TERRAIN AND ITS AFFECT ON THE USE OF ARTILLERY IN THE AMERICAN CIVIL WAR

The Battle of Perryville 8 October 1862

Judy Ehlen[1] and Robert J. Abrahart[2]
[1] *US Army Engineer Research and Development Center*
[2] *University of Nottingham, UK*

Abstract: The affects of dissected limestone terrain on the use and effectiveness of field artillery during the American Civil War are shown using examples taken from the Battle of Perryville, KY, a duel between artillery batteries and for artillery positions. Smoothbore weapons were more effective over shorter wavelength, more dissected terrain, whereas modern rifled cannon proved to be more advantageous over longer wavelength, more open terrain. Terrain-based optical illusions also had a significant affect on the outcome of the battle. These influences are illustrated using different methods of visualization and analysis based on a 30-ft-resolution raster digital elevation model.

Key words: American Civil War, Battle of Perryville, field artillery, battlefield terrain, military history

1. INTRODUCTION

The Battle of Perryville was fought between General Braxton Bragg's Confederate Army of the Mississippi and Major General Don Carlos Buell's Union Army of the Ohio. The two armies met on 7/8 October 1862 in the rolling hills just west of the small town of Perryville in the Bluegrass Region of Kentucky (Fig. 1). Kentucky, although a border state and neutral, was considered to have southern sympathies and it was believed that the presence of Bragg's army in the state would incite its citizens to rise in support of the Confederacy, that recruits would flock to join the Confederate armies, and that numerous weapons would be provided. To gain control of the state was thus the strategic objective of Bragg's Kentucky Campaign. For well-

D.R. Caldwell et al. (eds.), Studies in Military Geography and Geology, 155–172.
© 2004 *Kluwer Academic Publishers.*

documented accounts of the campaign see Connelly (1967) and McDonough (1994).

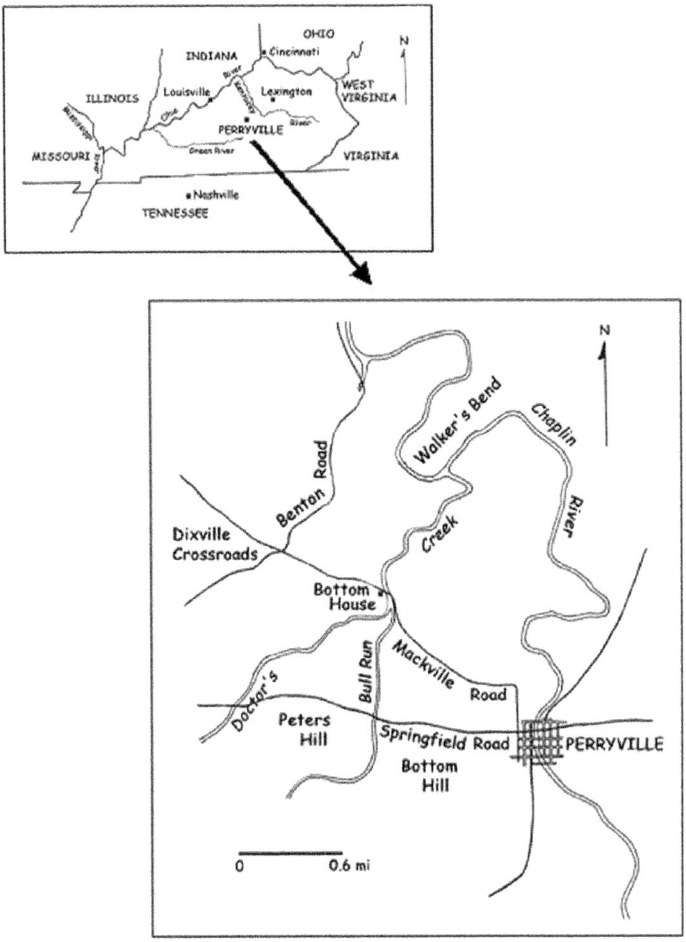

Figure 1. Location of Perryville, KY and details of the battlefield.

Perryville was a small market town with a population of about 500 (McDonough, 1994). The center of a rich agricultural region, the town sits astride the Chaplin River in Boyle County, KY (Fig. 1). Under normal conditions, there is an abundant supply of surface water with numerous creeks in addition to the Chaplin River. However, at the time of the battle, the region was suffering from a severe drought (OR, 1886). The Chaplin River and its major tributary, Doctor's Creek, were the only sources of water available over a large region and it was the need for water that attracted the two armies to this location (OR, 1886).

The ensuing battle was a duel between artillery batteries and for artillery firing positions, in which the nature of the terrain was influential in controlling the success of different types of weapons and, to a large extent, shaped the final outcome of the battle. The particular benefits and advantages of rifled cannon over smoothbores - longer ranges, greater accuracies, and higher striking velocities - had been demonstrated by the French and the British during the Crimean War (1854-56; Holmes, 2001). Yet their overall effectiveness in dissected, wooded terrain was brought into question during the American Civil War. The terrain over which the Battle of Perryville was fought exemplifies the limitations and advantages of the different types of field artillery in the mid-19th century and demonstrates how the relative effectiveness of such weapons could influence the outcome of a battle.

2. THE PERRYVILLE TERRAIN

Perryville is located in classic karst terrain - a landscape of green, gently rolling, limestone hills with abundant springs, sinkholes, swallow holes, caves, and dry valleys (see Day, 2004, this volume, for military operations in another type of karst terrain). Cressman (1974) indicates that the ridges are capped by interbedded limestones, shales, and siltstones of the Clays Ferry Fm., and the lower slopes and valleys are comprised of the Lexington Limestone, which is more susceptible to dissolution. The rounded hilltops and ridge crests are almost concordant, with a slight downward slope on the battlefield to the north. Slopes are convex and can be quite steep, particularly along the larger streams.

The main features of this terrain are the result of dissolution and differential weathering along joints. The major structural trend near Perryville is northeast, as exhibited by the course of Doctor's Creek (Fig. 1), with a lesser trend to the northwest. These patterns are reflected in the topography: The ridges and major stream valleys trend northeast parallel to the major joint set, whereas small tributaries that trend northwest, parallel to the minor joint set, form more dissected terrain between the major joints. The wider valleys and larger hills possess greater relief, but the slopes are longer, such that the landscape has a gentle, open, rolling appearance. However, southwest of Walker's Bend (Fig. 1), along the ridge where valleys parallel the northwest-trending joint set, relief is lower because the hills and ridges are closer together, and the topography appears to be more rugged and confined.

The topography had a significant effect on the employment of artillery during the battle, because of the different ranges of the weapons that were involved (Table 1) and variations in the distances between opposing artillery batteries. There are two types of topography at Perryville: More "open terrain" in the valley of Doctor's Creek parallel to the major northeast joint set, and more "dissected terrain" on the ridge southwest of Walker's Bend where valleys parallel the minor joint set (Fig. 1). These two terrain types

Table 1. Maximum ranges (yd) for typical Civil War cannon and the projectiles that were fired. Compiled from Gibbon (1863) and Ordnance Manual (1995).

	Range of projectiles at maximum elevation		
	Solid Shot[1]	Spherical Case Shot	Canister[2]
Smoothbores			
6-pounder	1523	1200	300-600
12-pounder (Napoleon)	1680	1300	300-600
Rifles			
3-inch	N/A	4180	N/A
10-pounder Parrott	N/A	6200	N/A

[1] Solid shot was not used for rifled cannon.

[2] Canister was also a smoothbore projectile, but on occasion, particularly when ammunition was low, gunners fired canister from rifled cannon.

can be differentiated according to "topographic wavelength" - the average distance from hillcrest to hillcrest or from valley to valley. Topographic wavelength is longer in the open terrain along Doctor's Creek such that the opposing batteries were 600-2300 yd apart (Figs. 2A, 2B, and 3; Table 2). Confederate Major General William J. Hardee described this terrain in his report of the battle:

> The country near Perryville is boldly undulating and varied with farm-houses, corn fields, and plantations, bordered by native forests. A creek called Chaplin Fork flows north-wardly though the village and unites 4 or 5 miles beyond with another little stream called Doctor's Fork. The space between the two from east to west is about 1½ miles. . . The position at Perryville is strong, and offered many tactical and strategical advantages (OR, 1886, 1120).

Table 2. Distances (yd) between batteries involved in the barrage based on their initial positions east and west of Doctor's Creek in longer wavelength terrain.

	Carnes	Lumsden	Semple	Darden
Loomis	1730	670	1080	1070
Simonson	1770	600	1200	1170
Harris	2330	1070	1970	1900

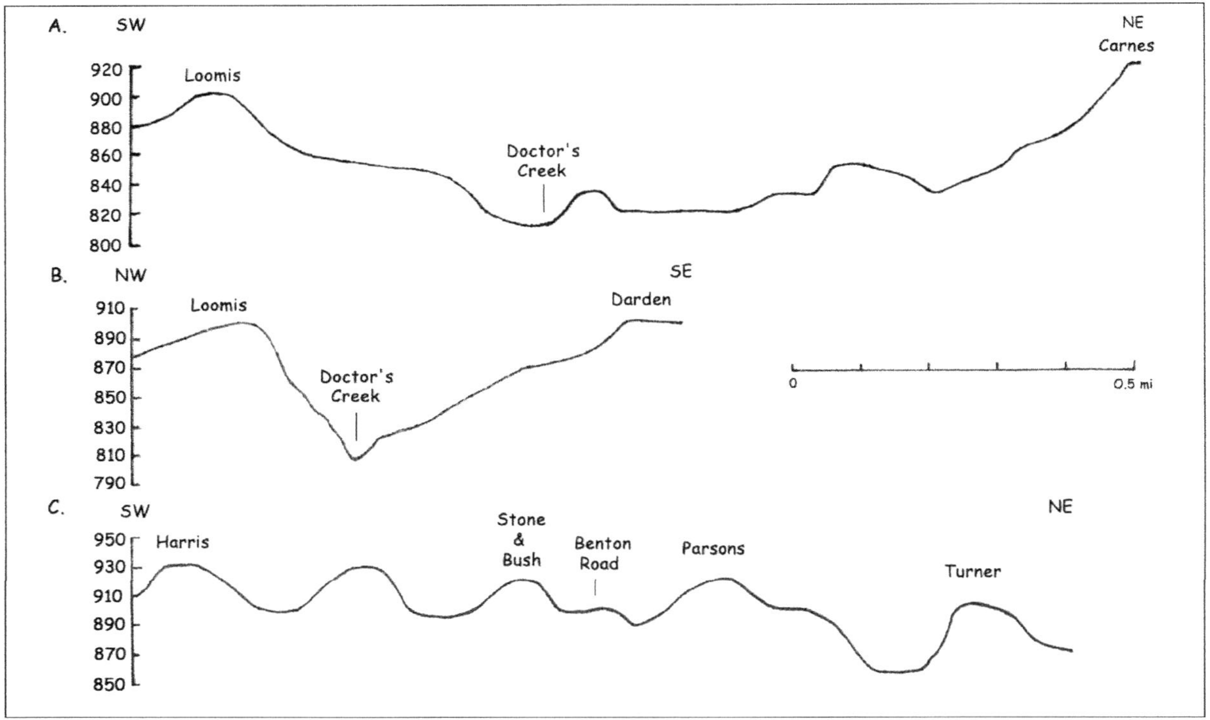

Figure 2. Topographic profiles of the battlefield showing differences in topographic wavelength. Vertical exaggeration is 13 times the horizontal distance. The vertical scale is in feet. A. Longer wavelength, open terrain parallel to the major northeast joint set along Doctor's Creek between Loomis' and Carnes' batteries. B. Across the longer wavelength terrain between Loomis' and Darden's batteries. C. Shorter wavelength, dissected terrain between Harris' and Turner's batteries southwest of Walker's Bend.

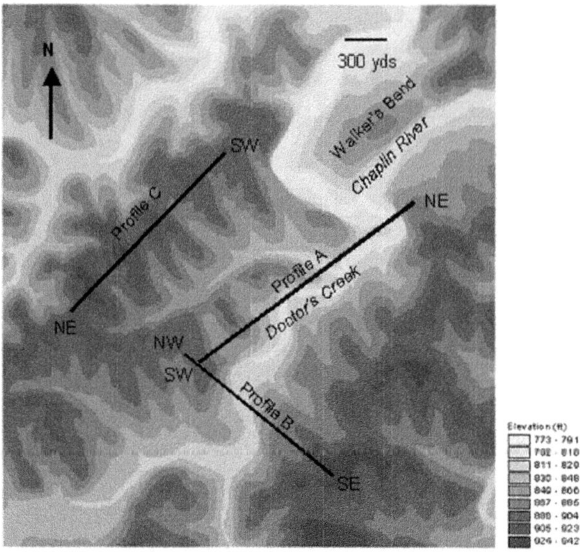

Figure 3. Locations of the terrain profiles shown in Fig. 2.

Topographic wavelength is much shorter southwest of Walker's Bend - some 300-1600 yd (Figs. 2C and 3; Table 3). Union Major General Alexander McDowell McCook described this terrain, that on his left flank, in his report of the battle:

> The battle-field was a chosen one for the enemy. . .The ground upon which the battle was fought was very much broken by hills and deep ravines, which afforded every facility to them for concealing their troops. . . The bluffs and dry channels of Chaplin River and Doctor's Fork afforded them every advantage for concealing and massing large bodies of troops (OR, 1886, 1042).

Table 3. Distances (yd) between batteries based on their initial positions southwest of Walker's Bend in shorter wavelength terrain.

	Parsons	Stone and Bush
Turner	330	700
Carnes	1030	1170

According to Kurt Holman (park manager, Perryville Battlefield State Historic Site, pers. comm., 2002), the battlefield looks today very much as it did in 1862. It is comprised of arable farmland with small patches of woodland, and narrow strips of mature trees along streams and fences. Trees on the north- sloping plateau surface are isolated or occur in widely scattered, small clumps. There were more fences and buildings on the battlefield in 1862 because some were destroyed during the battle and because of changes in farming practices. The road network is very much the same as it was in 1862; some roads that were mere farm tracks at the time of the battle, however, are now paved. One major difference in the terrain is that some of the larger sinkholes have been filled in and so no longer provide the formidable obstacles that existed in 1862.

3. THE BATTLE OF PERRYVILLE

Bragg's Army of the Mississippi, which arrived first on the battlefield, consisted of two wings, one commanded by Major General Leonidas Polk, and the other, by General Hardee. The two wings of Bragg's army had advanced together and thus deployed near Perryville at the same time. Both faced west, Hardee's on the left, south of Perryville, and Polk's on the right, north of Perryville. Buell's Army of the Ohio consisted of three corps: First Corps commanded by General McCook, Second Corps commanded by Major General Thomas L. Crittenden, and Third Corps commanded by Brigadier General (Acting Major General) Charles C. Gilbert. Third Corps arrived first on 7 October and took position west of Peters Hill (Fig. 1); Second Corps, which arrived next, deployed further to the south and west; and First Corps, which did not arrive until the morning of 8 October, deployed north and to the left of Third Corps. The total number of men involved was as follows: Bragg's two wings numbered just under 17,000; whereas Buell's three corps amounted to over 55,000 - although it was only the 20,000 men of Union First Corps who were engaged in the battle. For detailed descriptions of the battle see Hafendorfer (1991) or Noe (2001).

Noe (2001) provides the compositions of the artillery batteries in each army; those referred to herein are listed in Table 4. Although Bragg's Confederate army had 13 batteries made up of 56 guns, only 11 batteries took part in the battle. These batteries comprised 50 guns, of which 16% were rifles. Buell's Union army had 26 batteries made up of 147 guns. However, because Second Corps and most of Third Corps were not engaged in the battle, only 11 Union batteries participated. These batteries comprised 66 guns, of which 32% were rifles. The composition of individual batteries was also at odds: Only three of the 13 Confederate batteries, all of which were involved in the battle, contained rifles, whereas all of the Union batteries that were engaged had at least one rifle. Buell thus had two clear numerical advantages in terms of weapons during the battle: A 4:3 advantage with respect to total artillery pieces and an 8:3 advantage with respect to rifled cannon.

The main battle began with an artillery barrage at about 1230 on 8 October. Nine batteries were involved, four Union batteries and five Confederate batteries. About 1400 the Confederate infantry advanced on the

Union position southwest of Walker's Bend (the current location of Perryville Battlefield State Historic Site; Fig. 1). This part of the battle involved three Confederate infantry brigades and two artillery batteries; and three defending Union infantry brigades and four artillery batteries. Fighting in this sector stopped at sunset, about 1630, and culminated with the Union line forced back about 300 yd from its initial position.

Table 4. The batteries at Perryville and their guns. Compiled from Noe (2001).

	Smoothbores		Rifles			
	6-pdrs	12-pdrs	3-in	3.3-in	James	Parrott
Confederate Batteries						
Carnes	4					
Darden	2	2				
Semple		4			2	
Swett	6	2				
Lumsden		4				
Key		2				
Stanford			3			
Turner	2	2				
Slocomb	2	2		2		
Union Batteries						
Loomis						6
Simonson	2	2			2	
Hotchkiss		2				
Parsons		7				1
Bush	2	2			3	
Stone	2				2	2

Shortly after fighting began southwest of Walker's Bend, about 1430, the Confederates also attacked near the Bottom House (Fig. 1). This part of the battle, which occurred across the valleys of Doctor's Creek and Bull Run, involved five Confederate infantry brigades and nine artillery batteries against two defending Union infantry brigades and seven artillery batteries. Fighting here continued until after darkness fell; the Union line was pushed back about 1000 yd to the Dixville Crossroads (Fig. 1). Moreover, to avoid development of a highly vulnerable salient, the left flank of the Union line southwest of Walker's Bend moved to the vicinity of the Dixville Crossroads.

Even though the battle had ended with a tactical offensive victory for the Confederates, Bragg believed his army was too small to continue the fight - he had not found the reinforcements, arms, or supplies that had been expected in Kentucky - so he pulled back to Harrodsburg, to the east, and then followed his invasion route back into Tennessee. Bragg was correct in his assumption: 35,000 fresh Union troops still faced his tired, parched army on the evening of 8 October (OR, 1886). His Kentucky Campaign thus was a strategic loss - Bragg won the battle, but not the state (Griffith, 1989).

4. ARTILLERY AND TERRAIN

Two examples will be used to show how terrain affected the use of field artillery during the Battle of Perryville; first in the longer wavelength, open terrain along Doctor's Creek (Figs. 1, 2A, 2B, and 3); and second, in the shorter wavelength, dissected terrain southwest of Walker's Bend (Figs. 1, 2C and 3). The terrain profiles in Fig. 2 are drawn through battery positions; the names of the battery commanders are shown for reference.

4.1 Longer Wavelength, Open Terrain

The initial artillery deployments were on high ground on both sides of Doctor's Creek (Fig. 4). The Union batteries were on slightly higher ground, west of the creek, whereas the Confederate batteries were east of the creek, and closer to the town of Perryville. Union batteries commanded by Captains Cyrus Loomis, Peter Simonson, Samuel Harris, and William Hotchkiss opposed Confederate batteries commanded by Captains Putnam Darden, Henry Semple, William Carnes, and Charles Lumsden, and Lieutenant Thomas Key's section of Calvert's Arkansas Battery. The four Union batteries contained 18 guns, 11 of which were rifles (61%). The five Confederate batteries contained 20 guns, only two of which were rifles (10%). Table 2 shows the distances between the batteries that were involved

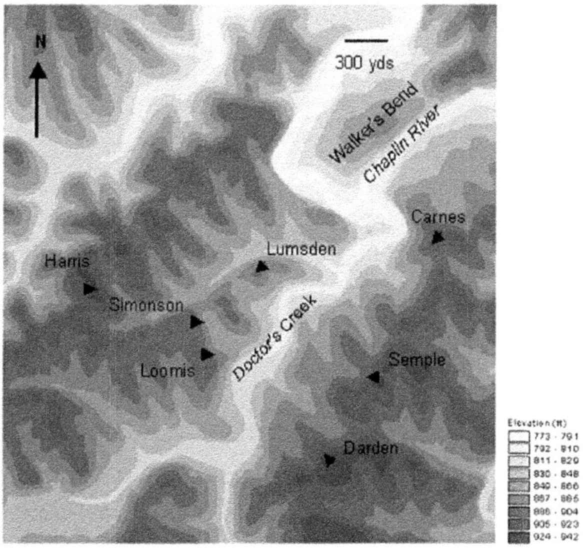

Figure 4. Initial deployment of artillery batteries during the barrage. Hotchkiss's position on Peters Hill is south of the area shown here (see Fig. 1). Key's position is unknown.

in the barrage, for which the exact positions have been identified. Fig. 2A shows the open terrain between Loomis' and Carnes' batteries that is parallel to the major joint set. Fig. 2B shows the dissected terrain between Loomis' and Darden's batteries that crosses this trend.

Both Loomis and Simonson, with their longer range rifles (Table 4), initially overshot Carnes' position whereas Hotchkiss' 12-pounder howitzers on Peters Hill (see Fig. 1) were unable to reach that far. The limited firepower of Carnes' 6-pounders was soon augmented by Lumsden's and Key's 12-pounders. At the height of the exchange, Carnes' battery was replaced by Captain T.J. Stanford's longer-range 3-in rifles, weapons that were more appropriate to the longer wavelength, open terrain in this area (Table 1).

Lumsden's fire against Simonson's battery was initially affected by the illusive nature of the terrain: From his position, Simonson's battery appeared to be located on the crest of the next ridge, and relic canister balls from Lumsden's guns have been found in that position (William Andrews, Kentucky Geological Survey, pers. comm., 2002; Fig 5A). Simonson's

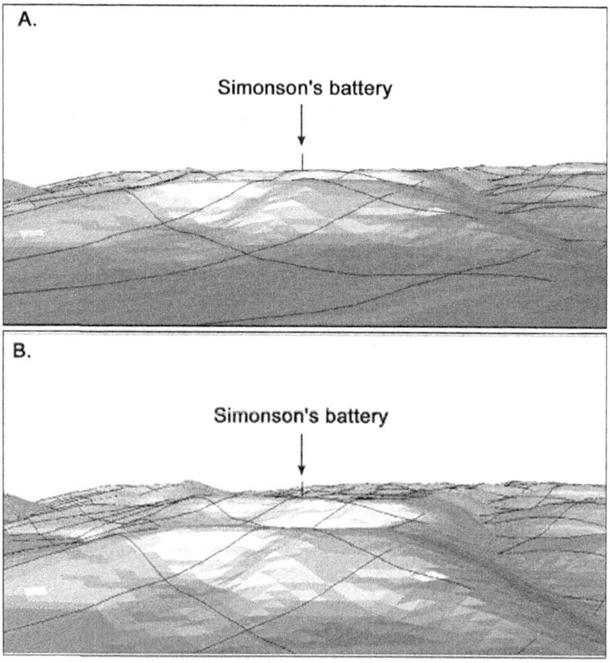

Figure 5. Terrain model looking south from Lumsden's position to Simonson's battery. A: The terrain viewed from Lumsden's position: Simonson's battery appears to be on the top of the next ridge. B. In this elevated view from the same point, the presence of an intervening valley can just be discerned showing that Simonson's battery was in fact positioned on the second, not the first, ridge.

battery was in fact positioned on the next ridge over (further southwest), but this was not visually apparent from Lumsden's position (Fig. 5B): The ridge upon which Simonson's defensive battery was deployed is slightly higher than the ridge upon which Lumsden thought it was located. Lumsden's location was also at a slightly lower elevation than the middle ridge, and from his position, the two ridges appear to be one. As the Union lines were pushed back by the combined efforts of infantry and artillery, Stanford, taking advantage of the open nature of the terrain, moved forward to replace Lumsden's Napoleons with his 3-in rifles (Tables 1 and 4). The fields of fire for Confederate batteries engaged in the barrage, for which exact positions have been identified, are shown in Fig. 6.

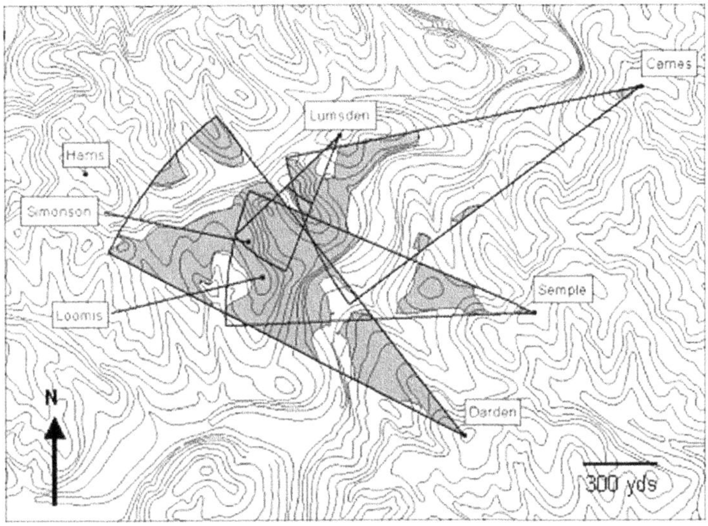

Figure 6. Weapons fan diagrams over 10-ft contour intervals for Confederate batteries during the barrage (see Guth, 2004, this volume, for information on weapons fans). The fields of fire for all batteries are set at 25°. The weapons, elevations, and projectiles used for each fan are: Carnes, 6-pounder, 5E, solid shot; and Darden, 12-pounder, 5E, solid shot; Semple, 12-pounder, 3E45', shell; Lumsden, 12-pounder, 5E, canister. Shaded section denotes visible region that could be fired upon using line-of-sight targeting from one or more positions. Hotchkiss's position on Peters Hill is south of the area shown here (see Fig. 1). Key's position is unknown. Data from Gibbon (1863), Ordnance Manual (1995), and Noe (2001).

4.2 Shorter Wavelength, Dissected Terrain

As noted above, the main battle began southwest of Walker's Bend about 1400 after the barrage was over. Due to faulty reconnaissance by Confederate cavalry commander Colonel John Wharton (about 1330), the initial Confederate attack was made against the strong Union center, in contrast to the intended and more vulnerable target, which was McCook's left flank (Connelly, 1967; Noe, 2001). Fig. 7 shows the initial positions of the batteries engaged. The shorter wavelength terrain forced the Confederates to attack with regiments, one attacking after another, rather than by brigades on line. Furthermore, the topography forced Confederate regiments to attack what they could see, artillery batteries on the high ground, not the Union infantry deployed lower on the slopes.

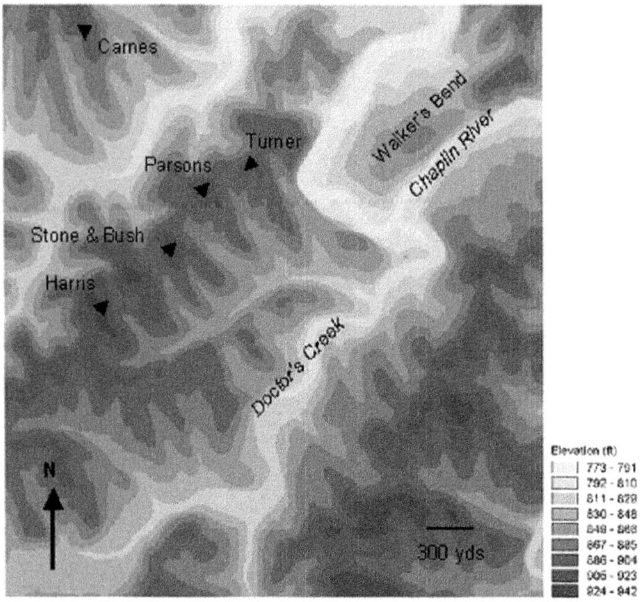

Figure 7. Initial deployment of artillery batteries at the beginning of the battle southwest of Walker's Bend.

The Union batteries involved in this area were commanded by Lieutenant Charles Parsons and Captains David Stone, Asahel Bush, and Samuel Harris. The Confederate batteries were commanded by Lieutenant William Turner and Captain Carnes (Fig. 7). Carnes moved to this area after retiring from his initial position during the barrage. The Union batteries contained 27 guns, 10 of which were rifles (37%), and the Confederate batteries, eight guns, none of which were rifles. The shorter wavelength, dissected terrain provided excellent artillery positions for both Union and Confederate batteries as well as good cover for advancing infantry (Fig. 2C). Table 3 shows the distances between batteries engaged in this part of the battle. The terrain on a line from Harris' battery to Turner's battery is shown in Fig. 2C. The fields of fire for the Confederate batteries engaged in this area are shown in Fig. 8.

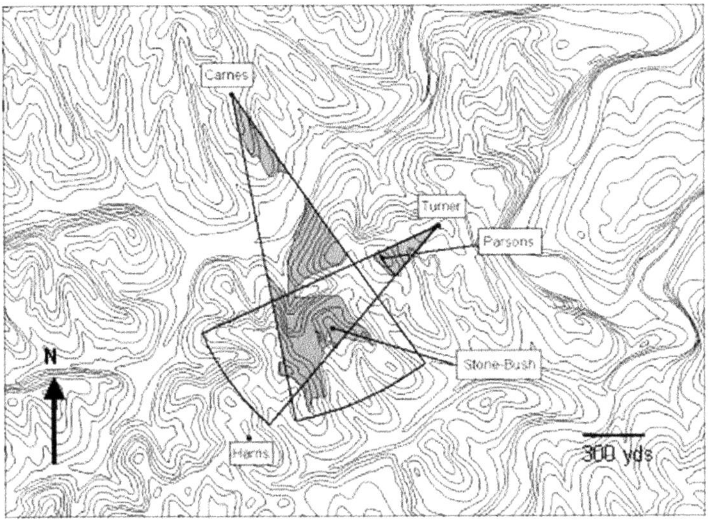

Figure 8. Weapons fan diagrams over 10-ft contour intervals for Confederate batteries southwest of Walker's Bend (see Guth, 2004, this volume, for information on weapons fans). The fields of fire for all batteries are set at 25°. The weapons, elevations, and projectiles used for each fan are: Carnes, 6-pounder, 5E, solid shot; and Turner, 12-pounder, canister (inner fan), spherical case shot, 4E (outer fan) (Table 1). Shaded section denotes visible region that could be fired upon using line-of-sight targeting from one or more positions. Data from Gibbon (1863), Ordnance Manual (1995), and Noe (2001).

Although the Confederate infantry attack was initially repulsed (it had no artillery support), a combined onslaught led by Confederate Brigadier General George Maney's brigade, supported by Turner's and Carnes' batteries, finally succeeded in pushing the Union infantry and artillery batteries back. Parsons' battery, deployed in front of the main Union line on an isolated knob, was forced back by the combined actions of the infantry advance and Turner's battery. Turner deployed his 6- and 12-pounders about 250-300 yd north of Parsons during the infantry advance: The shorter wavelength in this area allowed Turner's smoothbores to compete at least on an equal basis with the rifles of Parsons, Stone, and Bush (Tables 1 and 3).

The Union batteries of Stone and Bush were forced back by the Confederate 1st Tennessee, supported by Turner's and Carnes' batteries. Turner had moved forward to occupy Parsons' abandoned position, and Carnes' battery was now in his new position to the northwest (Fig. 7). The Tennessee regiment flanked the two Union batteries, approaching unseen from the north-northwest along a steep-sided valley on the Union left (Fig. 9A). The terrain masked the advance of the 1st Tennessee until they crested the slope, only tens of yards away from the two batteries (Fig. 9B; OR, 1886) - the gunners were focused on the Confederate infantry advance immediately to their front and right, and were under fire from Turner's battery, also to their front (Fig. 7). Bush, firing canister, was unable to depress the barrels of his guns sufficiently to repulse the Confederate infantry once they came into view. The short wavelength and dissected nature of the terrain in this area allowed Carnes' and Turner's smoothbores to be used with good affect, assisting the infantry attack.

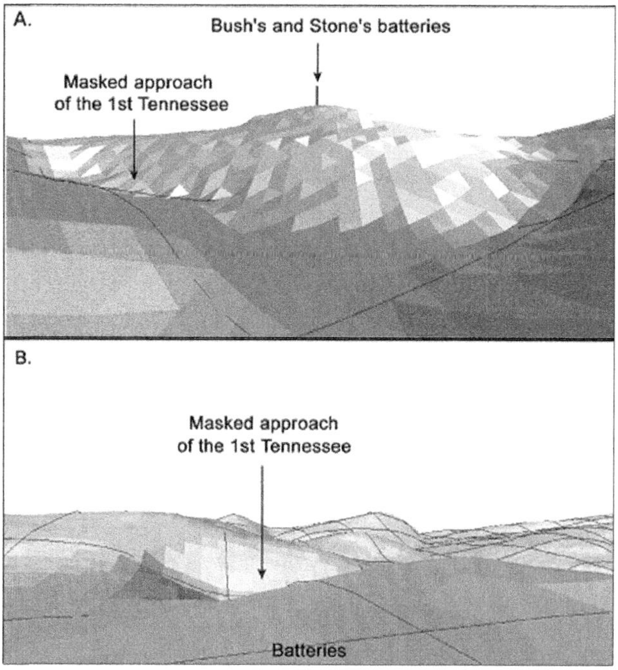

Figure 9. Terrain model showing the Confederate infantry attack on Bush's and Stone's batteries. A. View looking north showing the route taken by the 1st Tennessee as they approached Bush's and Stone's position. B. View from the batteries looking west showing how the terrain concealed the attacking force in the valley to the west. This view also makes clear the fact that the gun barrels could not be lowered sufficiently to repel the infantry attack.

5. DISCUSSION

Standard field pieces in the Civil War were the 6-pounder and light 12-pounder gun-howitzer or "Napoleon" smoothbores, and the relatively new 3-in ordnance rifles, 14-pounder James rifles, and 10-pounder Parrott rifles - effective rifled cannon for field use were only developed during the 1850s. Smoothbore weapons, usually bronze, were heavier, less accurate, and had shorter ranges than rifles (Table 1). Moreover, rifles fired pointed, cylindrical shells with striking velocities three times greater than smoothbore projectiles - being streamlined, rifle shells lost speed less rapidly. Rifled cannon also used the recently developed, highly effective, percussion fuses, but their shells were often defective. Furthermore, tube diameters were small so that they performed poorly with canister, the preferred short-range projectile (Manucy, 1949; Griffith, 1989). These factors resulted in enviable higher accuracies, but lamentably low killing power - the opposite qualities of those most needed to stop infantry attacks. Rifled field artillery pieces were thus considered experimental in terms of technological innovation and battlefield operations early in the American Civil War - Parrott rifles, for example, were not introduced until 1861 (see Schroeder, 2004, this volume for 19th century advancements in weapons).

Most field artillery pieces used in the Civil War were smoothbores, although most Union batteries contained at least one rifle. The smoothbore preferred by Union artillerymen was the highly reliable Napoleon, and the preferred rifle was the 3-in ordnance rifle (Katcher, 2001). The most common cannon in southern arsenals were 6-pounders (Katcher, 2001). Confederate batteries were forced to use whatever weapons were at hand, or could be captured, because the number of foundries capable of producing artillery pieces in the south was limited. Weapons capture was exemplified by Turner "trading" two of his 6-pounders for two of Parsons' Napoleons when he occupied Parsons' abandoned position (OR, 1886) in the fighting southwest of Walker's Bend.

Noe (2001) states that throughout the course of this battle the Confederate batteries outgunned the Union batteries at least in terms of being able to more effectively engage opposing targets. Manucy (1949), moreover, states that smoothbores were more effective weapons than rifles throughout much of the Civil War, because most battles occurred in wooded areas or in dissected terrain, where the longer ranges of modern weapons were of limited use and in which rifled projectiles were not as damaging as canister fired from smoothbores at close range. Griffith (1989, 169) has also noted the limitations of rifled cannon under these conditions, and states that:

> In this type of fighting there was actually not a very great role for the famous new long-range rifled cannon. The effect of artillery came mostly at 1,000 yards or less - in other words very much the same ranges as had been familiar to Napoleon - so the Civil War Commander was left with only a handful of specialist tasks for his longer-range pieces. . . .they could confer great advantages if a counter-battery duel was accepted by the enemy. But for stopping a massed attack the rifled cannon was much less useful than the smoothbore.

These arguments are supported in our analysis of the nature of the terrain over which the Battle of Perryville was fought. The battlefield was more or less devoid of woodland, but the contrast between open, longer wavelength terrain and more dissected, shorter wavelength terrain allowed muzzle-loading, bronze smoothbores to perform at least as effectively as modern rifles.

6. CONCLUSIONS

Effective utilization of the terrain has always been a key factor in gaining victory in land battles. The karst topography of the Perryville battlefield was instrumental in controlling the course of the battle and strongly affected the way in which the artillery batteries could be effectively utilized. The longer topographic wavelength of the terrain along Doctor's Creek - where fighting occurred across large, major dissolution features - restricted the use of shorter range, smoothbore weapons and enhanced the implementation of longer range, more accurate, rifled cannon. By contrast, the shorter topographic wavelength, dissected terrain southwest of Walker's Bend - where fighting occurred across smaller, less significant dissolution features - provided numerous favorable artillery positions within the shorter ranges of smoothbore weapons. The dissected nature and steepness of the terrain in this area allowed smoothbores to perform at least as well as the longer range rifled cannon.

The nature of the terrain, a dissected surface sloping very gently downward from the Union defensive positions toward the attacking Confederates, produced optical illusions that affected the outcome of the battle. For example, Lumsden thought Simonson was closer than he in fact was, so the range for his 12-pounders was set too short.

The Union and Confederate lines and the opposing batteries were much further apart in the open area along Doctor's Creek than in the dissected terrain southwest of Walker's Bend, and differences in topographic wavelength determined which types of weapons could be used most effectively in the different sectors of the battlefield. This is exemplified by the success of Turner's and Carnes' smoothbores against the rifles of Stone, Bush, Parsons, and Harris southwest of Walker's Bend. The shorter wavelength, more dissected terrain enabled the smoothbores to perform as highly effective, direct-support weapons for assaulting both Union infantry and artillery batteries. Civil War artillery fire was typically directed by battery commanders in the gun positions using binoculars or telescopes to observe the effects of direct fire. Short wavelength terrain allowed better control, resulting in more accurate fire and greater damage to opposing artillery batteries and infantry. The longer ranges of the rifled Union cannon in this type of terrain were thus not beneficial: Shorter wavelengths, steeper slopes, and dissected terrain did not allow their superior qualities to be used effectively. The success of Turner's and Carnes' smoothbores in the shorter wavelength terrain southwest of Walker's Bend can be contrasted against the lack of success of these weapons during the barrage in the more open, longer wavelength terrain along Doctor's Creek where Carnes' 6-pounders were ineffective against the rifles of Loomis and Simonson. Only one Union battery, that of Hotchkiss, was hindered in its performance at Perryville by the wavelength factor.

The terrain over which the Battle of Perryville was fought thus had two major influences: It controlled the use and relative effectiveness of both smoothbore and rifled cannon, and in this battle where the role of artillery was of paramount importance, it was a significant factor in the outcome. Union batteries were larger and better equipped than Confederate batteries (Table 4), but this proved to be of advantage only in the more open, longer wavelength terrain along Doctor's Creek. Shorter-range smoothbore weapons performed as well as or better than rifles in the dissected terrain where topographic wavelength was shorter.

ACKNOWLEDGEMENTS

We thank William Andrews, Kentucky Geological Survey, our co-author on an earlier paper on the Battle of Perryville that appeared in the proceedings of the 2003 International Military Geology and Geography Conference, for introducing us to the Battle of Perryville, for giving the senior author a guided tour of the battlefield, and for providing the digital elevation model and GPS locations for the positions of the batteries referred to herein. We also thank Allen Hatheway for an insightful and helpful review of this paper.

REFERENCES

Connelly, T.L. 1967. *Army of the Heartland, The Army of Tennessee 1861-1862.* Baton Rouge, LA: Louisiana State University Press.

Cressman, E.R. 1974. Geologic map of the Perryville quadrangle, Mercer and Boyle Counties, Kentucky. US Geological Survey Geologic Quadrangle GQ-11-85, 1 sheet, 1:24,000.

Day, M.J. 2004. Military campaigns in tropical karst - The Maroon Wars of Jamaica. In *Studies in Military Geography and Geology*, D.R. Caldwell, J. Ehlen, and R.S. Harmon, eds., Dordrecht, The Netherlands: Kluwer Academic Publishers, 79-88.

Griffith, P. 1989. *Battle Tactics of the Civil War.* New Haven, CT: Yale University Press.

Gibbon, J. 1863. *The Artillerist's Manual, Etc.*, 2nd edition. New York: D. Van Nostrand.

Guth, P. 2004. The geometry of line-of-site and weapons fan algorithms. In *Studies in Military Geography and Geology*, D.R. Caldwell, J. Ehlen, and R.S. Harmon, eds., Dordrecht, The Netherlands: Kluwer Academic Publishers, 271-285.

Hafendorfer, K.A. 1991. *Perryville: Battle for Kentucky,* 2nd edition. Louisville, KY: K.H. Press.

Holmes, R., ed. 2001. *The Oxford Companion to Military History*. Oxford: Oxford University Press.

Katcher, P. 2001. *American Civil War Artillery 1861-65 (1) Field Artillery*. Oxford: Osprey Publishing, Ltd.

Manucy, A. 1949. *Artillery through the Ages*. Washington, DC: US Government Printing Office.

McDonough, J.L. 1994. *War In Kentucky From Shiloh to Perryville*. Knoxville, TN: The University of Tennessee Press.

Noe, K.W. 2001. *Perryville, This Grand Havoc of Battle*. Lexington, KY: The University of Kentucky Press.

Schroeder, K.A. 2004. The development of tactical geography in the nineteenth century. In *Studies in Military Geography and Geology*, D.R. Caldwell, J. Ehlen, and R.S. Harmon, eds., Dordrecht, The Netherlands: Kluwer Academic Publishers, 39-51.

The Ordnance Manual for the Use of the Officers of the Confederate States Army, 1863 (Ordnance Manual). 1995. Dayton, OH: Morningside House, Inc.

The War of the Rebellion: A Compilation of the Official Records of the Union and Confederate Armies (OR). 1886. Series I, Volume 16, Part I, Reports. Washington, DC: US Government Printing Office.

Chapter 14

THE GEOLOGY OF THE CHICKAMAUGA CAMPAIGN, AMERICAN CIVIL WAR

Stephen W. Henderson
Oxford College, Emory Univeristy

Abstract: Regional and local geology exerted a very strong influence on the outcome of the Chickamauga Campaign of September 1863. The Union Army of the Cumberland met the Confederate Army of Tennessee at the Battle of Chickamauga in north Georgia. Topography of the Cumberland Plateau and Valley and Ridge Provinces of the Appalachian Mountains determined the movement of troops. General George Thomas was almost trapped in the confined anticline of McLemore's Cove. The opposing forces met near Chickamauga Creek in an area underlain by Ordovician carbonates. The Battle of Chickamauga can be related to differences in topography that are directly connected to differences in lithologies within these rocks. Thus, the maneuvers of armies prior to the Battle of Chickamauga and the battle itself were largely determined by both regional and local geology.

Key words: American Civil War, Valley and Ridge Province, Chickamauga, McLemore's Cove, differential weathering

1. INTRODUCTION

From summer through fall of 1863, the Chickamauga campaign of the American Civil War (1861-65), which culminated at the Battle of Chickamauga, was highly influenced by both large-scale regional and smaller-scale local geology. During this campaign, Union and Confederate armies moved through portions of the Cumberland Plateau and Valley and Ridge Provinces of the Appalachian Mountains (Fig.1).

The Cumberland Plateau is the southern portion of the Appalachian Plateaus Province, and the Allegheny Plateau is the northern part. Rocks of the Cumberland Plateau are generally horizontal late Paleozoic sedimentary rocks. Toward the eastern boundary with the Valley and Ridge Province,

D.R. Caldwell et al. (eds.), Studies in Military Geography and Geology, 173–184.
© 2004 *Kluwer Academic Publishers.*

their horizontal attitude grades into very broad open folds (Thornbury, 1965). The formation of the Appalachian Plateau resulted from uplifting brought about by the late Paleozoic Allegheny Orogeny. This mountain-building episode represents the collision of North America with Gondwanaland. The present topography of the Cumberland Plateau is highly dissected, in which the mountains and ridges are capped by resistant clastic rocks, especially Upper Carboniferous conglomerates and sandstones.

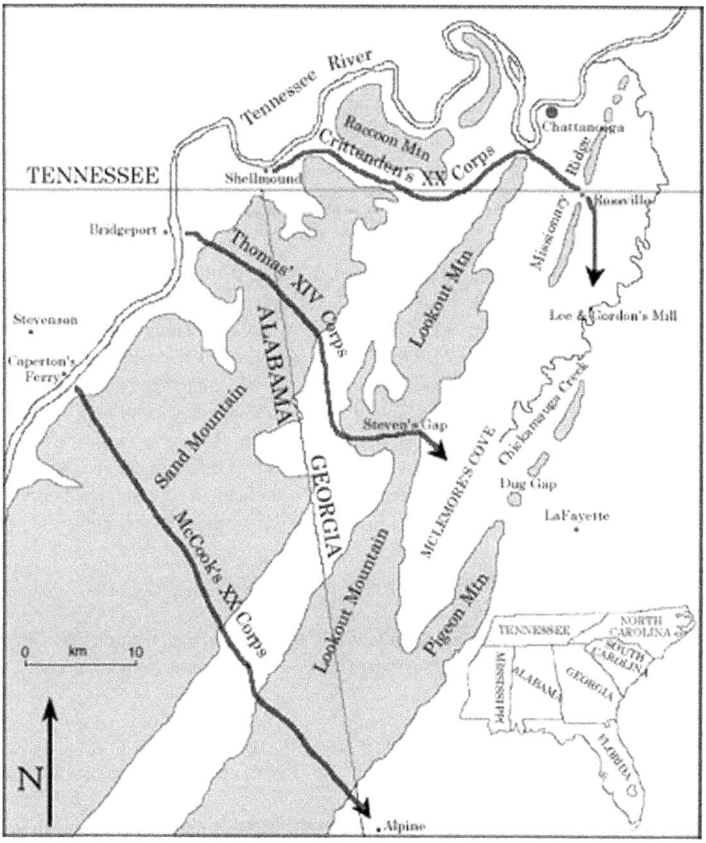

Figure 1. Regional locality map indicating major topographic features and Union Army of the Cumberland troop movements, with inset map of southeast US. After Lamont (1896).

The Valley and Ridge Province of the Appalachian Mountains consists of Paleozoic clastic and carbonate sedimentary rocks deformed through folding and faulting by the Allegheny Orogeny. In the early Paleozoic, the area was a passive, divergent continental margin. A rise in sea level is indicated by the transgressive depositional sequence, beginning with Lower Cambrian sandstones of the Chilhowee and Rome Fms. This is followed by Cambrian shales and limestones of the Conasauga Fm., continuing with the Upper Cambrian and Lower Ordovician dolomitic carbonates of the Knox Group. Limestones of the Middle Ordovician Stones River Group of the Chickamauga Supergroup represent the culmination of this transgression. By the Middle Ordovician, evidence of tectonic instability, including thick volcanic ash, indicates the nearby presence of a volcanic arc. At this time, the area became an active, convergent continental margin. From the Middle Ordovician to the Late Carboniferous, the Valley and Ridge in the southeast is dominated by two major clastic wedges. The first begins with westward migration of red-bed sedimentation of the Ordovician Greensport-Sequatchie Fms. and continues with deposition of the Silurian Red Mountain Fm. Subsequent downwarping of the basin, and resulting sediment-starved conditions, resulted in deposition of Upper Devonian black shale of the Chattanooga Fm. By the Early Carboniferous, a carbonate shelf developed and limestones, including the Monteagle and the Bangor, were deposited. In the Late Carboniferous, the basin was filled with deltaic sandstones and shales, such as occur in the Gizzard and Crab Orchard Mountains Fms. During the Allegheny Orogeny at the end of the Paleozoic, this active continental margin collided with Gondwanaland. Rocks of the Valley and Ridge were then detached from underlying rocks, folded, and thrust faulted. The geologic map and cross sections of northwest Georgia (Fig. 2) depict the resultant geologic structures. The present topography, shown on Fig. 1, consists of northeast-southwest oriented ridges and valleys, created through the combination of folding, thrust faulting, and differential weathering (Chowns, 1989a; Henderson, 1999).

2. BACKGROUND

The initial objective of the Union military campaign was capturing the important railroad town of Chattanooga. The influence of railroads on military strategy was extremely important. Chattanooga had railroad connections in all directions, which facilitated the movement of supplies and troops to all points of the Confederacy and theaters of war. The Western and Atlantic Railroad, opened in 1849, connected Chattanooga to Atlanta. Its route was influenced by Valley and Ridge structure and utilized natural gaps in the ridges, such as Ringgold Gap (Miles, 1991).

In late December of 1862, Major General William S. Rosecrans led his Union Army of the Cumberland from Nashville eastwards toward Murfreesboro, TN, where Confederate General Braxton Bragg's Army of Tennessee was located. Rosecrans' objective was to defeat Bragg's forces at Murfreesboro and capture Chattanooga. As Rosecrans' army made contact

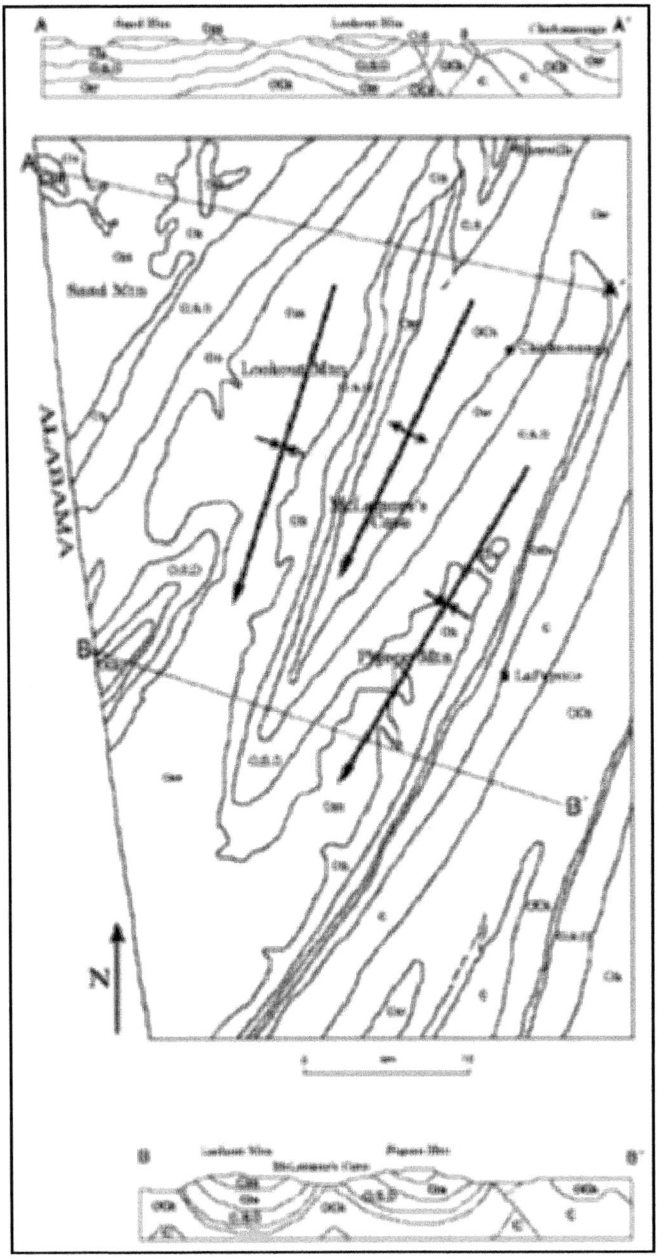

Figure 2. Geologic map and cross-sections of northwest Georgia (after Butts, 1948). Key: Css = Upper Carboniferous clastics; Cls = Lower Carboniferous carbonates; O,S,D = Ordovician, Silurian, and Devonian strata; Osr = Ordovician Stones River Group; OCk = Cambro-Ordovician Knox Group; C = Cambrian rocks.

with Bragg's forces, the Confederates attacked Rosecrans' right flank in the first phase of the Battle of Stones River on 31 December. The Army of Tennessee pushed the Army of the Cumberland back, leaving the Confederates in possession of the battlefield. The attack was renewed on 2 January 1863, when Bragg drove the Union army from the field. Massive Union artillery fire decimated the Confederates, ending the battle and forcing the withdrawal of Bragg's forces to Tullahoma, TN, some 65 km south along the Nashville and Chattanooga Railroad (Horn, 1987; Miles, 1991). Rosecrans went into winter camp in Murfreesboro, establishing a strong base of operations for his subsequent assault on Chattanooga (Miles, 1991).

Campaigning was renewed on 26 June 1863, when Rosecrans began moving southwards from Murfreesboro toward Tullahoma. The Tullahoma Campaign forced Bragg to withdraw from Tullahoma and fall back to Chattanooga. Rosecrans' casualties were minimal (Pittman, 2002). Rosecrans, an amateur geologist, was able to use the terrain to his advantage (Brown, 1963a; Brown, 1963b; Pittman, 2002). Brown (1963a) argues that Rosecrans used Safford's 1856 Geological Map of Tennessee during the campaign, although Pittman (2000; 2002) is skeptical as to exactly what sources Rosecrans used. However, Rosecrans developed an extensive mapping program with appointment of Captain William E. Merrill as "Engineer Officer in charge of the Topographical Department, Army of the Cumberland" (Brown, 1963a).

3. CHICKAMAUGA CAMPAIGN

The Chickamauga Campaign began on 16 August 1863, when Rosecrans' Army of the Cumberland began moving south toward Chattanooga and Bragg's Army of Tennessee. Rosecrans sent three brigades to shell Chattanooga as a feint. The shelling started on 21 August and over the next three weeks, Bragg became convinced that Rosecrans was preparing for an attack from the north across the Tennessee River. Meanwhile, the main force of Rosecrans' army went to the west and south, ending up southwest of Chattanooga, behind the cover of Raccoon and Sand Mountain at the eastern boundary of the Cumberland Plateau (Robertson, 1995). Raccoon and Sand Mountain are almost horizontal, very broad folds capped by conglomerates and sandstones of Upper Carboniferous Gizzard and Crab Orchard Mountains Fms. (Cressler, 1963; Chowns, 1989b; Figs.1 and 2).

The main force of the Army of the Cumberland consisted of General Thomas Crittenden's XXI Corps on the left, General George H. Thomas' XIV Corps in the center, and General Alexander McCook's XX Corps on the right. As can be seen in Fig. 1, Crittenden crossed the Tennessee River at Shellmound, TN; Thomas crossed at Bridgeport, AL, further to the west; and McCook crossed further to the southwest at Caperton's Ferry, AL (Scaife, 1983). Once the Union army was in position, they would then be able to flank the Army of Tennessee, cutting it off from Atlanta, and forcing Bragg to abandon Chattanooga without a fight. Brown (1964) concentrated on this portion of the Chickamauga Campaign.

At this point, Rosecrans further separated his three Union army corps and moved into the Valley and Ridge Province to isolate Bragg's Confederate forces. Due to differential weathering and erosion, the folded and faulted rocks of the Valley and Ridge have resulted in ideal terrain for movement of troops and defensive positions. Crittenden's XXI Corps crossed the syncline of Lookout Mountain at the northern end, moved through Rossville, and then south toward Lee and Gordon's Mill. Thomas' XIV Corps crossed Lookout Mountain through Steven's Gap and entered McLemore's Cove. McCook's XX Corps crossed Lookout Mountain at the southern end through Winston's Gap and Henderson's Gap, then moved on to Alpine (Miles, 1991). These gaps should not be considered to be true gaps where the mountains have been breached, but, instead, to be places where primitive roads were able to traverse steep inclines (Brown, 1964). Lookout Mountain extends from just southwest of Chattanooga in Tennessee, through northwest Georgia, and into northeast Alabama (Fig. 1). It is a broad, northeast-southwest oriented syncline at the western edge of the Valley and Ridge (Fig. 2) held up by resistant Upper Carboniferous clastics of the Gizzard and Crab Orchard Mountains Fms. (Cramer, 1986).

The ruse had worked. It was not until 7 September that Bragg realized that the Union army was attempting to flank him. The Confederates were forced to evacuate Chattanooga, where they withdrew into northwest Georgia, protecting their railroad supply line and Atlanta. Crittenden occupied Chattanooga and quickly headed south in pursuit of Bragg's Army of Tennessee. As the three army corps under Crittenden, Thomas, and McCook moved further into northwest Georgia, they bogged down and became separated, trying to negotiate the rough mountain roads (Robertson, 1995).

The Confederates halted their withdrawal on 9 September at LaFayette, GA, just east of Pigeon Mountain. Far from fleeing, Bragg prepared a trap for the Union forces. The Confederates planned to use the terrain to their advantage. Bragg's army was concealed behind the syncline of Pigeon Mountain (Robertson, 1995). Like Lookout Mountain, resistant sandstones of the Gizzard and Crab Orchard Mountains Fms. also cap Pigeon Mountain (GGS, 1976). Bragg knew that when Thomas' XIV Corps descended Lookout Mountain through Steven's Gap, it would end up in the confined anticlinal valley of McLemore's Cove, positioned between Lookout and Pigeon Mountains. Geologically, McLemore's Cove is an anticline plunging to the southwest (Fig. 2), floored by easily eroded Ordovician Knox and Chickamauga carbonate rocks (Milici and Smith, 1969).

A comparison of the locality map showing topography (Fig. 1) and the geologic map (Fig. 2) reveals the fidelity between the landforms and the geology. Bragg prepared to attack Thomas' Union forces from the northern opening to the cove and from the east via Dug Gap, cutting him off from escape. Narrow Dug Gap on Pigeon Mountain was held by Bragg, preventing Thomas from moving east (Robertson, 1995). Thomas would be bottled up at the nose of the anticline, where Lookout Mountain and Pigeon Mountain come together at the southwest end of McLemore's Cove.

Bragg organized the Army of Tennessee into four corps led by General Leonaidas Polk, General Daniel H. Hill, General Simon B. Buckner, and General William H. T. Walker. The lead division of Thomas' XIV Corps descended into McLemore's Cove on the night of 9 September. Bragg had 23,000 men and the artillery of Polk's corps at the northern open end of the anticline and one of Hill's divisions at Dug Gap, ready to attack in the morning. Polk waited until Hill's forces were in position before they began their attack. However, Hill never attacked. Thomas' lead division was alerted to their danger and escaped the trap by ascending Steven's Gap out of McLemore's Cove (Miles, 1991).

Rosecrans' Union Army corps were still seriously separated. On 12 September, Crittenden left Chattanooga, TN and advanced south toward LaFayette, GA. Two of his divisions were at Ringgold and the other one was dangerously isolated at Lee and Gordon's Mill, where McLemore's Cove opens out into a broad valley. Bragg ordered Polk to eliminate this isolated division and then destroy the rest of Crittenden's corps. By the time Polk got to the mill, Crittenden had united his forces and Polk decided not to attack. Rosecrans finally realized that Bragg was not retreating and ordered Thomas and McCook to move to Lee and Gordon's Mill and unite with Crittenden. McCook was located 50 km to the south. His corps took four days to traverse the mountain roads and descend into McLemore's Cove through Steven's Gap, where they followed Thomas north to the mill. Bragg did not attempt to attack while the Union forces were gathering (Miles, 1991).

4. BATTLE OF CHICKAMAUGA

The two armies were now concentrated along Chickamauga Creek near Lee and Gordon's Mill. This area lies near the boundary between the anticline that forms McLemore's Cove and the syncline of Pigeon Mountain.

The rocks underlying this broad valley are primarily easily weathered carbonates with local variations. They belong to the Lower Ordovician Knox Group and the Middle Ordovician Stones River Group of the Chickamauga Supergroup (Milici and Smith, 1969). Differences in lithology within these rock units played a significant role in the forthcoming Battle of Chickamauga.

Bragg ordered the attack to begin against the Union's left, held by Crittenden in the vicinity of the mill. His objective was to turn the left flank and push the Army of the Cumberland back into McLemore's Cove where it would be trapped. Rosecrans anticipated this plan of attack and had Crittenden extend his left flank and ordered Thomas to pass behind Crittenden and further extend the left. By the beginning of the Battle of Chickamauga on the morning of 19 September, the Union lines extended 5.5 km north of Lee and Gordon's Mill (Sullivan, 1956). Union forces were in possession of slightly higher ground than the Confederates. The Union line was just west of the north-south oriented topographic crest seen near Brotherton House on Fig. 3. This high ground is the result of resistant chert within the Middle Ordovician Murfreesboro Limestone of the Stones River Group. The Confederate line was located just east of the Union line in woods where the Middle Ordovician Lebanon and Ridley Limestones occur. These formations, which lie above the Murfreesboro Limestone, lack chert; thus, they are not resistant to weathering and produce a slightly lower topographic profile.

The first day's action was fierce but indecisive. Had Bragg initiated a full-scale assault on the Union lines, he probably would have broken through Thomas' line and turned the left flank, driving it into McLemore's Cove. Instead, Bragg responded to Thomas' advances throughout the day, never taking the initiative. As night fell, fighting stopped and the lines stabilized. There had been approximately 15,000 casualties that first day (Morris, 2000).

That night Bragg reorganized the Army of Tennessee into two wings. General James Longstreet had brought 12,000 reinforcements by train from the Army of Northern Virginia (Miles, 1991). Longstreet would command the left wing, including troops under the commands of Hood and Buckner. Polk commanded the right wing, consisting of Hill's and Walker's divisions. The Confederate attack on 20 September did not start until 0945. Polk had no written orders from Bragg and was waiting behind the lines for breakfast. Hill had no orders and was wondering what he should do. Longstreet had just arrived the previous night and was positioning his troops. This delay gave Union troops time to strengthen their lines (Morris, 2000). At first, the Union line held firm. At about 1030, a gap in the Union line's center was reported to Thomas by Captain Kellogg, one of his officers. Thomas reported the existence of the supposed gap to Rosecrans, who ordered General Thomas J. Wood to move his division to fill the gap. In reality, there was no gap. Kellogg probably didn't see the Union troops because trees shielded them. Wood also did not believe that there was a gap, but because he had been reprimanded earlier that day by Rosecrans for not moving quickly enough, he pulled his division out of the line and began to move them toward the non-existent gap (Robertson, 1995; Morris, 2000).

Figure 3. Topographic map of Chickamauga Military Park. From USGS (1982).

It was at that same moment that Longstreet ordered his three divisions to charge the Union line in the vicinity of Brotherton House. Unbeknown to Longstreet, this was the exact site of the real gap in the line that had been created when Wood pulled his troops out of position to fill the non-existent gap. Over 23,000 Confederate troops stormed through the Union position and sent them into headlong retreat. Most of McCook's corps and Crittenden's corps headed toward Chattanooga (Miles, 1991). Longstreet reported his victory to Bragg, "The enemy have fought their last man and he is running" (Scaife, 1983, 22). Longstreet's breakthrough was probably the most spectacular of the entire war. The right and center of the Union line fled toward Chattanooga through McFarlan's Gap in Missionary Ridge. Missionary Ridge is supported by resistant chert in the Knox Group where the Knox Group has been thrust-faulted over Ordovician and Silurian rocks (GGS, 1976).

Rosecrans' headquarters was also overrun and he and his chief of staff, General James A. Garfield, the future US President, rode to Rossville, where a road led back to Chickamauga. Rosecrans intended to return to the battle but Garfield convinced him that there would be greater need for him in Chattanooga, organizing the defense of the city. Rosecrans agreed and Garfield rode back to Chickamauga, where Thomas was still holding the left flank of the Union line. Garfield's harrowing ride back to Chickamauga was used years later during his presidential campaign (Miles, 1991).

Colonel John T. Wilder's "Lightning Brigade" (the 17th Indiana Infantry) was a unit of mounted infantry equipped with seven-shot repeating Spencer carbines. Their superior firepower allowed the troops to hold the line against Longstreet after the breakthrough. Wilder's brigade was located in the vicinity of Widow Glenn's cabin on high ground where the Wilder Tower now stands (Sullivan, 1956). This high ground and the farms and fields of the western portion of the battlefield are underlain by the Lower Ordovician Newala Limestone of the Knox Group (T.M. Chowns, pers. comm., 2002). The tower is composed of limestone of the Middle Ordovician Pond Spring Fm. of the Stones River Group (Chickamauga Supergroup), quarried near the town of Chickamauga (Maynard, 1912).

Wilder's stand gave Thomas time to organize defensive positions anchored on Snodgrass Hill. This position protected the road back to Chattanooga and had to be held. Snodgrass Hill is underlain by resistant cherty Knox, probably within the Chepultepec Fm. (T.M. Chowns, pers. comm., 2002). Interestingly, Longstreet was at the storming of Chepultepec during the Mexican War. Thomas' troops were running low on ammunition until fresh troops, commanded by General Gordon Granger from Rossville, were able to reinforce the lines. By early evening of 20 September, Longstreet sent his last fresh division up against Thomas. After 25 assaults, the Union line still held. Earlier, Garfield arrived with orders from Rosecrans to retreat. Thomas replied that retreat would be disastrous. He planned to hold his position until nightfall and then quietly withdraw to Chattanooga. Garfield informed Rosecrans that here is "Thomas standing like a rock" (Miles, 1991, 56). Thus, Union General George H. Thomas became the "Rock of Chickamauga." After dark, Confederates found themselves shooting at empty air as they closed in on Snodgrass Hill. Longstreet wrote that Thomas had gone "like magic" (Miles, 1991, 57).

The Confederates were now in complete possession of the battlefield. After two days of fighting, casualties numbered 34,000. Bragg lost 18,454 men and Rosecrans 16,179 (Miles, 1991). Chickamauga was the second bloodiest battle of the American Civil War. Although the Confederate victory was celebrated throughout the South, it was a costly victory and Bragg failed to follow it up with pursuit and an immediate assault on Chattanooga. Rosecrans' career was nearly ruined and the Union army was now under siege in Chattanooga.

5. CONCLUSIONS

Geology played a role in both the events leading up to the battle and the battle itself. From Raccoon Mountain in the Cumberland Plateau to McLemore's Cove and Pigeon Mountain in the Valley and Ridge, troop movements were shielded and traps were set. The resistant Upper Carboniferous clastic rocks that cap Sand, Raccoon, and Lookout Mountains created these ridges behind which the Union army was able to maneuver. These same clastics provided the cover for Confederate forces behind Pigeon Mountain. The anticlinal valley of McLemore's Cove, floored by non-resistant carbonates of the Knox and Stones River Groups, almost served as a trap for Thomas' corps. At Chickamauga, relative resistance of Ordovician rock units determined the high ground. The slightly higher Union line was created by resistant chert within the Murfreesboro Limestone, whereas the Confederate position was located within less resistant Lebanon and Ridley Limestones. Thomas' defense at Snodgrass Hill was made possible by resistant chert within the Knox Group. These geological circumstances significantly influenced the final outcome of the Battle of Chickamauga.

REFERENCES

Brown, A. 1963a. A geologist-General in the Civil War. Geotimes 7(7):8-11.

Brown, A., 1963b. Geology and the Tullahoma Campaign of 1863. Geotimes 8(1):20-22, 53.

Brown, A. 1964. The Chickamauga Campaign 1863 and geology. Geotimes 8(6):17-21.

Butts, C. 1948. Geologic map of Northwest Georgia. In *Geology and Mineral Resources of the Paleozoic Area in Northwest Georgia*. Georgia Geological Survey Bulletin 54.

Chowns, T.M. 1989a. Record of Paleozoic tectonism in the Valley and Ridge Province of Georgia and Alabama. In *Tectonostratigraphic Expression of Terrane Accretion in the*

Circum-Atlantic Paleozoic Orogen, Athens, GA: International Geological Correlation Project 233 Abstracts and Program 13-15.

Chowns, T.M. 1989b. Stratigraphy of major thrust sheets in the Valley and Ridge Province of Georgia. In *Georgia Geological Society Guidebook 9*, W.J. Fritz, ed., Geological Society of America SE Section Meeting Guidebook, 211-238.

Cramer, H.R. 1986. Geology of Lookout Mountain, Georgia. In *Geological Society of America Centennial Field Guide 6*, T.L. Neathery, ed., 153-157.

Cressler, C.W. 1963. Geology and ground-water resources of Catoosa County, Georgia. Georgia Geological Survey Information Circular 28.

Georgia Geological Survey (GGS), 1976, Geologic map of Georgia. Atlanta, GA: Georgia Department of Natural Resources.

Henderson, S.W. 1999. The geology of Civil War battlefields in the Chattanooga and Atlanta Campaigns in the Valley and Ridge of Georgia. In *Georgia Geological Society Guidebook 19*, T.M. Chowns, ed., 52-78.

Horn, S.F. 1987. The Battle of Stones River (Civil War Times Illustrated). In *The Battle of Stones River*, Conshohocken, PA: Eastern Acorn Press, 10-23.

Lamont, D.S. 1896. Military map showing the theater of operations in the Tullahoma, Chickamauga and Chattanooga Campaigns from the map of Col. William E. Merrill as published by Chief of Engineers U.S. Army, 1874.

Maynard, T.P. 1912. A report on the limestones and cement materials of North Georgia. Georgia Geological Survey Bulletin 27.

Miles, J. 1991. *Paths to Victory: a History and Tour Guide of the Stone's River, Chickamauga, Chattanooga, Knoxville, and Nashville Campaigns.* Nashville, TN: Rutledge Hill Press.

Milici, R.C. and Smith, J.W. 1969. Stratigraphy of the Chickamauga Supergroup in its type area. In *Precambrian-Paleozoic Appalachian Problems*, Georgia Geological Survey Bulletin 80:1-35.

Morris, R. Jr. 2000. Chickamauga: pyrrhic victory for the South. Military History 17:66-73.

Pittman, W.E. 2000. Geologists and the American Civil War. In *Geology and Warfare: Examples of the Influence of Terrain and Geologists on Military Operations*, E.P.F. Rose and C.P. Nathanail, eds., London: The Geological Society, 84-103.

Pittman, W.E. 2002. Tullahoma: Terrain and Tactics in the American Civil War. In *Fields of Battle - Terrain in Military History,* P. Doyle and M.R. Bennett, eds., Dordrecht, The Netherlands: Kluwer Academic Publishers, 99-115.

Robertson, W.G. 1995. *The Battle of Chickamauga.* Fort Washington, PA: Eastern National Press.

Scaife, W.R. 1983. *Atlas of Chickamauga and Chattanooga Area Civil War Battles.* Cartersville, GA: Civil War Publications.

Sullivan, J.R. 1956. *Chickamauga and Chattanooga Battlefields.* Washington, DC: US National Park Service, Historical Handbook Series No. 25.

Thornbury, W.D. 1965. *Regional Geomorphology of the United States.* New York: John Wiley and Sons, Inc.

US Geological Survey (USGS). 1982. Fort Oglethorpe, GA-TN 7½ minute topographic quadrangle map.

Chapter 15

MILITARY GEOLOGY AND GEOGRAPY IN THE AMERICAN CIVIL WAR

The Cloyds Mountain/New River Campaign, May 1864

Robert C. Whisonant
Radford University

Abstract: During the American Civil War, geology and physiography figured prominently in military activities in southwestern Virginia. In this region, strategic mineral operations and the Virginia and Tennessee Railroad, all crucial to the Confederate war effort, drew the interest of Union commanders throughout the war. In the spring of 1864, Union Major General George Crook and his Army of the Kanawha conducted a campaign against the railroad that resulted in the Battles of Cloyds Mountain and the New River Bridge. Terrain profoundly impacted this expedition, both at the operational and tactical levels. Simply to reach and return from its targets, Crook's army had to cross the rugged mountains of the Appalachian Plateaus and Valley and Ridge. At the tactical scale, geologic formations and structures created the topography over which the troops fought, and local terrain elements such as ridges and valleys, drainage, and karst features were important factors in the battles.

Key words: American Civil War, military geography, military geology, Southwestern Virginia, terrain analysis

1. INTRODUCTION

On Monday, 9 May 1864 - a beautiful sun-splashed day in the mountains of southwestern Virginia - the largest battle ever fought in that sector of the state erupted at the base of Cloyds Mountain (Fig. 1). Both Union and Confederate veterans of that engagement, many of whom had fought in larger and more important battles elsewhere, claimed ". . . that for fierceness and intensity, Cloyds Mountain exceeded them all . . ." (Humphreys, 1924, cited in McManus, 1989, 41). Of the roughly 9000 soldiers engaged, 1226

D.R. Caldwell et al. (eds.), Studies in Military Geography and Geology, 185–198.
© 2004 *Kluwer Academic Publishers.*

became casualties. Union killed, wounded, and missing were approximately 10% of their force and Confederate losses reached an appalling 23%.

The next day, 10 May, another splendid spring day, Union and Confederate soldiers clashed again when a sharp three hour cannon duel broke out at nearby Central Depot, now the City of Radford. This battle determined the fate of the Virginia and Tennessee Railroad's crucial New River Bridge (Fig. 1). The little-known actions at Cloyds Mountain and the New River Bridge were fought over classical Appalachian Valley and Ridge topography, which significantly impacted the events of the campaign. This paper examines the linkages between the May 1864 Federal expedition against the Virginia and Tennessee Railroad and the geology and geomorphology of the region.

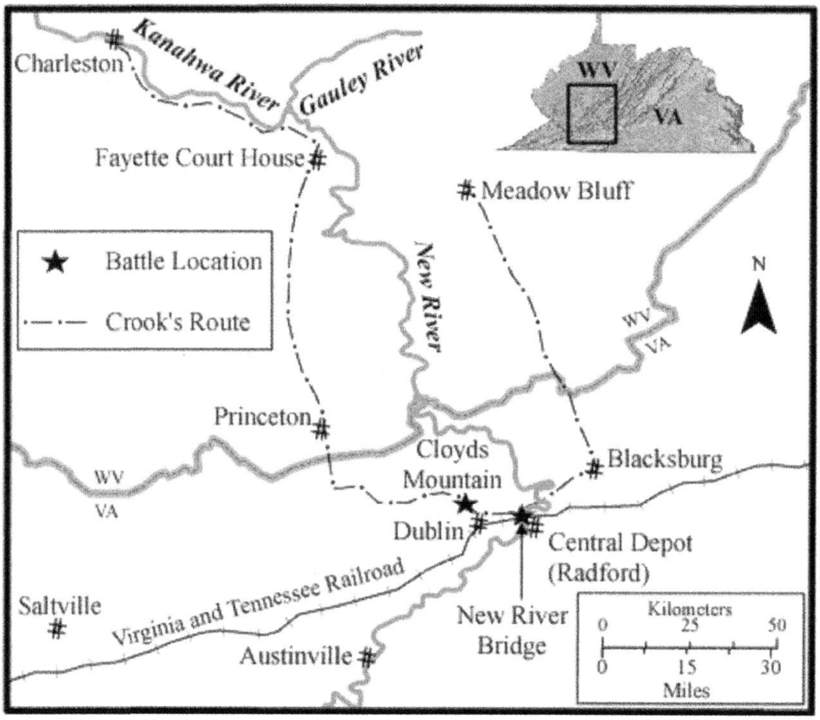

Figure 1. Map showing the region of the Union Cloyds Mountain - New River Valley expedition of May 1864. Note the battle sites at Cloyds Mountain and New River Bridge, the Virginia and Tennessee Railroad, and the mineral operations at Saltville and Austinville. Modified from McManus (1989) and US Geological Survey Bluefield sheet, scale 1:250,000.

2. SALT, LEAD, AND RAILS: THE SIGNIFICANCE OF SOUTHWESTERN VIRGINIA TO THE CONFEDERATE WAR EFFORT

When the American Civil War broke out in 1861, Virginia was by far the leading mineral-producing state in the entire Confederacy (Boyle, 1936; Dietrich, 1970). Among the principal mined materials needed to fight a war in the 1860s were salt, lead, iron, niter (saltpeter), and coal (Whisonant, 1997). Of all southern states, the Old Dominion ranked first in the production of each of these mineral resources except iron, where she was a close second to Alabama. In terms of national importance to the Confederate war effort, the value of southwestern Virginia's salt and lead operations cannot be overemphasized. Salt is essential in the human diet and during the American Civil War, every soldier's rations included it. Salt is also necessary for livestock; a hoof and mouth disease that appeared among the cavalry horses of Lee's army in 1862 was attributed possibly to a lack of salt (Lonn, 1933). In the 1860s, salt was by far the primary means of preserving meat. Additional uses included packing certain foodstuffs, particularly eggs and cheese, and preserving hides during leather making. In addition, salt was employed in numerous chemical processes and various medications (Holmes, 1993). Therefore, the Saltville salt works were of national importance to the Confederacy, ultimately producing two-thirds of the total southern salt supply.

Lead was also an extremely valuable mineral resource because of its use in manufacturing bullets. In the later years of the conflict, as pre-war stockpiles and smuggled quantities of lead became increasingly scarce, the Confederacy came to rely almost exclusively on the one significant lead mining operation in the entire South - the lead mines at Austinville (Donnelly, 1959; Robertson, 1993a). By war's end, Austinville had produced approximately one-third of all the lead consumed by Confederate forces. Both Austinville and Saltville are located in the remote mountains of southwestern Virginia; therefore, in the 1860s they depended heavily on the region's only railroad, the Virginia and Tennessee, to get the desperately needed lead and salt to manufacturing and distribution centers in the east and west.

The Virginia and Tennessee Railroad, built in the 1850s, ran completely through southwestern Virginia along the length of the Great Valley, then cut through the Blue Ridge and extended to Lynchburg (Noe, 1994). From there, other lines continued to Richmond, thus providing the Confederate capital with its most direct connection to the mineral riches of southwestern Virginia as well as other resources farther west. United States President Abraham Lincoln himself called the Virginia and Tennessee the "gut of the

Confederacy" (Noe, 1994, 112) as indeed the supplies shipped to Richmond and Petersburg were essential to fueling the voracious southern war machine in eastern Virginia. The weakest parts in military terms of this indispensable artery were the bridges, the most important of which "stood long and inviting" (Marvel, 1992, 19) across the New River at Central Depot.

3. PHYSIOGRAPHY OF SOUTHWESTERN VIRGINIA

Southwestern Virginia lies primarily in the Valley and Ridge province of the Appalachian Mountains. The Valley and Ridge, an archetypal "folded mountain belt," is characterized by a strong northwest-southeast alignment of linear, parallel ridges and valleys. Here, late Paleozoic mountain building folded and faulted a thick series of sedimentary strata of varying resistances to erosional processes. Easily decomposed lower Paleozoic carbonate rocks predominate in the outcrop belt of the eastern Valley and Ridge, thus forming the extensive lowland from Alabama to Pennsylvania known as the Great Valley. In contrast, the western Valley and Ridge has more erosionally resistant middle and upper Paleozoic clastic sequences exposed; hence, this area is dominated by high sandstone- and conglomerate-capped ridges separated by narrow limestone- and shale-floored valleys.

During the American Civil War, two physiographic attributes of southwestern Virginia exerted a tremendous influence on the strategy and tactics of military campaigns. The first of these, the basic topography of the region, presented nightmarish problems to commanders on both sides, but particularly to the Federals, who had to press their attacks against defenders skilled in using mountainous terrain to their advantage. The steep ridges, narrow valleys, numerous streams, and poor roads in these southern mountains made it virtually impossible to supply a sizeable army. In such difficult country, a relatively small Confederate force could cut supply lines at will and starve a large army into submission (Walker, 1985). Moreover, until very late in the war, Union forces had to attack southwestern Virginia across the trend of the mountains from bases in West Virginia and Kentucky, the only Union-controlled areas close enough to the salt, lead, and railroad targets from which expeditions could be launched. This meant that Union troops and artillery had to be moved first across the treacherous, deeply dissected Appalachian Plateaus and then through the high relief western part of the Valley and Ridge. At last, in December 1864, Union Major General George Stoneman was able to leave his base in Knoxville, TN, and advance directly into southwestern Virginia following an easy invasion route, the Great Valley lowland.

In addition to the fundamental landforms of the region, the New River itself is the second great physiographic factor figuring prominently in American Civil War military operations in southwestern Virginia. The New River is very distinctive geomorphologically; it is the only stream in the Appalachians to flow more or less directly northwestward across three major provinces - the Blue Ridge, Valley and Ridge, and Plateaus. Owing to this, the New River is generally considered to be one of the oldest, if not the oldest, streams in the Appalachians (Bartholomew and Mills, 1991).

Because it flows northwest across the Valley and Ridge, the New River constitutes a major obstacle to movement southwest or northeast along southwestern Virginia's natural transportation corridor, the Great Valley. Builders of the Virginia and Tennessee Railroad in the 1850s dealt with the New River barrier by crossing it only once; a single bridge over the stream was built at Central Depot. This so-called "Long Bridge" had a 700 ft long wooden superstructure covered by a tin roof, and was supported by metal piers anchored in the river bottom (McManus, 1989). For Union raiders, destruction of the vital span at Central Depot, the "Achilles tendon" of the Virginia and Tennessee (McManus, 1989), was the most effective way to cut the railroad. Of great interest also to Union strategists was nearby Dublin Depot, headquarters of the Confederate Department of Southwest Virginia. Significant military personnel and provisions, in addition to important railroad facilities, were located here. Determination to strike at Dublin and, above all, to demolish the New River Bridge eventually precipitated the expedition against the Virginia and Tennessee Railroad in May 1864.

4. UNION RAIDERS IN THE NEW RIVER VALLEY, MAY 1864

In March 1864, Lieutenant General Ulysses S. Grant became General-in-Chief of the Armies of the United States. Grant intended to crush Confederate General Robert E. Lee in Virginia by force of arms and by denying him supplies from the west and Deep South (Johnson, 1986). To execute his grand strategy, Grant had Union forces on the march throughout Virginia in spring of 1864. Major General Benjamin Butler moved his Army of the James toward Richmond from Fortress Monroe in the Hampton Roads area. Major General Franz Sigel, commander of the Department of West Virginia, advanced southward in the Shenandoah Valley toward Staunton. Grant joined the Army of the Potomac in the field and pushed down from Washington, DC toward the Confederate capital at Richmond.

The fourth major component of Grant's grand strategy in Virginia was to strike with Major General George C. Crook's Army of the Kanawha from

Charleston, WV, through the Allegheny Mountains, and sever the Virginia and Tennessee Railroad in the New River Valley. If Crook was successful, he would continue to Lynchburg, there join Sigel's forces advancing from the Shenandoah Valley, and thus isolate Lee completely from desperately needed resources coming from the west.

On 29 April, Crook left Charleston with 6155 soldiers, intending to attack Confederate headquarters at Dublin first, then destroy the New River Bridge (McManus, 1989). Among his three brigade commanders was Colonel Rutherford B. Hayes, an Ohioan who had served in West Virginia since the early days of the war. Within his command, Hayes had a young lieutenant in the 23rd Ohio named William McKinley. These two future presidents campaigned together throughout the New River Valley expedition; ironically, McKinley would survive the carnage at Cloyds Mountain only to fall to an assassin's bullet during his presidency.

Crook's men first moved up the Kanawha Valley, then left the river bottom to advance southward across the precipitous Allegheny Mountains of the Plateaus. On 6 May, as Lee and Grant continued a second day of bloody battle in the Wilderness, Crook's column reached Princeton and quickly overwhelmed a small Confederate command there. The next day, Union forces entered the Valley and Ridge of Virginia along the Rocky Gap Road, present-day Interstate 77, pushing Confederate skirmishers before them. By the evening of 8 May, another day of marching and skirmishing, Crook's army was encamped just northwest of Cloyds Mountain. Only 2 mi away, across the mountain top to the southeast, Confederate Brigadier General Alfred Jenkins, in temporary command of the troops in southwestern Virginia, waited expectantly with his hastily assembled men, determined to stand fast in defense of Dublin and the railroad just to their rear.

The local geology and topography of the Cloyds Mountain area played a prominent role in the bloody events of 9 May 1864 (Fig. 2). Cloyds Mountain is a typical ridgeform of this region, supported by erosionally resistant upper Paleozoic sandstones dipping toward the southeast. The ridge crest is formed by the Mississippian Cloyd Conglomerate; the dip (southeastern) slope consists of Mississippian sandstones and shales overlying the Cloyd. These strata are overthrust by Cambrian carbonate and shale formations along the Pulaski fault at the base of the mountain. Back Creek, a New River tributary, flows near and in some places along the trace of the Pulaski fault at the battlefield. Southeast of Back Creek and the fault trace toward Dublin lies low, rolling terrain characterized by karst features developed on the Cambrian carbonates. The small knolls lying athwart the Dublin-Pearisburg Turnpike, present-day Route 100, are developed on resistant units in the limestones and dolomites just southeast of the Pulaski thrust fault zone. Atop these low hills sat the Confederates, entrenched behind barricades of logs, fence rails, and earth (Fig. 3).

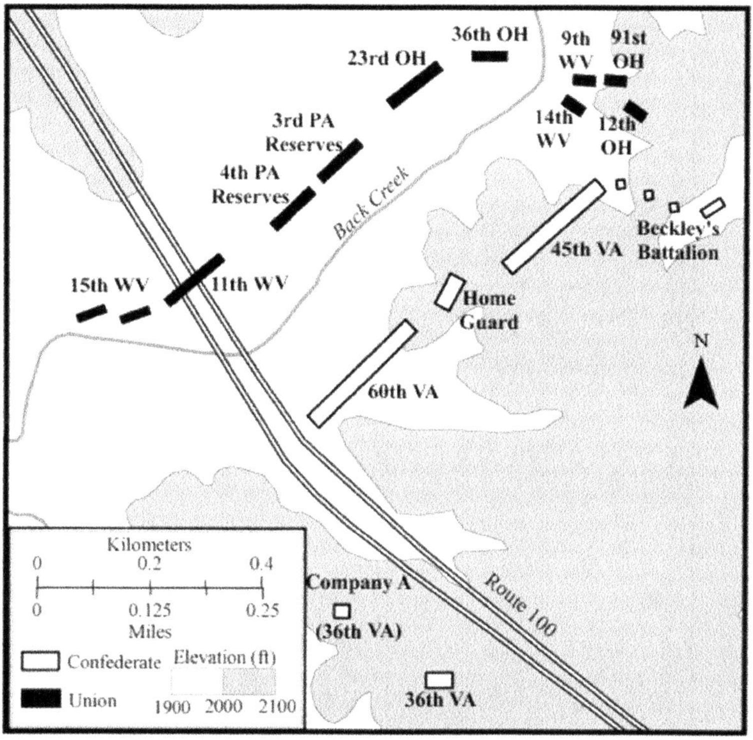

Figure 2. Map of the Cloyds Mountain battle site showing the locations of the Union and Confederate forces at the beginning of the battle. Modified from McManus (1989) and US Geological Survey Staffordsville Quadrangle, scale 1:24,000.

Crook's soldiers moved out of camp at dawn on 9 May. Expecting to meet the main Confederate forces along the mountain top, Union troops advanced anxiously through thick brush and woods up the northwestern slope of Cloyds Mountain. But only skirmishers awaited them at the ridge crest and these were soon driven off. Jenkins had chosen not to place his Confederate defenders on the mountain top, where dense forest and steep ground reduced visibility and the effectiveness of infantry and artillery fire. Rather, he selected the low ridge near the base of the mountain because its frontal slopes dropped off sharply into an open meadow through which Back Creek flowed, with banks steep enough to hinder advancing Federals. Moreover, the open ground provided excellent fields of fire for Confederate cannon and soldiers behind their log-and-rail breastworks. Crook, upon reaching the mountain's summit, observed the formidable half-mile long southern positions before him at the foot of the mountain just beyond Back Creek and opined: "The enemy is in force and in a strong position. He may whip us but I guess not" (Reader, 1890; cited in McManus, 1989, 26).

Figure 3. Cloyds Mountain Battlefield. A. View looking south across Back Creek valley to low hills defended by Confederates. B. View looking east along Back Creek; note Cloyds Mountain in background on left, low hills occupied by Confederates on right.

The Union army by now numbered 6555 men with 12 pieces of artillery (McManus, 1989). The Confederates had assembled 2400 men with ten cannon to meet the Union onslaught. The battle proper began at about 1100 when a brief artillery duel broke out. Soon, the entire Confederate right flank was under assault; musketry, grape, and canister swept the field, tearing huge holes in the Union lines. Back Creek ran red with blood. Dry leaves covering the ground caught fire and cremated an unknown number of men (Robertson, 1993b). For a few moments, the Confederates sensed victory, but at the critical time, Hayes' Ohioans smashed into the Confederate right-center. The battle reached a murderous peak as frenzied soldiers fought hand-to-hand with clubbed muskets, bayonets, knives, and fists. Jenkins went down, later to die in a Union field hospital, when a musket ball shattered his left arm; command passed to Colonel John McCausland, whose brigade included two infantry regiments, an infantry battalion, and a battery of light artillery (McManus, 1989). The thin Confederate line began to crack and suddenly, the entire right wing collapsed. Some men began fleeing toward Dublin in panic, but McCausland rallied his troops and fought a skillful rear guard action. Shortly, with much of his artillery and walking wounded safely off the field, the Confederate commander ordered retreat.

The Battle of Cloyds Mountain was brief, lasting only about an hour, but extremely vicious (McManus, 1989). Casualty rates attest to the ferocity of the struggle. The Union lost 688 men, about 10% of their force. Confederate casualties were 538 soldiers, nearly 23% of their command.

McCausland's men abandoned Dublin following the battle that afternoon, taking many of the supplies with them, but also leaving behind large quantities of commissary, quartermaster, and ordnance stores (McManus, 1989). Moving eastward about 10 mi with his small army, the Confederate colonel determined to make a stand at the New River Bridge (Fig. 4). During the evening of 9 May, McCausland positioned his men and artillery on the southeast side of the river, concentrating his forces in fortifications near the end of the railroad bridge (Fig. 5). Meanwhile, Crook's forces occupied Dublin, and early next morning, Tuesday, 10 May, began the destruction of militarily important supplies and facilities (McManus, 1989). These included the depot, telegraph office, a water tank, some woodsheds and wood, and a number of other buildings and properties belonging to the railroad. In addition, Confederate government warehouses, buildings, and military provisions were put to the torch. Finally, the Union soldiers came upon stores of Confederate ammunition, some of which they took for their own use and destroyed the rest.

That same morning, having completed the devastation of military and railroad facilities and supplies in Dublin, Crook's army abandoned the town and advanced toward the New River Bridge at Central Depot, where McCausland and his men lay in wait. At around 1000, Union forward skirmishers reached the rocky bluffs along the river near the north end of the bridge. Confederate batteries immediately opened fire as the Federals came up quickly in force. Fourteen Confederate cannons opposed 12 Union guns, but the Union weapons had the advantage of higher ground (McManus, 1989). Over most of the next three hours, cannons roared as the two sides fought for control of the bridge.

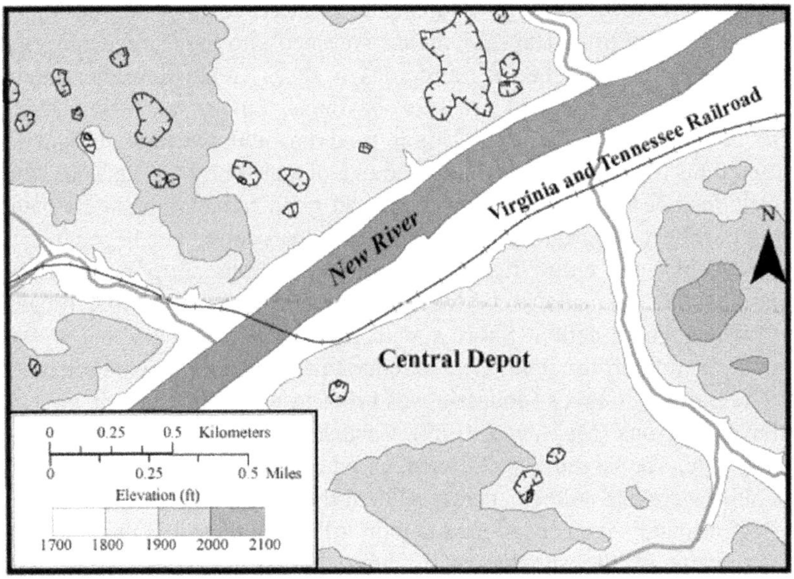

Figure 4. Map of the New River Bridge battle site. Note the abundant sinkholes on the north side of the New River. Modified from US Geological Survey Radford North Quadrangle, scale 1:24,000.

Nestled on the south side of a large meander loop of the river, Central Depot, like Dublin, is located on the Pulaski thrust sheet and underlain principally by carbonate-dominated rocks of Cambrian age; karst features are therefore very abundant (Fig. 4). Steep dolomite bluffs rise some 200 ft above the river on the north (cutbank) side. South of the river (slip-off slope), elevations are generally lower where fluvial processes have formed a flood plain and river terraces. Several interesting local geological connections with the 10 May Battle of New River Bridge exist. First, because Union artillerists occupied the higher ground created by the dolomite bluffs, they easily observed where their shells struck. Confederate cannoneers firing against the horizon saw only those shell bursts that were fired too low, a decided disadvantage in an artillery exchange (McManus, 1989). Second, one of the few casualties incurred by either side during the battle took place when a trooper under Hayes refused to take refuge in one of the sinkholes near the north end of the railroad bridge. Johnson (1986, 42) quotes Hayes as telling the following story:

Figure 5. Modern New River railroad bridge looking north. Note high ground on right and left where Union cannon were emplaced.

> There was a large lime stone sink hole," says Hayes, "in which I ordered the men to lie down. All obeyed promptly except one dismounted cavalryman who in a pert and saucy way turned to me and said, "Why don't you get off your horse and hide too?" On my repeating the order, the cavalryman replied, "I'll get down when you do." Just as I was insisting on his obeying the order a shell burst near us - the cavalryman was fatally and shockingly wounded and was then discovered to be a woman. She died almost instantly.

Shortly after noon, Confederate fire slackened as ammunition stores ran low. Crook ordered the bridge burned, and soon flames raced rapidly through its dry timbers (McManus, 1989). The regimental band of the 23rd Ohio played a stirring tune as the troops, waving flags and cheering triumphantly, lined up along the dolomite bluffs to watch the collapsing bridge plunge into the New River. At about 1300, as burning pieces of timbers floated away downstream, McCausland broke off the fight. With his ammunition now virtually exhausted, the Confederate commander withdrew toward Christiansburg, planning to defend the railroad there.

But Crook had other ideas. The day before in Dublin, his soldiers had discovered Confederate telegraph dispatches indicating that Lee had repulsed Grant in the Wilderness. This was erroneous information. Lee had indeed inflicted fearful losses on the Union army, but its tough-minded new commander wasn't retreating. Rather, Grant was sidestepping toward

Richmond and final denouement with the Army of Northern Virginia. Concerned about Grant's supposed defeat and knowing nothing of Sigel's fate, Crook believed that Confederate forces would now be free to trap him deep in hostile territory. Thus, having accomplished his major goals of destroying Dublin and the New River Bridge, the Union general decided to turn his command back toward the northwest and the safety of West Virginia.

Withdrawing from Central Depot after the battle, Crook crossed the New River a few miles downstream from the burned-out bridge and spent the night of 10 May on the eastern side. Next day, the Union commander advanced into Blacksburg where his army bivouacked for the evening. Early on the morning of 12 May, in a driving rain, the Union army began crossing a succession of sandstone-capped ridges, some of which towered over 4000 ft. For the next several days, Crook's troopers, skirmishing along the way, continued their tortuous journey in drenching rain through the forbidding high country of the western Valley and Ridge. Finally, the bedraggled Army of the Kanawha descended the last mountain slopes into the Greenbrier Valley and the relative sanctuary of West Virginia. On 19 May, Crook's veterans reached their home base of Meadow Bluff, whereupon the first sunshine they had seen in nine days broke through the clouds. The Virginia and Tennessee Railroad campaign was over.

5. SUMMARY AND CONCLUSIONS

What was the significance of the Virginia and Tennessee Railroad raid of 1864? For 21 days, Crook's army had marched 270 mi through some of the most difficult terrain in the eastern United States. Seventeen rugged mountain ridges and countless streams had to be crossed (McManus, 1989). For 16 days, heavy storms plagued the troops. Crook considered the expedition a success because he destroyed facilities and supplies at Dublin and burned the New River Bridge. Less kind critics, pointing to the campaign's failure to achieve the greater strategic goals of linking up with other Union armies to choke off Lee, concluded that the results were "hardly worth the powder" (Williams, 1971, cited in McManus, 1989, 78).

Ironically, the centerpiece of the raid, the seemingly successful demolition of the bridge at Central Depot, was of no long term importance. Only the wooden superstructure had burned, leaving the metal piers intact in the river. Crook had failed to bring explosives to blow up the foundation; solid shot cannon balls had bounced off harmlessly. Within five weeks of the Battle of New River Bridge, Confederate workmen had the span completely rebuilt, this time using fire-resistant green timber. Rail traffic again rumbled unimpeded across the New River. Despite a later attempt to fire the bridge once more, this new structure would not ignite (Walker, 1985). It survived to the very end of the war.

ACKNOWLEDGMENTS

I wish to thank Howard McManus and Linda Killen, local historians and experts on the New River Valley campaign, for information about the battles. Radford University graduate students John Surber and Claire Stull helped to draft the manuscript and maps, respectively. I particularly appreciate the help of Sharon Hollaway who prepared the manuscript.

REFERENCES

Boyle, R.S. 1936. Virginia's mineral contributions to the Confederacy. Virginia Division of Mineral Resources Bulletin 46:119-123.

Bartholomew, M.J. and Mills, H.H. 1991. Old courses of the New River: Its late Cenozoic migration and bedrock control inferred from high-level stream gravels, southwestern Virginia. Geological Society of America Bulletin 103:73-81.

Dietrich, R.V. 1970. *Geology and Virginia*. Charlottesville, VA: Virginia Division of Mineral Resources.

Donnelly, R.W. The Confederate lead mines of Wythe County, Va. Civil War History 5:402-414.

Holmes, M.E. 1993. Salt. In *Encyclopedia of the Confederacy*, R.N. Current, ed., New York: Simon and Schuster.

Humphreys, M.W. 1924. *A History of the Lynchburg Campaign*. Charlottesville, VA: The Michie Co.

Johnson, P.G. 1986. *The United States Army Invades the New River Valley, May 1864*. Christiansburg, VA: Walpa Publishing.

Lonn, E. 1993. *Salt as a Factor in the Confederacy*. New York: Walter Neale.

Marvel, W. 1992. *The Battles for Saltville*. Lynchburg, VA: H.E. Howard, Inc.

McManus, H.R. 1989. *The Battle of Cloyds Mountain*. Lynchburg, VA: H.E. Howard, Inc.

Noe, K.W. 1994. *Southwest Virginia's Railroad*. Chicago, IL: University of Illinois Press.

Reader, F.S. 1890. History of the Fifth West Virginia Cavalry, Formerly the Second Virginia Infantry, and of Battery G, First West Virginia Light Artillery. New Brighton, PA: Daily News.

Robertson, J.I., Jr. 1993a. Lead. In *Encyclopedia of the Confederacy*, R.N. Current, ed., New York: Simon and Schuster.

Robertson, J.I., Jr. 1993b. Cloyds Mountain, Virginia. In *Encyclopedia of the Confederacy*, R.N. Current, ed., New York: Simon and Schuster.

Walker, G.C. 1985. *The War in Southwest Virginia, 1861-1865*. Roanoke, VA: Gurtner Graphics and Printing Co.

Whisonant, R.C. 1997. Geology and the Civil War in southwestern Virginia: Union Raiders in the New River Valley. Virginia Minerals 43:29-40.

Williams, C.R., ed. 1971. *The Diary and Letters of Rutherford Birchard Hayes, v. 2, 1861-65.* New York: Reprinted by Kraus Reprint Co.

Chapter 16

GERMAN MILITARY GEOLOGISTS AND GEOGRAPHERS IN WORLD WAR II

Roles in Planning for Operation Sealion - The Invasion of England Scheduled for September 1940

Edward P.F. Rose[1] and Dierk Willig[2]

[1] *Royal Holloway, University of London*

[2] *Amt für Geoinformationswesen der Bundeswehr*

Abstract: During World War II the German army developed the largest organization to be used by any nation to contribute military applications of earth science in wartime. In the summer of 1940 two of its military geologist "groups" as well as units of military geographers focused their activities on preparations for Operation Sealion - the cross-Channel invasion of England, planned to be the greatest amphibious assault to that time in world history. The German military geographic service generated topographic maps and target appraisals, and the military geology units produced specialist geotechnical maps which analyzed the terrain of southeast England in terms of coastal geomorphology, water supply, construction materials, and cross-country trafficability. British victory in the aerial Battle of Britain led to cancellation of Sealion in late September 1940, but the maps (similar in category to those prepared for the Allied cross-Channel invasion of Normandy in June 1944) were preserved amongst archive documents seized by American forces at the end of the war.

Key words: England, German army, military geography, military geology maps, Operation Sealion, World War II

1. INTRODUCTION

From 10 May 1940, early in World War II, German forces advanced westward through the Low Countries and initiated the Battle of France, advancing with such speed that the British Expeditionary Force was compelled to evacuate via Dunkirk between 26 May and 3 June. An armistice was imposed on the French on 22 June. Yet the British were

D.R. Caldwell et al. (eds.), Studies in Military Geography and Geology, 199–214.

© 2004 *Kluwer Academic Publishers.*

unwilling to sue for a speedy peace. Accordingly, on 16 July the *Führer* Adolph Hitler issued a directive "On the preparation of a landing operation against England" (Kieser, 1999). Plans were quickly developed to deploy two German armies (the 9th and 16th) into southeast England (Fig. 1) during mid or late September 1940.

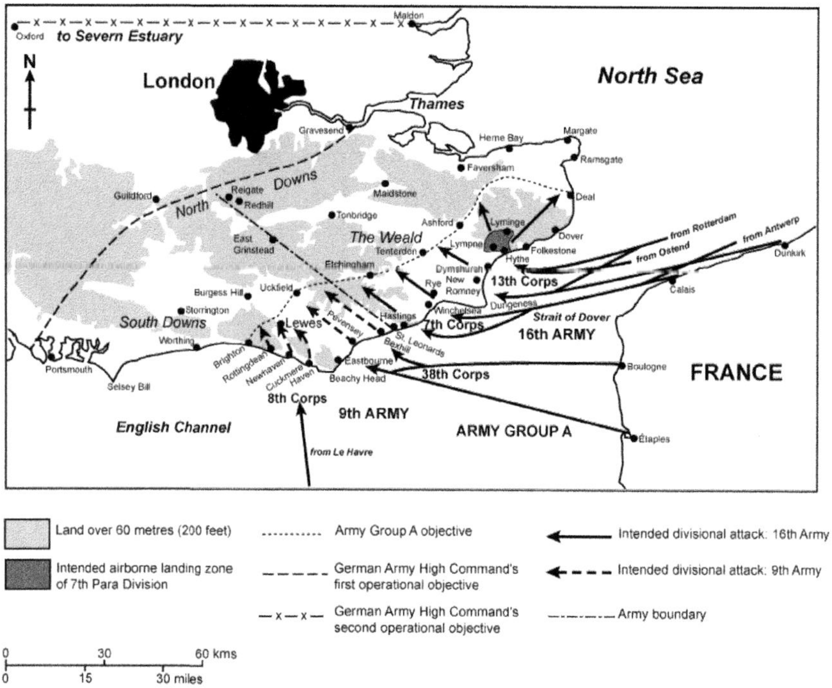

Figure 1. Map of southern England and northwest France showing deployment of German troops planned for Operation Sealion. Redrawn after Wheatley (1958) and Maier (1995); from Rose et al. (2002), by permission of Oxford University Press and the Geologists' Association.

Well-known preparations included the recording of all available sea and river craft in Germany and the countries already occupied by German forces; embarkation and disembarkation exercises; and the formation of occupation authorities that planned, among other tasks, the arrest of certain prominent British citizens (see, e.g., Wheatley, 1958; Schenk, 1990; Kieser, 1999). By mid-September the plan was to land ten divisions (seven infantry, two mountain, and one airborne) in the first wave of the attack, in total about 138,000 men in two days, followed by more divisions in the second and third waves to build up to a total of some 248,000 to 300,000 troops within two weeks. Yet it is largely unknown that to guide planning, military geology units were deployed in support of the two armies (Rose et al., 2002; Rose and Willig, 2002a, b) to generate specialist maps that analyzed the terrain in terms of coastal geomorphology, water supply, construction materials, and cross-country trafficability.

By late September the German air force had failed to achieve the air supremacy deemed essential for the success of a cross-Channel amphibious assault: The (British) Royal Air Force was victorious in the aerial Battle of Britain. German plans for Sealion were therefore largely set aside. However, copies or drafts of the specialist maps were subsequently transferred to the Military Geology Staff (*Wehrgeologenstab*) of the Army High Command in Berlin. Late in the war, the archives and libraries of the *Wehrgeologenstab*, the National Geologic Service (*Reichsamt für Bodenforschung*) and the German Patent Office were stored underground for safety, in a salt mine at Heringen in the Hessen province of Germany. This "Heringen Collection" was captured by American troops at the end of the war. Parts were transferred to the United States, some eventually to the National Archives and Records Administration (NARA) at College Park, MD. There the relevant documents are now catalogued within flats 20 and 23 of Record Group 57, line sequences 72 and 73, "Mil Geol Branch: German geologic maps of Europe 1917-45."

2. GERMAN MILITARY GEOLOGISTS

A military geological service, founded in the German army early in World War I but disbanded at its end, was re-established from 1937 (Häusler and Willig, 2000; Rose et al., 2000). The German armies that occupied northern France and the Low Countries in May and June 1940 were therefore aided by considerable geologic expertise. Five military geology groups (*Wehrgeologengruppen*) were deployed in their support, based on Brussels, Lille, Paris, Nancy, and Baden-Baden, but soon with 20 reconnaissance units or out-stations (and therefore a total of some 20-50 geologists) deployed much more widely (Häusler and Willig, 2000; Rose and Willig, 2002a; Fig. 2). German armed forces were ultimately to employ about 400 geologists during World War II, mostly in the military geological service - the largest organization developed by any nation to contribute military applications of earth science in wartime (Häusler, 1995a, b; Willig, 2003) - and smaller but still significant numbers in the air force, navy, *Waffen-SS*, and paramilitary construction agency *Organisation Todt*.

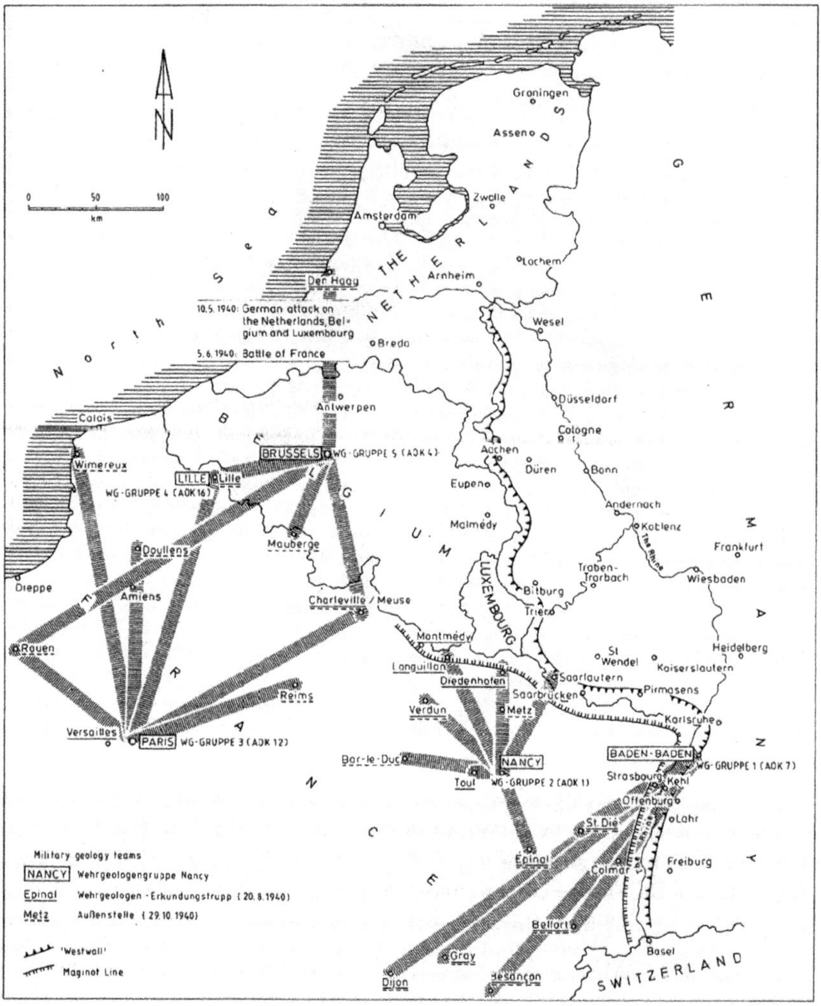

Figure 2. Map showing deployment of the five German military geology groups on the Western Front, with headquarters at Brussels, Lille, Paris, Nancy, and Baden-Baden, and all with subordinate reconnaissance units (*Erkundungstrupps*) by 20 August or out-stations (*Aussenstellen*) by October 1940, following the Battle of France. Fortification positions of the German West Wall and French Maginot Line are also shown. From Häusler (2000), by permission of the author and the Geological Society, London.

The 16th Army, part of Army Group "A" that had led the occupation of northern France (Häusler, 2000), was supported by the military geology group based at Lille (Häusler and Willig, 2000), directed by Otto Burre (Rose and Willig, 2002b). None of the 9th Army specialist maps bears either the name(s) of the geologist(s) responsible or the place name of preparation, and differences in style and scale compared with the 16th Army maps imply that the 9th Army geologists functioned quite separately. Some of the maps show preparation by a military geology unit belonging to 8th Corps of the 9th Army. The Corps and Army headquarters were located near Rouen, and by October 1940 Rouen was also the base for a military geology out-station of *Wehrgeologengruppe Brussels* (Häusler, 1995a), with links to *Wehrgeologengruppe Paris* (Häusler, 2000; Fig. 2), so it seems likely that the 9th Army's specialist maps were generated by German military geologists working at or near Rouen (Rose and Willig, 2002b).

3. BASE MAPS

Colored topographic maps were annotated by the geologists to produce specialist geotechnical maps. The base maps were those produced for general operational use by the German army's Mapping and Survey Agency from British (Ordnance Survey) maps at the most similar scale. Maps for Sealion were mostly prepared at metric scales of 1:50,000, 1:100,000, and 1:250,000 (Fig. 3), derived from British maps scaled in Imperial units of one-inch-to-one-mile (1:63,360), half-inch, and quarter-inch, respectively. Sheet numbers correspond with those of the then most recent Ordnance Survey map series. For the 1:50,000 scale maps, major differences in style between some sheets reflect preparation from different editions (4th, Popular, or 5th) of 1:63,360 scale Survey maps then current in particular areas. Minor differences in typography and layout between the German maps suggest that speed of preparation was of greater importance than standardization of format, and more than one map production unit may have been involved.

4. MAPS OF COASTAL GEOMORPHOLOGY

NARA contains a 1:500,000 scale map of southern England in which coastal areas are color coded to indicate zones of decreasing suitability for amphibious landing: Coasts with wide beach and insignificant cliffs, wide beach but high cliffs, narrow beach and high cliffs, no beach but high cliffs. Marginal data show that the map was prepared for the Army General Staff by its "War" Mapping and Survey Agency, Branch 4: Military Geography, in August 1940 (*Generalstab des Heeres, Abt. für Kriegskarten u. Vermessungswesen, [IV Mil-Geo], VIII. 1940*).

More detailed maps were prepared by geologists rather than geographers, by annotating 1:50,000 scale topographic base maps. Two were produced by geologists supporting the German 16th Army (sheets 117 and 135 of Fig. 3),

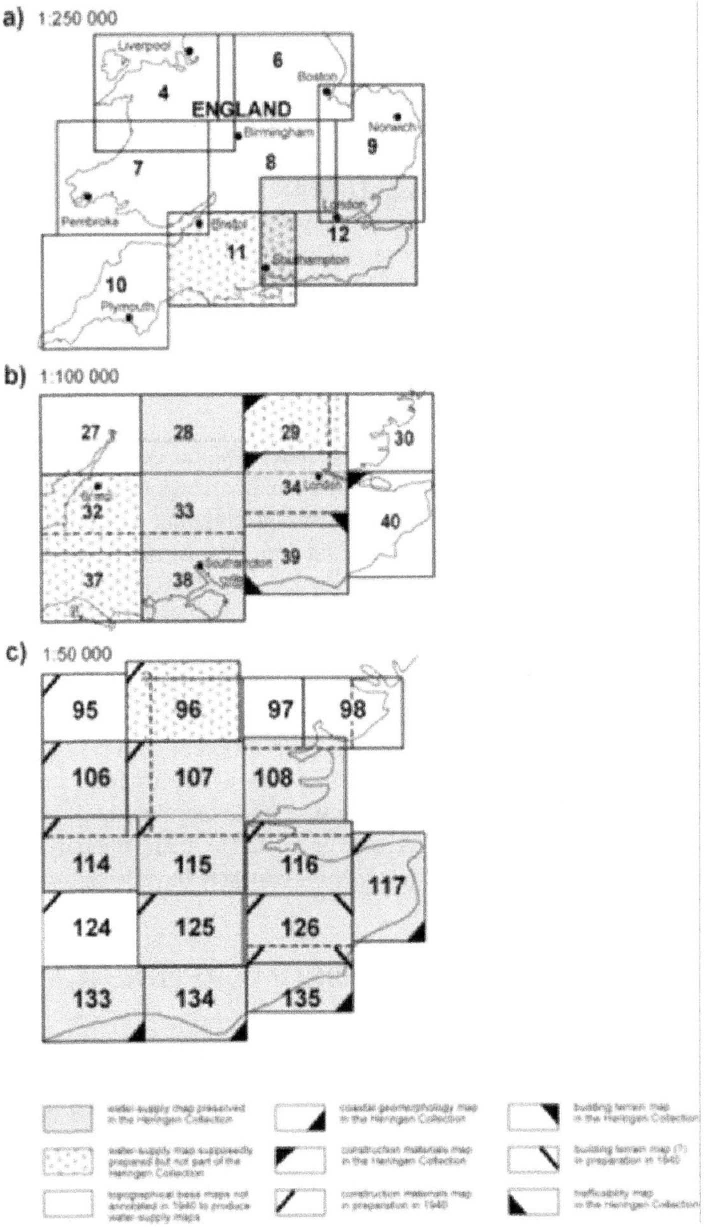

Figure 3. Area of southeast England covered by specialist maps prepared by German military geologists to facilitate the invasion of England in 1940, with sheet numbers indicated. Dashed lines indicate sheet overlap. Compiled from maps in the Heringen Collection, courtesy of NARA. Amended from Rose et al. (2002), courtesy of the Geologists' Association.

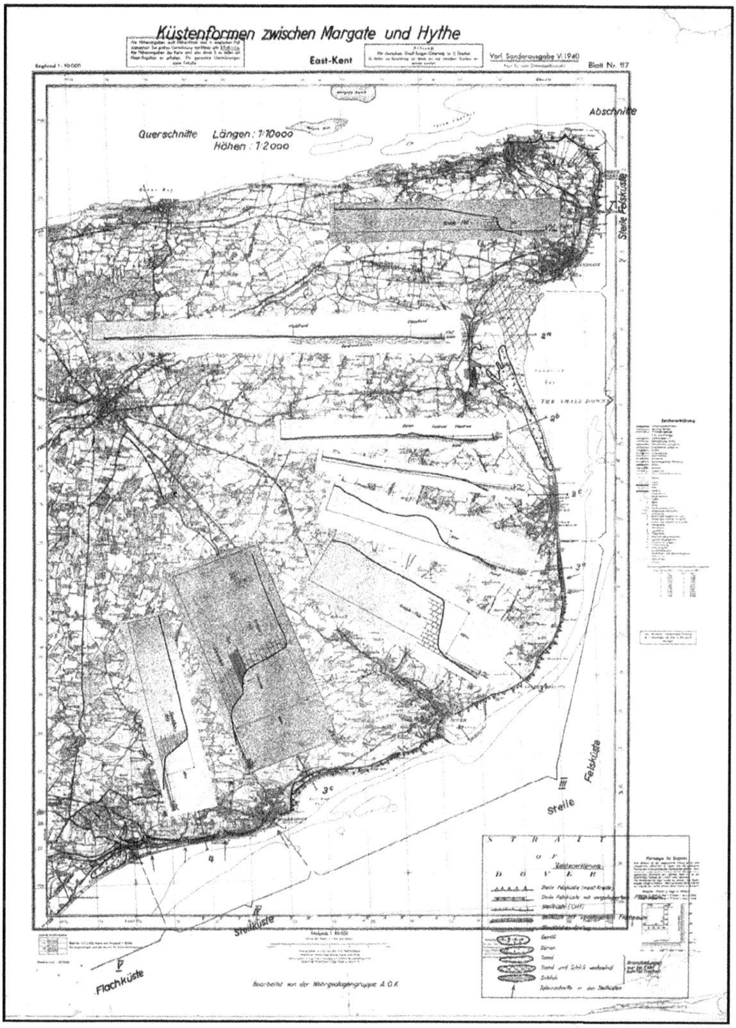

Figure 4. German 16th Army coastal geomorphology map, sheet 117, one of the maps in the Heringen Collection (see Fig. 3). Eight geological sections glued to the map show cliff profiles and lithologies. The five coastal regions sidelined right comprise: I steep rocky coastland, II flat coastland, III steep rocky coastland, IV steep coastland, V flat coastland. Arrows indicate the positions of major valleys that notch the coastal cliffs. The key (bottom right) distinguishes five types of shore morphology (steep rocky coastland - mainly chalk, steep rocky coastland with flat shore, steep coastland, steep coastland with flat shore, coastland with gradual inland ascent) and five types of beach sediment (gravel, dunes, sand, alternating sand and silt, silt). To amplify the maps, the geologists provided two additional sheets, of face-view coastal diagrams at scale of 1:2000, plus one sheet of block diagrams. Reproduced courtesy of NARA.

and two by those of the 9th Army (sheets 133 and 134). 16th Army maps divide the coastal region into sectors that are described in an accompanying leaflet. The maps themselves (e.g., Fig. 4) distinguish five types of shore morphology and five categories of beach sediment. Geological sections glued to the map surface inland against the line of section indicate the topography of the cliff and the adjacent beach areas; the underlying lithostratigraphy; and the position of high and low water levels relative to the cliff and beach, hence the width and nature of the intertidal ground. 9th Army maps were produced at the same scale as the 16th Army maps, and also by red ink annotation of a topographic base map, but the style is somewhat different. Only two types of coastal geomorphology are distinguished; one category of beach sediment; and fewer geological profiles are drawn. Yet by drawing the profiles in marine areas that would otherwise be largely blank space, the compilers have not obscured the inland topography in the manner of the 16th Army maps - and have been able to annotate key features, e.g., major scarps. High, middle, and low tide levels are indicated, but there is no accompanying leaflet to amplify map data.

5. WATER SUPPLY MAPS

Geologists attached to the 16th Army generated 1:50,000 scale water supply maps, an explanatory leaflet to accompany each sheet, and a two page overall summary for ten areas of southeast England, which together represented the Army's initial ground objective (Rose et al., 2002). Each map comprised a base map overprinted with annotation in black ink to show water supply data (e.g., Fig. 5). Sheets are divided into between two and seven numbered regions of different geology that are ornamented to distinguish only three categories of suitability for installing shallow (drive) wells. They thus illustrate a pattern adopted for specialist military maps from at least World War I, irrespective of nationality, in which ground for a specific military activity is mapped in terms of: "go" (suitable), "no go" (unsuitable), and "slow go" (intermediate between suitable and unsuitable). A leaflet providing an explanation for the numbered regions in terms of topography, subsoil, and potential water supply accompanied each map. The overall summary outlined general principles for obtaining groundwater under military field conditions, guidelines for the installation of field wells, and procedures to be adopted in coastal areas.

Maps for the 9th Army were prepared in a different general style, and primarily at 1:100,000 rather than at 1:50,000 scale, together with a 1:250,000 scale overview. Geological annotation, although again in black ink on a base map, provided less information than shown on the 16th Army maps, and arguably less clearly. Regional data were printed on the map itself, rather than more extensively in an accompanying leaflet.

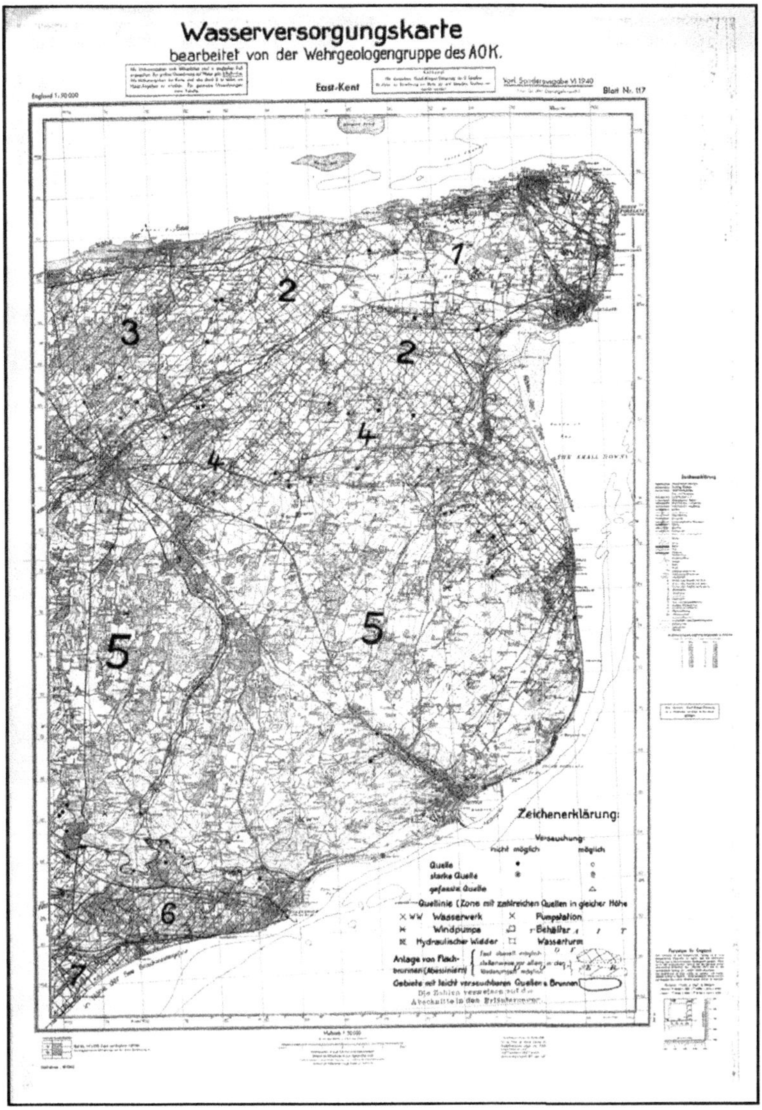

Figure 5. German 16th Army water supply map, sheet 117, one of the maps in the Heringen Collection (see Fig. 3). Numbers shown on the map refer to regions whose hydrogeology was described in an accompanying leaflet. Ornament is used to indicate regions unsuitable for shallow wells (blank), suitable in places (diagonal lines), generally suitable (crossed diagonals). See Rose et al. (2002) for translation of the key and accompanying leaflet, and further discussion. Reproduced courtesy of NARA.

6. MAPS SHOWING QUARRY LOCATIONS

NARA contains four maps completed at 1:100,000 scale, plus 12 seemingly in-course-of-preparation at 1:50,000 scale, which plot sites of quarries or pits for extraction of raw materials in southeast England.

Three 1:100,000 scale maps (sheets 29, 34, 40 of Fig. 3), each headed (in German) "construction materials map," indicate compilation by the military geology group of the 16th Army. Sites for rock extraction are plotted by a point within a colored circle on the topographic base map. The key distinguishes quarry/pit localities by color, with an indication of product use, e.g., for sheet 29: Blue (limestone, suitable for calcining), purple (dolomite and limestone, suitable for road-making, drainage, strengthening embankments), yellow (gravel and sand, for road making and route surfacing), brown (building sand and filtration sand, i.e., clay free), green (clay and loam, i.e., basic material for dike construction and brick manufacture), colorless (no longer in use).

By contrast, sheet 39 is headed "building terrain and construction materials map (engineering map)." Although set to the same scale, this map was produced by a different (9th Army) unit, and probably later than the others. It differs from the 16th Army maps in depicting regions differing in potential engineering use as well as extraction sites - both features of the local geology. Lithological boundaries as well as quarry and pit sites are plotted on the topographic base map, and the different terrain regions as well as these sites are color coded, to produce in effect a simplified geologic map, with the distribution of six main rock types depicted. Each of these is classified in the key according to its engineering properties (Table 1).

The 1:50,000 scale maps are in various stages of preparation, with sites commonly marked by pencil rather than ink. In general, geologists of the 16th Army seem to have been more advanced than those of the 9th Army in this aspect of their work.

7. CROSS-COUNTRY TRAFFICABILITY MAP

A single trafficability map is preserved at NARA, at 1:100,000 scale: Sheet 39 of Fig. 3. The map is annotated to show preparation by the same unit as the corresponding "building terrain and construction materials map (engineering map)" and so was presumably generated at about the same time. The trafficability map is complete except for its key. It is important in so far as it demonstrates the German intention and ability to produce maps of this particular type, but the lack of a key by which to interpret the complex of areas depicted on the map would make comment too speculative for convenient summary here.

Table 1. Rock description and characteristics

[Map color]	Rock	Workability	Stability	Permeability	Remarks
Blue	Chalk	By hand tools	Very good	Very permeable	Excellent for trenches and dugouts
Green	Sandstone (often with sandy and clay layers)	By hand tools, or sometimes explosives	Good to slight (gabions unnecessary)	Permeable to poorly permeable	
Grey	Clay	By hand tools and diggable	Poor	Impermeable	
Brown	Loam, also loam on the high chalk plateau	Mostly diggable	Slight	Poorly permeable	
Red	Loamy gravel	By hand tools to diggable	Poor (gabion mesh necessary)	Permeable	
	Sandy gravel		Poor (gabion mesh always necessary	Very permeable	
Blank	Alluvium and peat				Not land for construction!

Translation of part of key to German 1:100,000 scale sheet 39 Brighton "building terrain and construction materials map (engineering map)" preserved in the Heringen Collection.

8. SOURCE MATERIAL USED BY GEOLOGISTS

Rose et al. (2002) have demonstrated that the information depicted by the geologists of the 16th and 9th Armies on their water supply maps of southeast England could all have been derived from the geologic maps and memoirs published prior to 1939 by the Geological Survey of Great Britain, used together with the 10,560 and smaller scale topographic maps published by the Ordnance Survey, without recourse to significant covert military intelligence, aerial photography, or other geologic literature published by that time. British Geological Survey publications purchased pre-war would have been accessible either through the well-resourced National Geologic Service (*Reichsamt für Bodenforschung*) or geologic institutions in German-occupied France and Belgium (notably those in Lille, Paris, and Brussels). These primary sources, together with published tidal data for the coastal maps, would seem to have been sufficient for the German geologists to generate the other specialist maps - for coastal geomorphology, quarry sites, and trafficability.

9. MILITARY GEOGRAPHY

The appraisals of coastal geomorphology, water supply, inland quarry sites, and cross-country trafficability described above were all generated by military geologists. Military geographers provided a quite separate service (Smith and Black, 1946). This large organization (*Mil-Geo*) comprised Group IV of the "Military" Mapping and Survey Agency (*Abteilung für Kriegskarten und Vermessungswesen*) for the General Staff of the Army, and so supported the Army High Command. Already established at the outbreak of war, *Mil-Geo* was developed to provide a large central office plus branch units attached to military headquarters in German-occupied countries. Elaborate files of information were compiled, and used to prepare two main categories of publication: Handbooks and maps. Examples of these are preserved at NARA as part of the Heringen Collection, and in Germany, in the *Bundesarchiv-Militärarchiv* at Freiburg-im-Breisgau. Those relating to southeast England comprise: A military geographic account of England, of which the "London" and "South-coast" handbooks were completed in August 1940; and military geographic objective maps and pictures, of which the "SE England and London" volume was completed only in October 1941. (In addition to a 1:250,000 scale topographic map of the region, 1:10,000 scale town plans for Reading, Brighton, Dover, and Folkestone, and maps of the London area at 1:50,000 and 1:20,000 scales, this provided a booklet of 66 photographs of bridges, railway lines, factories, etc., which had been identified as potential military or air force targets.) Only the handbooks were published in time to be relevant to Operation Sealion; neither handbooks nor objective maps and pictures are geologic in scope.

In time, handbooks were prepared for all the countries likely to be occupied by the German army, and later, for the coastal regions of western and southern Europe that it intended to defend, reaching a peak output in 1943. By the end of 1943, *Mil-Geo* had produced 102 volumes covering 39 areas (Smith and Black, 1946). Specialist military maps were also produced, designed to provide military information in a form more compact than the handbooks. By the end of 1943, *Mil-Geo* had issued 244 sheets. Some specialist maps were prepared for southeast England (notably those showing known fortifications), but seemingly without geologist input - although *Mil-Geo* in general did have access to geologic expertise (Rose et al., 2000).

10. CONCLUSION

About 8500 years ago, or so it is generally believed, water broke through the neck of land that then connected England to France, and Britain became an island. Since then, although colonized by successive waves of immigrants from mainland Europe and harried by foreign fleets, England has been actually conquered by amphibious invasion only twice: By the Romans under Emperor Claudius in AD 43, and by the Normans under Duke William in 1066. The Germans under Adolph Hitler planned, for a third success, the greatest amphibious assault to that time in world history.

Maps of the coastal geomorphology were a primary requirement to guide the landings. About 67,000 troops were to land in the first echelon, on S-Day; a further 71,000 in the second echelon (probably S plus 1), followed by 77,000 more by S plus 10 (or 14). The large number of troops in the first two days but smaller number thereafter was a function of the long turnaround time required for the landing ships (Schenk, 1990). Landings were apparently to be made on sectors of coastline with insignificant cliffs, or where the coastline was notched by major valleys. The close correlation between favorable areas indicated on the coastal geomorphology maps and the sites actually scheduled for troop landings (Fig. 1) indicates their perceived value in operational planning.

Once the landings had been effected and the initial ground objectives achieved, water supply to sustain the advancing troops might have become a problem. It is not surprising, therefore, that after completion of the coastal geomorphology maps geologist priority seems to have been completion of those for water supply. At least 19 such maps were ready for printing (Rose et al., 2002), amply covering the area set by the Armed Forces High Command as the operational objectives.

As the campaign and occupation progressed, supply route maintenance and enhancement in the southeast would potentially have become important, together with the construction or repair of military installations. To obtain the necessary raw materials, the four 1:100,000 scale construction materials maps which covered southeast England provided an appropriate guide to key quarrying sites. Had the invasion succeeded, it seems likely that the more detailed 1:50,000 scale series of construction materials and also building terrain maps would have been completed to facilitate military engineering works by the army of occupation.

Possibly because southeast England was well-served by a network of roads, unlike some countries of German occupational interest, trafficability or "going" maps were not given a high priority. The Allies gave them similarly low priority at the start of the Normandy campaign (Rose and

Pareyn, 1996b). This contrasts with their much higher priority given in terrain evaluation (terrain analysis) for military purposes in later years.

In using geologists to generate specialist maps to guide beach landings, obtain adequate potable water supplies, win raw materials for route maintenance and military engineering works, and to guide assessment of off-road trafficability, the Germans adopted much the same roles as the British for the Allied cross-Channel amphibious invasion of Normandy in June 1944. However, the scales were somewhat different. The Germans planned to land about 138,000 troops in two days, building up to 248,000 to 300,000 within about two weeks, across a front of some 140 km (narrowed to some 100 km as planning developed) (Schenk, 1990). The Allies actually landed over 132,000 troops across a front of 80 km in a single day (Rose and Pareyn, 2003), building up more quickly to 326,000 in six days, 929,000 in less than four weeks (Schenk, 1990). But the opposing forces were smaller: Britain had only 29 divisions and eight independent brigades, all under strength, dispersed across the UK to oppose Operation Sealion (Schenk, 1990), whereas the Allies in Normandy faced the might of some 30 German divisions already stationed in northern France.

A striking difference in the two situations is that whereas the Allies gave greatest geological priority to selection of candidate areas for construction of temporary airfields (Rose and Pareyn, 1995, 1996a, 1996b, 1998, 2003), correctly deeming command of the air over the battlefield to be of major importance, the Germans do not seem to have used their army geologists in this way at this time. It was German failure to obtain command of the air as much as of the sea lanes that brought cancellation of the operation despite much careful preparation. However, such wartime military geologic and geographic studies by both sides effectively pioneered the development of terrain evaluation (terrain analysis) as a distinct discipline in the years that followed World War II. With the benefit of hindsight and ground access unavailable to the German planning staffs, it seems likely that the specialist maps would have been useful but imperfect aids to facilitate invasion (Rose et al., 2002). German geotechnical mapping skills and resources were significantly enhanced as the war progressed (Häusler, 1995a; Willig, 2003).

ACKNOWLEDGMENTS

We thank staff at the *Bundesarchiv-Militärarchiv*, Freiburg-im-Breisgau, the *Amt für Wehrgeophysik*, Traben-Trarbach, Germany, and especially at NARA, for access to archive documents; also copyright owners as acknowledged in figure captions for permission to reproduce illustrations.

REFERENCES

Häusler, H. 1995a. *Die Wehrgeologie im Rahmen der Deutschen Wehrmacht und Kriegswirtschaft Teil 1: Entwicklung und Organisation.* Vienna: Informationen des Militärischen Geo-Dienstes 47.

Häusler, H. 1995b. *Die Wehrgeologie im Rahmen der Deutschen Wehrmacht und Kriegswirtschaft Teil 2: Verzeichnis der Wehrgeologen.* Vienna: Informationen des Militärischen Geo-Dienstes 48.

Häusler, H. 2000. Deployment and role of military geology teams in the German army 1941-45. In *Geology and Warfare: Examples of the Influence of Terrain and Geologists on Military Operations*, E.P.F. Rose and C.P. Nathanail, eds., London: Geological Society, 159-175.

Häusler, H. and Willig, D. 2000. Development of military geology in the German Wehrmacht 1939-45. In *Geology and Warfare: Examples of the Influence of Terrain and Geologists on Military Operations*, E.P.F. Rose and C.P. Nathanail, eds., London: Geological Society, 141-158.

Kieser, E. 1999. *Operation "Sea Lion": The German Plan to Invade Britain, 1940.* London: Cassell.

Maier, K.A. 1995. Sealion. In *Oxford Companion to the Second World War*, I.C.B. Dear, ed., Oxford: Oxford University Press, 988-989.

Rose, E.P.F. and Pareyn, C. 1995. Geology and the liberation of Normandy, France, 1944. Geology Today 11:58-63.

Rose, E.P.F. and Pareyn, C. 1996a. Roles of sapper geologists in the liberation of Normandy, 1944. Part 1 Operational planning, beaches and airfields. Royal Engineers Journal 110:36-42.

Rose, E.P.F. and Pareyn, C. 1996b. Roles of sapper geologists in the liberation of Normandy, 1944. Part 2 Quarries, water supply, bombing and cross-country movement. Royal Engineers Journal 110:138-144.

Rose, E.P.F. and Pareyn, C. 1998. British applications of military geology for "Operation Overlord" and the battle in Normandy, France, 1944. In *Military Geology in War and Peace,* J.R. Underwood, Jr. and P.L. Guth, eds., Boulder, CO: Geological Society of America Reviews in Engineering Geology XIII, 55-66.

Rose, E.P.F. and Pareyn, C. 2003. *Geology of the D-Day Landings in Normandy, 1944.* Geologists' Association Guide No. 64.

Rose, E.P.F. and Willig, D. 2002a. British contrasted with German military geologists and the Battle of France, 1940: No comparison? Royal Engineers Journal 116:154-160.

Rose, E.P.F. and Willig, D. 2002b. German military geologists and terrain analysis for Operation "Sealion": The invasion of England scheduled for September 1940. Royal Engineers Journal 116:265-273.

Rose, E.P.F., Häusler, H., and Willig, D. 2000. A comparison of British and German military applications of geology in world war. In *Geology and Warfare: Examples of the Influence of Terrain and Geologists on Military Operations*, E.P.F. Rose and C.P. Nathanail, eds., London: Geological Society, 107-140.

Rose, E.P.F., Mather, J.D., and Willig, D. 2002. German hydrogeological maps prepared for Operation "Sealion": the invasion of England planned for 1940. Proceedings of the Geologists' Association 113:363-379.

Schenk, P. 1990. *Invasion of England 1940: The Planning of Operation Sealion.* London: Conway Maritime.

Smith, T.R. and Black, L.D. 1946. German geography: War work and present. Geographical Review 36:398-408.

Wheatley, R. 1958. *Operation Sea Lion: German Plans for the Invasion of England, 1939-1942*. Oxford: Clarendon Press.

Willig, D. 2003. *Entwicklung der Wehrgeologie: Aufgabenspektrum und Beispiele: II, von 1918 bis 1945*. Traben-Trarbach: Fachliche Mitteilungen des Amtes für Wehrgeophysik 226.

Chapter 17

WAR IN THE HEARTLAND

The Role of Geography in Operation Barbarossa 1941-1942

Burl E. Self

Southwest Missouri State University

Abstract: The goal of this paper is to identify and analyze the significance of Russia's terrain and geography during Operation Barbarossa in 1941. Like Napoleon's army before them, the Germans were unable to overcome geography and destroy the Russian army. Russia's geography created delaying actions, depriving the attacker of the effective application of mass, economy, concentration, and movement. Many serious errors were committed by the German command: Among the most serious of them was the supposition that the entire structure of the Soviet regime would topple after the first defeats of the Red Army. The Russian colossus could afford to withdraw troops and sacrifice entire geographic regions. Without destruction of Russia's army, a victory was not possible. In the end, geography mounted an effective and conclusive defense of Fortress Russia.

Key words: Operation Barbarossa, geostrategic, military geography, Sino-Western conflict, Eurasia

1. INTRODUCTION

The greatest land surface battle in military history began when the German army entered Russia on the early morning of 22 June 1941. Operation Barbarossa was the German code name given to this blitzkrieg attack. Initially victorious, practically all strategic initiative was seized from Germany during the ensuing winter because of the geographic challenges of Russia's vast steppe and Great Northern Forest. By 6 December 1941, the German army was forced to begin its slow, inexorable withdrawal from Russia, no longer able to dictate combat conditions and terms of battle. Table 1 shows the time line for these events and the German units involved.

D.R. Caldwell et al. (eds.), Studies in Military Geography and Geology, 215–226.

© 2004 *Kluwer Academic Publishers.*

Table 1. Chronology of Operation Barbarossa 1940-41

Timeline/Action	Location	Units Involved
Early 1940: Planning began		
22 June 1941: Invasion launched		
	Germany	14 motorized divisions
		120 infantry divisions
		1 cavalry division
		19 panzer divisions
		2 motorized brigades
		1 motorized regiment
	Romania	12 infantry divisions
		4 cavalry brigades
		1 tank brigade
		3 divisions
	Slovakia	1 brigade
		2 small infantry divisions
	Spain	1 division
November-December 1941: Soviet winter defensive		
6 December 1941: Tactical withdrawal begins		

Through careful review of combat accounts of Russian-German forces from June 1941 to late 1944, critical geographical principles that preserved Russia from military defeat have been identified. Not surprisingly, these same principles play a direct, strategically significant role in protecting Russia today.

2. RUSSIA'S GEOSTRATEGIC POSITION

With its physical geography and sheer immensity, modern Russia is the Eurasian Heartland. It is the most geostrategically positioned country on the Eurasian land mass and is key to the flow of energy resources to world markets.

In early 1940, the occupation of Russia was the geopolitical focus of German foreign policy. Germany's eastward push was largely motivated by a desire to access desperately needed Eurasian resources vital to continental domination. Germany's control of Russia and Eurasia would have threatened Africa, and rendered subordinate North and South America and the Pacific (Brzezinski, 2002).

Germany's failure to occupy Russia is a testimony to the importance of physical geography in defending Russia. Since attackers must fight through

Russia's natural obstacles, an assault from the west has never succeeded. Russia's geography literally creates delaying actions, depriving attackers of the effective application of mass, economy, concentration, and movement.

While planning their 1941 invasion, the Germans committed serious errors, including their supposition that the Russian regime would topple after the first defeats of Russia's army. Germany believed it could gain freedom to maneuver by breaking through the Russian front and occupying all strategic terrain features. However, sweeping around major geographical obstacles was tactically impossible because Russian forces always occupied the most advantageous positions. Further, the huge Russian army could not be out-flanked. Geography compelled the Germans to fight through natural obstacles to access Eurasia, while the Russian colossus used the immense land and terrain to its advantage. In effect, Russia's fortress-like geography simplified its effective and conclusive defense (Fig. 1).

Russia's 38,000 mile, vast, land frontier is nine times the size of that of the United States. Russia contains vast areas of boreal forest and steppes, and an inhospitable climate.

In 1941, Russia's geography influenced every large scale and small unit operation of the German military campaign, whether on land, over water, or in the air. At the time, the German army did not fully recognize the fundamental importance of Russia's geography and was not prepared to withstand its effects. Only the ability of German soldiers to bear up under tremendous adversity prevented earlier total disaster. On 6 December 1941, one day before the United States declared war on all Axis Powers, Germany began withdrawal from Russia (Jacobsen and Rohner, 1965). As later battles in 1943-44 proved, the German Army never recovered from its first hard blow in the winter of 1941-42.

The German Army faced geographic conditions significantly different than those to which it was accustomed. German soldiers crossing into Russian territory often felt they had entered a different world, opposed not only by Russian forces but also by the forces of nature. Conquering the raging elements of nature was extremely difficult because their fury and effect had not been fully recognized. Soldiers were neither trained nor equipped to counter the cold and the inhospitable terrain. Military planners wrongly assumed Russia would be defeated west of the Dnepr River (Fig. 1) and there would be no need for conducting operations in snow and mud (Anon., 1986a).

Figure 1. General reference map for Russia west of the Ural Mountains. AZER. - Azerbaijan; BUL. - Bulgaria; EST. – Estonia; LAT. – Latvia; LITH. – Lithuania; ROM. – Romania.

2.1 Combat on the Russian Taiga

War in Russia can be waged only with a great deal of fighting in taiga, which Russians refer to as "the Great Northern Forest." This vast sub-Arctic forest interlaced with bogs, swamps, and small drainages covers over one-half of Russia. Fig. 1 shows areas of taiga west of the Ural Mountains. By the time German troops had traveled over 1000 mi in this foreboding terrain, they were greatly reduced in numbers, ill-supplied, often starving and freezing, and scarcely mobile in dense forest.

To conduct a successful campaign against an opponent able to strike from Russia's peculiar forests and swamps, Germany needed well-trained troops with high morale and physical endurance (Anon., 1986b). Not surprisingly, German units attempting to penetrate Russia's forested areas without preparation usually were defeated. Many German offensives resulted in troops losing their bearings, deviating from their attack objective, and firing on friendly forces. Attacking units quickly became intermingled, difficult to control, and lost much of their striking power (Anon., 1986b).

Where possible during their 1941 advance, Germans showed a tendency to bypass large forests and swamps. As a result, large Russian formations in the taiga remained untouched and undetected. After initial contact, Russian units would simply retire into the forest depths and continue to fight.

In addition to defensive cover, the taiga provided Russia's crucial offensive advantage at strategic moments. For example, in the latter half of December 1941, Russians encircled in a dense forest were joined by reinforcements, who then attacked the Germans across Lake Ilmen to split the German Sixteenth Army in two (Fig. 1; Anon., 1986b). Throughout the war, undetected Russian combat units in the taiga continued to attack vulnerable supply lines and rear areas until Germany's retreat into Eastern Europe.

2.2 Combat on the Russian Steppe

The term "steppe" implies a flatland, or prairie environment, and includes large areas of modern Russia, especially Russian Europe (Fig. 1). This vast flatland is accessible in all parts, with the exception of the one natural obstacle called "balkas." Balkas are deep ravines, typically with high, steep slopes overgrown with brush and thickets. Since the balkas can be reconnoitered and avoided, the steppe is the ideal battleground for armored and motorized units during late spring, summer, and early fall.

During the war, lack of cover placed withdrawing German forces at a decided disadvantage. For example, in its retreat through the steppes of the Kerch Peninsula in the spring of 1944 (Fig. 1), Germany's V Infantry Corps was left with no tanks. Air cover was not available since remnants of German aviation were already committed against the Russian breakthrough on the narrow isthmus joining the northern Crimea with the mainland. In completely open terrain, every German truck, horse- or tractor-drawn gun was plainly visible over a considerable distance. Soviet air and armored forces arrived at a very simple division of labor: While the tanks proceeded to put German infantry out of action, Russian fighters swooped down on vehicles, especially heavy weapons and artillery. Horses and motorized equipment were blown to bits by strafing aircraft, and the few that remained succeeded only by towing away under cover of darkness. The rest were destroyed or left abandoned.

3. FAULTY STRATEGY AND FLANK PROTECTION

The Germans' invading force comprised three army groups: Central, with Moscow as its objective; North, whose objective was St. Petersburg (Leningrad); and South, whose objective was the Caucasus oil fields (Fig. 1). When Germany began its invasion campaign in June 1941, the main effort and highest priority was Army Group Central's Moscow objective. In August, Germany's General staff made a dramatic shift in strategy and focused its main effort on Army Group South and, on a lesser extent, to Army Group North.

This change in strategy was a crucial mistake in timing. Germany lost the opportunity to knock the Russians out of the war when it did not concentrate its efforts on a direct thrust on Moscow. With forces now diverted south and north, there were not enough German troops to provide the Army Group Center adequate flank security. As a consequence, the southeast flank was exposed and extremely vulnerable to attack. Russian units simply used geography to their advantage, waited to be bypassed, and then struck the German army from the rear.

The vastness of Russia's space made German efforts for all-round strategic security for its attacking forces untenable (Dunnigan, 1977). Furthermore, weather, stiffening resistance, natural obstacles, and miscalculations of time and crucial resources destroyed Germany's strategic advantage and shifted it to the still largely intact Russian army.

Adolf Hitler did not plan on a winter war in Russia. Lieutenant General Alfred Jodl, chief of the Armed Forces Operations Staff, had told his staff "In the autumn of 1941, after the consummation of the Russian defeat, our Luftwaffe will appear in the skies of western Europe in greater strength than ever before" (Anon., 1986a, 5). Hitler and his General staff concluded that "Therefore, it was best to wait until May 1941, to invade Russia and then bring the campaign to a successful conclusion within five months" (Anon., 1986a, 5). Germany believed the total time required to attain all military objectives would vary from 9-17 weeks at maximum (Anon., 1986a).

Miscalculations of time, and the subsequent changing seasons, became increasingly important factors once the invasion began. Inclement weather and great distances significantly contributed to defeat by slowing down every element of the operation. Furthermore, Germany's initial delay in launching the offensive, a month lost vacillating over the continuation of offensive, and, finally, diversion of forces to the south and north, away from Moscow, cost the Germans time, equipment, and irreplaceable manpower.

Hitler's overconfidence immeasurably compounded the inevitable hardships of the winter campaign. Because a victory was anticipated by autumn, he intended to withdraw two-thirds of the divisions and leave the remainder as an occupation army. Subsequently none of the German troops arrived with winter coats and boots. The amount of winter clothing procured months in advance was inadequate since it was planned for a much reduced occupation force. The grim situation became even worse when clothing arrived much too late because of transportation breakdowns.

4. INADEQUATE MOBILITY AND VASTNESS OF SPACE

From the outset, the Germans lacked sufficient mechanized vehicles, aircraft, supplies, and fuel operations in Russia. Many German infantry divisions were without transportation. Instead of using full maneuverability when conditions allowed, tank divisions were frequently forced to halt to permit the infantry to catch up.

On the Russian plains, German equipment was much too heavy for warfare, which created even more problems. The premier King Tiger tank was super heavy, weighing 75 tons. During inclement weather, it frequently was mired in mud or buried in snow. When disabled, it could be towed only by another King Tiger. Conversely, the Russians' smaller, lighter weight tanks with wider tracks could maneuver easily and were not dependent on a road- or bridge-rich environment (Anon., 1986a).

Russia's difficult terrain imposed performance limits on tanks and motor vehicles, requiring frequent maintenance or replacement. Germany was clearly disadvantaged by its scarcity of equipment and maintenance supplies. The harsh terrain took its toll, not only on motorized equipment, but on horses as well. During the first year, over one million horses were lost. Thereafter, Germany was forced to depend on Russian horses and horse-drawn equipment.

The vastness of Russia greatly exacerbated Germany's problems of limited resources, few supplies, and immobile troops. As the weeks passed, the German army lost more and more of its mobility - the essence of its striking power.

5. THE EFFECT OF WEATHER

Adverse weather began to hamper German mobility a mere four months after the invasion of Russia began. Momentum gained by major German victories at Vyazma and Bryansk (Fig. 1) in October 1941 was significantly dampened when offensive operations were essentially stopped due to heavy rains that began on 7 October and continued through the month. Germans relied on trucks and horse-drawn vehicles to deliver supplies from distant rearward railheads. Impassable road conditions brought these efforts to a halt until 14 November, when operations resumed. By then, attack routes to Moscow were barred by an in-depth and expanded Russian defensive line (Wray, 1986).

Sudden thunderstorms often changed easily passable dirt roads and open terrain into mud traps. In August 1941, such a storm was almost fatal to a regiment of a German motorized infantry division near Kiev (Fig. 1). Ordered to block the last escape route of Russian forces encircled north of Cherkassy (Fig. 1), the division moved over dry roads and reached the area of encirclement to accomplish its mission. Relieved from the blocking position, the division was then ordered to join Germany's Second Panzer Group for the drive on Bryansk (Anon., 1986a).

Forward elements hardly had moved when heavy rain began, and the roads became so slippery that the last regiment was completely mired. At this critical moment, Russian tank forces attempting a relief thrust on Kiev hit the rear of the regiment. Russian armor with its wide tracks still could move, but the German motorized infantry was anchored by its weight and wheels. Lacking the firepower to mount a defense, the German troops set fire to their vehicles and scurried on foot to join the remainder of the division, which also was bogged down to the north.

Muddy roads stalled critical German offensive efforts on numerous occasions. During another incident, after the 22nd Panzer Division broke through the Parpach battle positions in preparation for seizing the Kerch Peninsula in the summer of 1942 (Fig. 1), a sudden cloudburst mired the road making movement impossible. A perimeter defense was thrown up; the division sat until the storm was over and the summer sun had dried the road to a passable condition. In the meantime, Russian units withdrew intact to regroup and fight another day (Anon., 1986a)

During 1941-42, "only one German tank in ten survived" the mud and the snow (Anon., 1986a, 4-7), and those that did could not move through the snow because of their narrow tracks. When fair weather returned, the steppe's dirt roads dried out rapidly and were usable, provided undisciplined, over-eager drivers did not plow them up while the roads were still soft.

Russia's inclement weather slowed, and often stopped, the construction of German defensive positions. Neither trenches nor bunkers could be properly excavated in the muddy ground cased by thawing snow. Winter snow trenches and ice parapets quickly dissolved into slush. In late March, spring rains and mud halted almost all large scale operations. Only in late May or early June had the ground dried enough to construct effective positions (Wray, 1986).

To compound matters, the severe Russian winter combined with extreme distances created massive disruptions in the flow of German supplies. With great difficulty, only the most urgently needed rations, ammunition, and gasoline could be moved forward. Because of Russia's poorly developed road system and lack of suitable airfields, logistical support was almost completely dependent on the country's damaged railroads. Conditions were worsened by the scarcity of supplies and materials for repairs. By contrast, Russia's strategic ground forces were well equipped and able to continue offensive operations throughout the war.

Severe cold weather played a dramatic role in combat efficiency of German troops and their weapons as well. Since relatively short period of suitable fighting weather could be expected immediately following Russia's fall muddy season, it is surprising that German forces were not redistributed and billeted to withstand the rigors of a steppe winter: Hitler and his military advisers assumed that the campaign would end before the onset of bad weather (Winters et al., 1998). German troops were caught unprepared for winter. Furthermore, winter clothing had been ordered for only the 60 divisions that advisors planned to retain in Russia to form the military occupation force after victory (Anon., 1983).

On 30 November 1941, the temperature plummeted to -45°F. Germany's General von Bock informed Field Marshal von Brauchitach, Chief of Staff, that his men still had not received winter coats (Anon., 1986a). Nearly three weeks later, an angry General Heinz Guderian, commander of the Second Panzer Army, confronted Hitler with the stark fact that no winter clothing had arrived in the forward areas. Troops were reduced to removing clothes from enemy corpses, improvising straw boots, and taking other emergency measures. During the first five months of the invasion, Germany's eastern army suffered more than 734,000 casualties (nearly one-quarter of its average strength of 3,200,000 troops). Twice as many men died from cold weather than from enemy action (Winters et al., 1998).

The frigid conditions shattered the reliability of German combat equipment. At the beginning of December, the Sixth Panzer Division was 15 km from Moscow (Fig. 1) and 25 km from the Kremlin when temperatures fell to -30°F (Wray, 1986). The severe cold, coupled with a surprise attack by fresh Siberian troops, stopped their drive. Troops could not aim their small arms fire. Bolt mechanisms jammed, and firing pins shattered. Machine guns iced up, recoil fluid froze in heavy guns, and the ammunition supply failed. Mortar shells exploded harmlessly in deep snow, and land mines were no longer reliable. On 5 December 1941, Germany's Colonel General Hans Reinhardt reported that his troops were exhausted; he could hold his front only if the Russians did not attack (Wray, 1986). On the same day, Guderian recommended that the German offensive be halted; the cold had become too severe for troops and vehicles. On 6 December 1941, the first retreat order was given (Wray, 1986).

Upon retreat much of Germany's artillery and heavy equipment was abandoned due to lack of transport. Weak in men and firepower, and lacking Russia's winter cross-country mobility, the Germans found it increasingly difficult to break contact with the Russians and move across frozen terrain with unit integrity (Wray, 1986). Hitler ordered a large scale withdrawal from Russia on 15 January 1942 (Jacobsen and Rohner, 1965).

6. CONDITION OF GERMAN COMBAT FORCES

By the end of the 1941-42 winter, the German infantry divisions' combat efficiency had dropped by 35%, and the motor armored divisions had fallen 40-50%. The tank attrition rate was estimated to be at 65-75%, reaching 90% in some divisions (Dunnigan, 1977).

Nearly one-third of the German officers were killed in action, decimating the leadership ranks. Based on these figures, it is estimated that the real combat value of the 136 German divisions employed in the Russian theater was equivalent to that of only 83 full strength divisions - hardly a strategic advantage (Anon., 1988).

Physical and psychological effects of Russia's weather and geography on the average German infantryman were horrendous as evidenced in the following statements by infrantrymen from Fritz (1995):

> Everything has become a sodden black mush. (105)
> God save us from a winter campaign in the east. (109)
> Because of the cold our machine guns would not work at all. (110)
> Our fingers froze in our gloves, we could not have fired our rifles. (110)
> Heavy snow fall buried the road we needed a compass to dig it out again. (111)
> Life in Russia was a perpetual shivering fit. (113)
> Any desire to piss was announced so that hands could be held out under the warm urine. (117)
> A landscape characterized by death. (123)
> The immensity of Russia caused one to lose the feeling of time. (124)

> The spaces seemed endless, we became very depressed. (125)
> Russia is like a cold iron coffin. (138)

7. ODDS AGAINST GERMANY'S BLITZKRIEG

Victory in Russia was essential to realizing Hitler's dream of a vast, German-controlled Eurasian Heartland - the world's largest storehouse of minerals, rich farmland, cheap labor, and an unlimited oil supply. These resources were vital to Germany's long term success and to its most powerful military weapon, the "Blitzkrieg" (lightning war). Using coordinated and concentrated air and ground forces to deal shattering blows against opposing armies, the lightning war's effectiveness was also marked by its vulnerability. Mobility and speed provided the momentum for surprise, yet success of the Blitzkriegs was utterly dependent upon adequate resources and constant support.

Although the German plan to quickly defeat Russia did not allow for the escape of the Russian army, this is precisely what happened. Driving ever deeper into the Russian abyss, the German army tried for one last decisive encirclement to defeat the Russian opposition, but the opportunity never came. Russia's sheer size, difficult terrain, and inhospitable climate presented insurmountable supply and mobility problems that have never been experienced in modern warfare.

As early as October 1941, Russia's weather literally slowed German mobility and supplies to a standstill. Only one in ten Panzers survived beyond the first spring thaw on the Russian steppe. In the first five months of the invasion, three-quarters of a million German soldiers died. During this time, twice that many soldiers died from the cold than from battlefield action (Anon., 1986a).

8. SUMMARY

Russia used geography to its distinct advantage during the 1941 German invasion. The taiga was difficult to navigate, allowing large groups of Russian combat units to remain undetected and to attack aggressors at will. Conversely, the open steppes left the German army fully exposed to Russian forces, with every tank, truck, and gun plainly visible. Few Germans survived.

Geography preserved Russia from military defeat. Russia's geography created delaying actions, depriving attackers of the effective application of mass, economy, concentration, and movement. Russia's fortress-like geography solidified its effective and conclusive defense.

REFERENCES

Anonymous (Anon.). 1983. Military improvisations during the Russian Campaign. Washington, DC: US Army Center of Military History Publication 104-1.

Anonymous (Anon.). 1986a. Effects of climate on combat in European Russia. Washington, DC: Army Center of Military History Publication 104-6.

Anonymous (Anon.). 1986b. Terrain factors in the Russian Campaign. Washington, DC: US Army Center of Military History Publication 104-5.

Anonymous (Anon.). 1988. The German campaign in Russia: Planning and operations (1940-1942). Washington, DC: US Army Center of Military History Publication 104-7.

Brzezinski, Z. 2002. *The Eurasian Chessboard.* New York: The Caspian Sea Library, Basic Books.

Dunnigan, J.F. 1977. *War in the east: The Russian-German conflict 1941-1945.* New York: Strategy and Tactics Staff Study, M.I. Simulations Publications.

Fritz, S.G. 1995. *Front-Soldaten. The German Soldier in World War II.* Lexington, KY: University of Kentucky Press.

Jacobsen, H. and Rohner, J., eds. 1965. *Decisive Battles of World War II: The German View.* New York: Putnam.

Winters, H.A, Galloway, G.E., Jr., Reynolds, W.J., and Rhyne, D.W. 1998. *Battling the Elements, Weather and Terrain in the Conduct of War.* Baltimore, MD: Johns Hopkins Press.

Wray, T.A. 1986. Standing fast: German defensive doctrine on the Russian Front during World War II. Fort Leavenworth, KS: US Army Command and General Staff College Combat Studies Institute Research Survey 5.

PART III

TECHNOLOGIES OF THE TWENTYFIRST CENTURY

Military geography must fall within the context of a particular time frame. Some abstract notions of the subject may retain their validity over a long period of time, but for the most part, it derives relevance from the association between conditions of the total environment, technological capabilities of the time, and the mission to be accomplished. Military geography, then, depends upon technological capability, and varies with it.

Louis C. Peltier and G. Etzel Pearcy
Military Geography
1966

Chapter 18

MILITARY FOOT TRAFFIC IMPACT ON SOIL COMPACTION PROPERTIES

Kenneth W. McDonald
US Military Academy

Abstract: The study of military training on Army installations focuses extensively on vehicle impact, and foot traffic impact is not well documented. At the US Military Academy, West Point, NY, foot traffic impact was modeled using three functions (logistical growth, Gamma, and Wiebull) and field and laboratory data. The modeled data were compared to the actual condition of the soil in the Bataan Bayonet Assault Course, indicating a moderate level of compaction at the end of summer training. Bulk density, mean infiltration rates, and soil resistance were also compared, also indicating a moderate impact on soil compaction. The average infiltration rate decreased while the bearing capacity increased during the training cycle. The results indicate that the soil recovers moderately during the subsequent freeze-thaw cycle. This model has potential for greater application for predicting soil conditions.

Key words: bulk density, compaction, infiltration, foot traffic, military training, modeling

1. INTRODUCTION

Surface disturbance caused by off-road vehicles and heavy foot traffic often leads to disruption of soil structure, reduced plant cover, degradation of biological and physical soil crusts, soil compaction, reduced water infiltration, increased runoff, and accelerated erosion (Doe, 1993; McCarthy, 1996; Kade and Warren, 2002; Gilewitch, 2004, this volume). By its nature, military training constitutes an extreme form of landuse. Erosion caused by military training becomes a great concern to the military not only because of the limited amount of land available for training, but also because of the desire of the military to be good stewards of the land entrusted to them, which often conflicts with established military training practices. The military must ensure it does everything possible to protect and maintain its

D.R. Caldwell et al. (eds.), Studies in Military Geography and Geology, 229–242.
© 2004 *Kluwer Academic Publishers.*

training lands (e.g., Patrick et al., 2004, this volume). The impact of military vehicle traffic on soil is well known; however, little is known concerning the impact of foot traffic on soil. This study analyzed the effect of military training conducted by dismounted soldiers (i.e., foot traffic) on soil compaction at the Bataan Bayonet Assault Course (BBAC), the most heavily trafficked training area within the West Point Military Reservation, NY.

Two important components in determining soil compaction on these lands are: (1) the type of training conducted and (2) the soil response to the training. The soil response to different types of training can be quantified in terms of increases in bulk density. Determining the impact and response interaction of training and soil response to training can provide a tool for training area managers to develop improved management techniques.

Current land condition assessment models focus on erosion characterization as the basis for determining the "land condition" and compaction is overlooked (USAEC, 1999). The US Army Training and Testing Area Carrying Capacity (ATTACC) model, which is a modified version of the Revised Universal Soil Loss Equation (RUSLE) computer program, is used to assist in managing the impact of military training on land degradation (USAEC, 1999). The ATTACC model provides for the input of a training factor that estimates an impact assessment for a specific military training activity. These training factors are based on historical data from numerous experiments involving military vehicle maneuvers. Currently the ATTACC model uses an estimated foot impact factor, which does not have a scientific basis. Therefore, this research project was an attempt to validate foot traffic impact and develop protocols to establish foot traffic impacts applicable for use at other installations.

2. BACKGROUND

Changes in landuse management commonly involve adjusting a particular force on an ecosystem and the imposition of another (Milchunas et al., 1999). Removal of an existing management regimen may have a synergistic or an antagonistic affect, depending on the type and intensity of the new management practice. For example, the landuse practices of the West Point Military Reservation have dramatically changed over the past 200 years, ranging from agriculture, to industrial smelting, and ultimately, to military training. Each of these practices impacts the West Point landscape in different ways. From 1944-60, military training was vehicle intensive, but over the last 40 years, the training impact on the land is exclusively from foot traffic (M. Anderson, US Military Academy (USMA), pers. comm., 2002).

The West Point Military Reservation consists of a cantonment area of 2250 acres and a training area of 13,830 acres. Prior to World War II, it consisted of approximately 2000 acres. The government then deemed it necessary to expand the training environment for cadets to include realistic training with contemporary weapon systems (M. Anderson, USMA, pers. comm., 2002). The selected training land, northwest of the original cantonment area, was rural farmland consisting of several family-operated farms or orchards and an iron ore smelting operation. Following acquisition, the army converted the area to training land characterized by open fields, forests, training areas, weapon ranges, and impact areas. Additionally, several ponds and lakes became public recreation areas and reservoirs. The acquired area includes the watershed for West Point Military Reservation and the local community of Highland Falls, NY (M. Anderson, USMA, pers. comm., 2002).

3. METHODS TO DETERMINE SOIL CONDITION

Common methods for determining soil condition caused by military training, in general, include the Integrated Training Area Management system (ITAM; US Army, 1998) and the US Army Training and ATTACC models. These systems qualify a soil condition based on soil loss (erosion) and nothing more. The ITAM system also uses Land Condition Trend Analysis (LCTA) that consists of ten years of historical data and uses a visual and subjective assessment of land condition based on 100 m transects throughout a study area. Existing research and data supporting dismounted impact is limited, and accepted techniques for estimating soil condition are too general to be highly effective.

Anderson et al. (2002) analyzed the modified RUSLE model support practice factor (P factor) to test the uncertainty in the model's ability to predict land disturbance at the Fort Hood, TX training facility. Their results indicated that the use of a high quality, periodically updated vegetation map can reduce the prediction uncertainty. Likewise, Bartsch et al. (2002) found that erosion models were designed to estimate the amount of sediment based on agricultural applications at the farm-field scale in humid climates. When these empirical models were applied at the landscape scale, it was difficult to accurately assess erosion potential in a semi-arid environment. The Bartsch et al. (2002) work at Camp Williams, UT used the RUSLE model with a Geographic Information System to identify erosion potential areas, but the lack of quantitative accuracy was too general to be highly effective.

4. DATA COLLECTION

Field experimentation and data collection (ground truthing) are the most accurate means of determining soil compaction. Ground validation is a labor- and time-intensive process and, in most cases, is not feasible due to accessibility to training lands and time constraints. Military land managers must manage large, expansive areas with minimal staffs so that ground truthing is often not a realistic option. Using computer-based methods for assessing soil erosion is more practical.

Soils are complex and have varying chemical and physical properties associated with specific regions based on climate, topography, parent material, organisms, and time (Jenny, 1941). Soil texture (i.e., the proportions of sand, silt, and clay) contributes to the soil structure. Sand provides strength to the soil structure and resists compaction. Silt also provides structural strength and has little nutritional value, whereas clay plays an important role in nutrient supply and structural support. On average, these three components provide approximately 45% of a soil's volume with pore space taking up as much as 53% of the remaining volume, and organic matter filling the remainder (Dunn, 1984). The soils in this study area are classified by the American Association of State Highway Officials as A-7-5 or A-7-6, silty and clayey soils (Liu and Evett, 2003).

The most noted impact from foot traffic is compaction and loss of vegetation. Within military training areas, the impact is noticeably more severe than for common foot traffic elsewhere due to persistent use of these areas. The impact from foot traffic reduces soil volume by compressing void spaces and "wearing out" the vegetation. The loss of void space reduces the infiltration rate from the surface, whereas the loss of vegetation reduces the surface retention time, allowing rainwater to soak into the ground more rapidly.

Understanding soil response to foot traffic requires soil compaction analysis. Common methods for determining compaction include determination of bulk density (mass per unit volume), surface soil infiltration rates (absorption of water per unit of time), and penetrability (resistance to penetration).

5. STUDY AREA AND METHODOLOGY

The BBAC was chosen as the study site because it consistently has a high degree of training use. Every year approximately 1000 cadets traverse the training course during the summer. The course is rectangular in shape, oriented southwest to northeast (Fig. 1), and approximately 40 m wide by

300 m long. The course consists of nine lanes, each lane containing identical "stationary obstacles" that a foot solider might commonly encounter in a wartime environment. Examples of the stationary obstacles include: Ladder walls, parry structures, bunkers, and culverts (Fig. 1). Each lane is identical and all obstacles are positioned uniformly in each lane. The BBAC exterior was surveyed, staked, and logged using a Trimble GPS Model 4000SE. Data collection transects (1-7) were established approximately 65 m apart across the BBAC. Data collected from transects 8 and 9 were not used because the lanes merged, and impacts on each lane could not be separately determined. The transect end points were connected using nylon line to ensure data was gathered at the same location throughout the field data collection period (Fig. 1). The study consisted of three phases: (1) the foot traffic experiment, (2) the BBAC data gathering, and (3) numerical modeling.

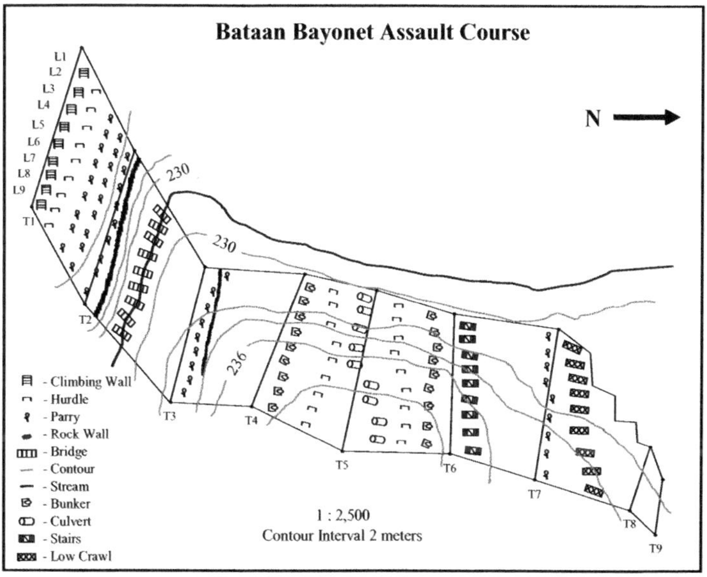

Figure 1. The Bataan Bayonet Assault Course on the West Point Military Reservation, NY.

5.1 Foot Traffic Experiment

The foot traffic experiment was an impact response experiment on an undisturbed piece of terrain similar to the BBAC. Data gathering included soil bulk density readings, infiltration rates, and soil penetrometer measurements taken before, during, and after the summer training cycle to determine the BBAC soil response and condition change caused by the training.

The induced foot traffic experiment consisted of taking periodic bulk density samples as cadets walked through the specified piece of terrain. Cadets walked through the experiment site throughout the summer training cycle for a total of 3088 passes. Plotting the bulk density against the number of soldier passes produced a distinctive response (Fig. 2). Fig. 2 also shows the maximum bulk density (1.30 gm/cm^3) the BBAC should experience based on the load cadets exert on the soil.

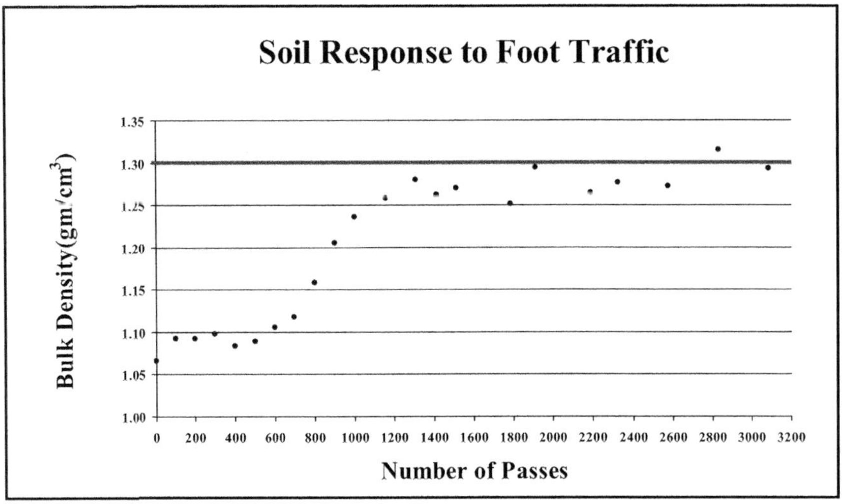

Figure 2. Induced foot traffic experiment at the BBAC, with optimum compaction identified by the solid line at 1.30 gm/cm^3.

5.2 Data Gathering at the BBAC

The BBAC data gathering phase also consisted of taking before, during, and after measurements for bulk density, soil infiltration, and soil penetrometer resistance. Transects were used to identify locations for measurements throughout the training cycle. The "before" samples were taken prior to cadets arriving on site, the "during" samples were taken approximately mid-way through the training cycle, and the "after" samples were collected at the completion of training. The data was tabulated and graphed (Figs. 3, 4, and 5) followed by statistical analysis of the average values. Samples for bulk density determination were taken to include the surface organic layer in order to better understand rate and effect of the loss of organics during the training cycle.

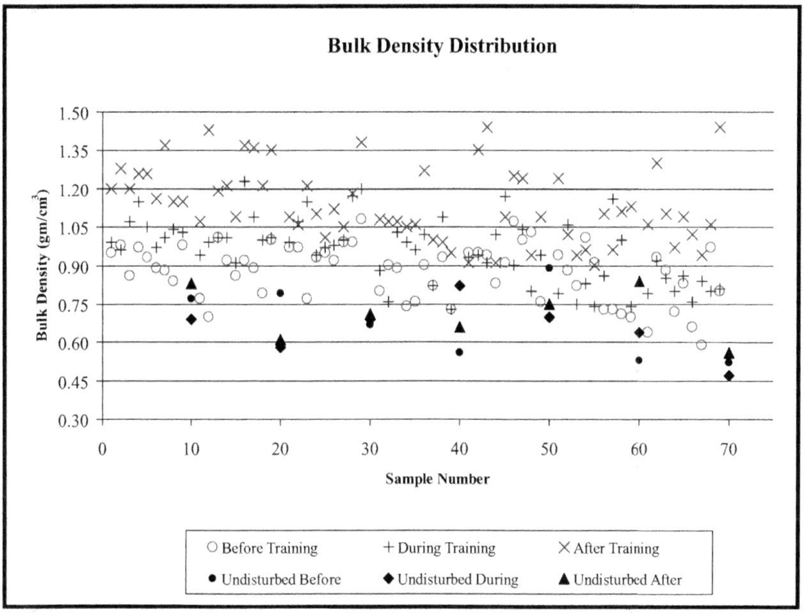

Figure 3. Bulk density distribution of soil samples collected before, during, and after testing at the BBAC.

5.3 Numerical Modeling

The modeling phase combined laboratory experimental data and the field data into mathematical models that replicated the induced foot traffic response curve. The general shape of the soil bulk density response curve is an increasing function that is bounded below at an initial bulk density and above at a maximum bulk density based on a specified applied load. Cumulative distribution functions (CDF) from probability and statistics theory also carry the same above and below bounding characteristics (Devore, 2000). Two CDFs, the Wiebull function, and the Gamma function can be used to portray a soil's response to foot traffic. By shifting the CDFs, models were developed to replicate the soil's response to foot traffic. Additionally, a logistical growth curve can also replicate this type of bounding characteristic and, by adjusting the rate of growth, a soil response curve can replicate the induced foot traffic effect on the soil (Polking et al., 2002). These three approaches were adjusted to fit the natural curve obtained at the BBAC and compared statistically to determine the "best fit" model.

Laboratory experiments and the field collection data mentioned earlier were required to determine the bounded limits for the soil bulk density. Characterization of the soil based on set experimental procedures and quantitative analysis is essential for understanding the soil and its natural reaction to induced foot traffic. Therefore, a series of geotechnical engineering tests (determination of specific gravity, Atterberg limits, and modified standard compaction) assisted in determining the soil's maximum bulk density. Combining the data from the laboratory experiments (maximum density) with the field data (minimum density and natural curve), the bounded upper and lower portions of the curve were established and applied to the three mathematical functions to identify an accurate model for soil response to induced foot traffic. With these two bounding values, the three models were populated and compared statistically to the actual curve generated from the foot traffic experiment.

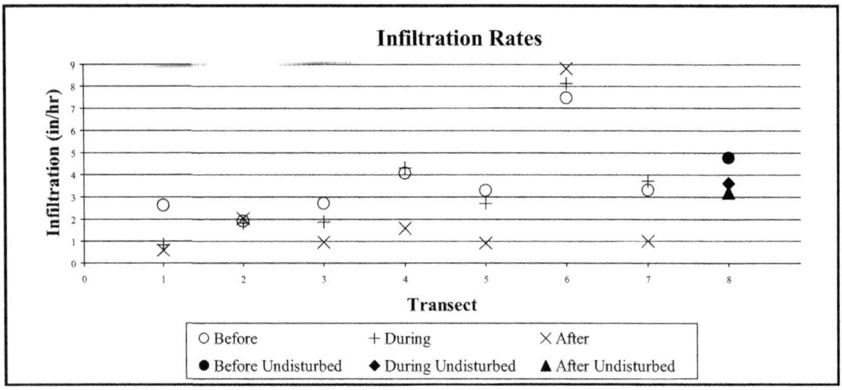

Figure 4. Infiltration rate in in/hr for soil samples collected before, during, and after testing at the BBAC.

6. RESULTS

6.1 Foot Traffic Experiment

Initial bulk density and infiltration rate measurements and penetrometer readings were made and recorded prior to the traffic experiment. Personnel then walked over the test site, and bulk density measurements were recorded approximately every 100-150 passes, for a total of 3088 passes. The traffic experiment revealed measurable changes in soil bulk density in response to foot traffic. Initial bulk density values remained relatively stable until 700-800 passes. At 700-800 passes (Fig. 2), soil compaction occurred at a higher

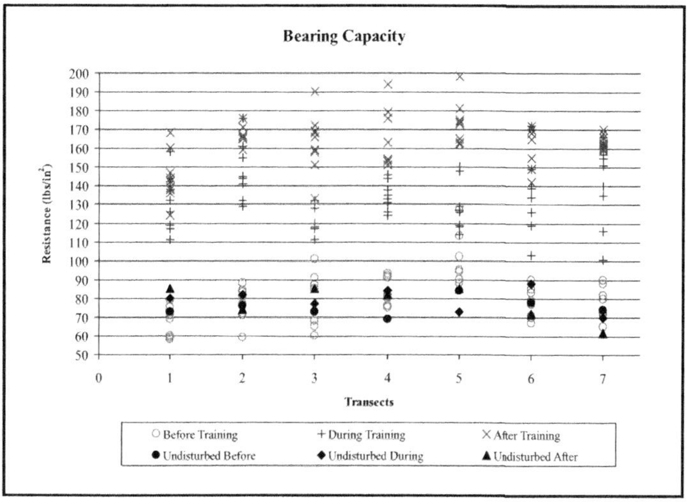

Figure 5. Soil penetrometer resistance readings in lb/in^2 for soil samples collected before, during, and after testing at the BBAC.

rate and a marked increase in the soil bulk density occurred; bulk density continued to increase through 900 and 1400 passes, then started to level out, remaining below 1.30 gm/cm^3 with the exception of one outlier. During the traffic experiment, loss of vegetative cover and ground litter occurred within the first 400-500 passes.

6.2 Compaction Study

The results of the bulk density measurements indicate a distinct increase in the average bulk density as cadets progress through the training cycle. A visual inspection of Fig. 3 shows the bulk density distribution for the "undisturbed," "before," "during," and "after" samples. As the training continued during the summer, the bulk density measurements increased accordingly. To verify the change in bulk density, a One-way ANOVA test was run: There are statistically significant differences between the "before," "during," and "after" bulk density means when compared to the "undisturbed" mean that are definitely not the result of a natural response. Likewise, differences between the "before" and the "during" and "after" means could not have occurred naturally.

Comparing the "after" and "before" density means shows a 24% decrease in bulk density over the 10 month period between training cycles. A comparison of the "before" and the "undisturbed" soil densities shows a lesser 22% difference in soil densities.

6.3 Infiltration Study

The results of the infiltration study indicate a decrease in the average infiltration rates as the training cycle progresses. A visual inspection of Fig. 4 shows the infiltration rate distribution for the "undisturbed," "before," "during," and "after" samples. The infiltration analysis indicates an overall drop in infiltration rates over the training cycle. A One-way ANOVA test determined that there are statistically significant differences between the "before" and "after" means, whereas comparing the "before" and "during" means and the "undisturbed" and "before" means produced no statistical differences.

The difference between the average "undisturbed" and "before" infiltration rates was only 6%. The difference between the average "before" and "during" infiltration rates was 8%, whereas the overall decrease for the "before" and "after" rates was 38%.

6.4 Penetrometer Study

The results of the bearing capacity measurements indicate an increase in the average bearing capacity as training progressed. A visual inspection of Fig. 5 shows the bearing capacity distribution for the "undisturbed," "before," "during," and "after" samples. As the training progressed the bearing capacity of the soil increased accordingly. A two-sample t-test ($P = 0.135$) was used to verify the change in bearing capacity. There are no statistically significant differences between the "undisturbed" and "before" bearing capacity means. Conversely, comparison of the "during" and "after" bulk density means to the "undisturbed" and "before" bearing capacity means shows strong statistical differences (One-way ANOVA test); the average resistance for each set of readings could not have occurred naturally.

Comparing the "after" and "before" bearing capacity means shows only a 4.2% increase. Large increases of 40% and 50% occur during the training cycle when comparing the "before" bearing capacity mean to the "during" and "after" bearing capacity means.

6.5 Numerical Modeling

Applying the results of the geotechnical engineering laboratory experiments and the field experiment to the three numerical models (the logistical growth model and the Gamma and Weibull function models) produced distinctive curves (Fig. 6). A regression analysis produced the following R^2 values for each model: Logistical growth = 0.85, Gamma =

0.95, and Weibull = 0.94. The corresponding R^2 values indicate that the Gamma model is the best-fit model, accounting for 95% of the variation observed in the test data.

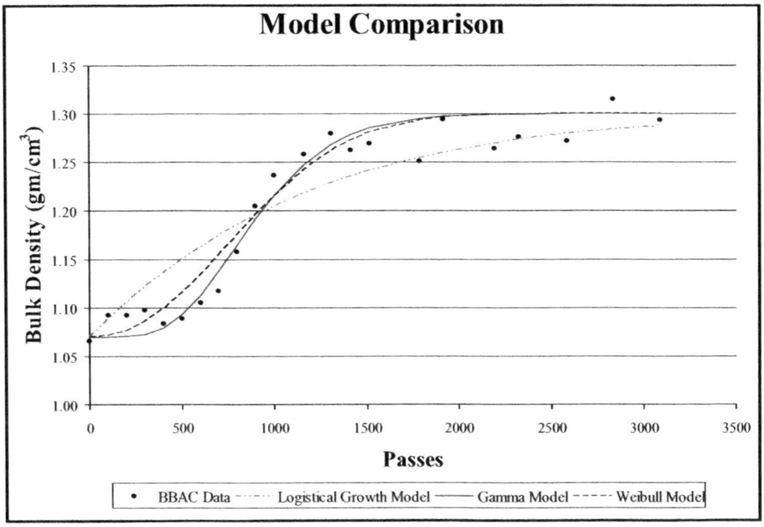

Figure 6. Comparison of soil bulk density measurements at the BBAC with data from the numerical models.

7. DISCUSSION AND CONCLUSION

The traffic experiment revealed measurable changes in soil bulk density in response to foot traffic (Fig. 2). The initial 700 passes had minimal effect on soil compaction. Initial bulk density values remained relatively stable until 700-800 passes. At 700-800 passes, soil compaction occurred at an exponential rate and a marked increase in the soil bulk density occurred. Soil compaction continued to increase through 1400 passes. A possible explanation for this bulk density change after 700 passes is vegetation cover loss. During the foot traffic experiment, the visible vegetation layer disappeared at approximately 400-500 passes. The organic layer and vegetation may act as a cushion to lessen the impact of the traffic and assist in preventing a quick compaction of the soil.

The increase in bulk density indicates increased compaction due to foot traffic (Fig. 6). Initial "before" bulk density values show an average compaction level of approximately 0.87 g/cm^3. The "during" and "after" values increase to 0.96 g/cm^3 and 1.14 g/cm^3, respectively. This increase in bulk density is the result of cadet training on the BBAC and the loss of the organic layer, which provides cushioning against impact. Visual inspection of the BBAC lanes shows a "wearing out" of the organic layer, which can exaggerate the impact effect of the foot traffic.

Comparison of the bulk density mean distributions for the "undisturbed" and "before" sample means indicates a return of the soil to a lower bulk density after the 2002 summer training cycle. Although the soil does not return to the "undisturbed" bulk density level, the recovery does provide some benefit for the soil to endure another training cycle. Several of the "before" readings actually fall within some of the "undisturbed" readings. This was unexpected and may indicate a quick soil recovery. Furthermore, the unexpected low "before" bulk density readings may be attributed to the thick organic layer, which acts as a cushion to lessen impact on the soil. A comparison of the traffic experiment results and the BBAC bulk density measurements indicate initial compaction at the BBAC is below the compaction measured initially at the undisturbed site. This is important considering that the compaction reading represents the initial compaction at which the BBAC begins a new season of training.

The pronounced difference in infiltration rates between the undisturbed site and the BBAC sites indicates a direct relationship between infiltration rate and foot traffic (Fig. 4). As foot traffic increased over the summer the infiltration rates decreased. Visual inspection of the soil profiles at the infiltration sites revealed a 2-5 cm thick, distinctive, platy structure layer directly below the organic layer, which overlies a relatively undisturbed blocky structure layer. The platy structure developed over the course of the summer, thus affecting the measured infiltration rates. The soil in the undisturbed area displayed no signs of such a platy structure. Soil peds were predominately blocky, indicative of a soil structure conducive to infiltration and plant regrowth. The similarity between the average "undisturbed" and "before" infiltration rates indicates that the soil has recovered almost completely from the previous training cycle in 2002.

The surface bearing capacity increased dramatically over the course of the training cycle (Fig. 5). A dramatic jump in "during" bearing capacity indicates that the surface soil layer becomes compacted quickly and continues to compact until training is complete. The "undisturbed" and "before" bearing capacities are statistically the same. The "undisturbed" bearing capacity is nearly identical to the "before" training bearing capacity, indicating that the soil's bearing capacity has recovered almost completely from the 2002 summer training cycle.

The modeling results provide a good foundation for future research. A visual inspection of the results (Fig. 6) shows the foot traffic experiment data displaying a distinctive exponential growth curve, which generally can be duplicated, although uniquely different (different growth slope, initial density and maximum density), in other soils. The curves generated from the models illustrate their ability to replicate the observed results and have the flexibility to be applied to different soils, whereas statistical comparison proved that the Gamma CDF to be the "best fit" model. With this new model, a similar type of application is possible at other military installations to assist in characterizing soil condition based on field and laboratory analysis of soil variables.

The results of this study strongly suggest that inclusion of the compaction model into the ATTACC and ITAM assessments of land condition would be a positive improvement and, therefore, should be addressed and developed. Erosion, as a lone condition qualifier, is no longer a viable solution to proper land management protocols. Additional research is required to ensure the accuracy of the model and its applicability to military land management techniques. Replication of this technique at other military posts and using different soils will ensure the validity of the approach and may have long-lasting benefits to training land sustainment.

REFERENCES

Anderson, A., Fang, S., Wente, S., Gertner, G.Z., and Wang, G. 2002. Uncertainty analysis of predicted disturbance from off-road vehicular traffic in complex landscapes at Fort Hood. Environmental Management 30:199-208.

Bartsch, K.P., Van Miegroet, H., Boettinger, J., and Dobrowolski, J.P. 2002. Using empirical erosion models and GIS to determine erosion risk at Camp Williams, Utah. Journal of Soil and Water Conservation 57:29-37.

Devore, J.L. 2000. *Probability and Statistics for Engineering and the Sciences.* Pacific Grove, CA: Duxbury Press.

Doe, W.W., III. 1993. Simulation of the spatial and temporal effects of maneuvers on watershed response (spatial effects) (unpublished Ph.D. dissertation). Boulder, CO: University of Colorado.

Dunn, B.A. 1984. Recreation effects on forest soil and vegetation. Clemson, SC: Clemson University, Department of Forestry Technical Paper 16.

Gilewitch, D. 2004. The effect of military operations on desert pavement - A case study from Butler Pass, AZ (USA). In *Studies in Military Geography and Geology*, D.R. Caldwell, J. Ehlen, and R.S. Harmon, eds., Dordrecht, The Netherlands: Kluwer Academic Publishers, 243-258.

Jenny, H. 1941. *Factors of Soil Formation.* New York: McGraw-Hill, Inc.

Kade, A. and Warren, S.D. 2002. Soil and plant recovery after historic military disturbances in the Sonoran Desert, USA. Arid Land Research Management 16:231-243.

Liu, C. and Evett, J.B. 2003. *Soil Properties - Testing, Measure, and Evaluation.* Upper Saddle River, NJ: Prentice Hall, Inc.

McCarthy, L.E. 1996. Impact of military maneuvers on Mojave Desert surfaces: A multiscale analysis (unpublished Ph.D. dissertation). Tucson, AZ: University of Arizona.

Milchunas, D.G., Schulz, K.A., and Shaw, R.B. 1999. Plant community responses to disturbances by mechanized military maneuvers. Journal of Environmental Quality 28:151533-151547.

Patrick, D.M., Roth, K.M., and Lemire, R.A. 2004. Managing ground water resources at Camp Shelby Training Site, MS (USA). In *Studies in Military Geography and Geology*, D.R. Caldwell, J. Ehlen, and R.S. Harmon, eds., Dordrecht, The Netherlands: Kluwer Academic Publishers, 321-332.

Polking, J., Boggess, A., and Arnold, D. 2002. *Differential equations with boundary value properties*. Upper Saddle River, NJ: Prentice Hall Inc.

US Army. 1998. Integrated Training Area Management (ITAM). Washington, DC: US Government Printing Office, Army Regulation 350-4.

US Army Environmental Center (USAEC). 1999. US Army Training and Testing Area Carrying Capacity (ATTACC). Aberdeen Proving Ground, MD: US Department Army Environmental Center.

Chapter 19

THE EFFECT OF MILITARY OPERATIONS ON DESERT PAVEMENT

A Case Study from Butler Pass, AZ (USA)

Daniel A. Gilewitch
US Military Academy

Abstract: Military activities cause physical alteration of landscapes that can provide geomorphologists unique opportunities to better understand natural processes. Today's desert pavement in western Arizona exhibits scars from 1940s era tracked vehicle maneuvers. This study examines the impact of these military maneuvers with particular attention to an unexpected field observation of deeper soil moisture penetration beneath scarred pavement than under undisturbed pavement, a condition that appears contradictory to results of previous work. Field work and backscatter electron microscopy images reveal destruction of the Av horizon and fractured soil plasma near the surface under track scars; this fracturing promotes soil moisture penetration relative to the naturally indurated, compact, surface material of undisturbed pavement. Increasing moisture penetration of the subsurface may encourage plant growth and have consequences to the ecological functioning of a largely barren landform.

Key words: desert pavement, vehicle track scars, soil compaction, sustainable landuse

1. INTRODUCTION

Most military operations cause physical alterations to natural landscapes regardless of mitigation strategies (Fig. 1). Although these alterations can provide geomorphologists unique opportunities to better understand natural processes (Prose, 1985; Marston, 1986; Ayers, 1994; Koch and El-Baz, 1998; Watts, 1998; Gatziolis et al., 2000; Prose and Wilshire, 2000; Prosser et al., 2000), they are often considered to be destructive to the environment. Arid regions of the world are particularly susceptible to long term

D.R. Caldwell et al. (eds.), Studies in Military Geography and Geology, 243–258.
© 2004 *Kluwer Academic Publishers.*

anthropogenic alteration and damage because of the prevalence of stable landforms such as desert pavement (Mabbutt, 1977).

Figure 1. World War II era track scars are clearly discernable on the well-developed desert pavement at the study site in Butler Pass, AZ.

Desert pavement occurs extensively in deserts and has earned regional names such as *reg* in North Africa and *gibber* in Australia; similar surfaces also occur in mountain, arctic, and periglacial regions (Cooke, 1970). They are commonly found atop alluvial deposits in deserts, including alluvial fans and fluvial terraces worldwide (Cooke, 1970). Desert pavement consists of an armored surface of abundant, closely packed stone fragments of pebble to cobble size that is only one or two stones thick, often set on finer sediment several centimeters to meters thick (Mabbutt, 1979; Elvidge and Iverson, 1983; McFadden et al., 1987; Thomas, 1989; Cooke et al., 1993; Thomas and Goudie, 2000). The closely spaced surface particles protect underlying fine material from further erosional forces, thus earning the name of *desert pavement* or *desert armor*. The sediment layer is usually characterized by low infiltration rates arising from surface crust formation under raindrop impact and washing of fine sediments into near-surface pores. Occasionally, salts act as a bonding agent (Cooke, 1970; Cooke et al., 1993).

Whereas desert pavements are indicative of surface stability that can last tens of thousands of years (Mabbutt, 1979), the pavement surface itself is exceptionally fragile and easily destroyed. The armored layer, although resistant to further erosion by wind and water, can be easily broken when weight is applied to the surface. A human walking on the surface can break the layer of accreted pebbles and sediment matrix and reveal the loosely held fine particles beneath (see McDonald, 2004, this volume). This results in a scar in the pavement.

Alterations to desert environments by human activity are common. The most damaging impacts are related to soil compaction and subsequent increases in erosion rates. Wilshire and Nakata (1976), for example, reported that off-road motorcycle use in the Mojave Desert caused significant soil compaction and increased soil erosion. Iverson et al. (1981) noted a decrease in desert soil porosity and infiltration capacity after off-road vehicle use. Cross-country maneuver accelerates water and wind erosion (Gillette and Adams, 1983; Hinckley et al., 1983; see also Marston, 1986), has a negative impact on desert soil stabilizers (Wilshire, 1983), and a negative effect on desert vegetation and wildlife (Brattstrom and Bondello, 1983; Bury and Luckenbach, 1983; Lathrop, 1983). One of the most significant and long lasting impacts of vehicle use in the desert is compaction of the underlying sediment, which changes density, porosity characteristics and infiltration rates (Webb, 1983). Soil compaction is a widely cited undesirable alteration (Wilshire and Nakata, 1976; Iverson et al., 1981; Braunack, 1986a, b; Ayers et al., 1990; Lovich and Bainbridge, 1999; Prose and Wilshire, 2000).

The US Geological Survey recently conducted a detailed investigation of the lasting effects of military movement across desert pavement (Prose and Wilshire, 2000). This study investigated relic tank track scars from the same military maneuvers that this research considers (ca. 1942-44), and track scars from later exercises conducted in 1964. Prose and Wilshire (2000) compared soil compaction (through use of a penetrometer), soil bulk density, surface reflection, surface clast size, infiltration rates and plant cover and density. They concluded that soil density is greater under the 1940s era track scars despite the heavier weight of tanks in 1964, and that water infiltration rates under track scars are up to 55% lower than under undisturbed pavement at the Ward Valley, CA study site.

The purpose of this research is to assess the impact of 60 yr old tank track scars on desert pavement in western Arizona with particular emphasis on an unexpected observation of deeper soil moisture penetration beneath track scarred pavement, than under undisturbed pavement. An examination of the soil profile after a winter rain event in January 2002 revealed that the depth to which moisture penetrated the subsurface of desert pavement in track scars was greater than the depth of penetration in adjacent, undisturbed pavement. Prior observations in the literature on infiltration capacity after off-road vehicle (ORV) activity (Iverson et al., 1981; Webb, 1983; Webb and Wilshire, 1983; Wilshire, 1983; Prose and Wilshire, 2000) suggest that these initial field observations are anomalous.

Information from this research has implications for military maneuver area management as well as civilian recreational ORV control in the fragile arid lands of the southwestern United States and elsewhere. Desert pavements are ubiquitous and see frequent archaeological use in the form of trails, cleared patterns, and stone alignments collectively known as *geoglyphs* or *Earthen Art* (Von Werlhof, 1987). The growing popularity of civilian off-road recreational vehicle traffic and the preponderance of military activity in arid regions will likely lead to repeated incursions on this fragile landform. This study may be useful in assessing management strategies for archaeological, military, and civilian use of these areas.

2. STUDY SITE

Between April 1942 and April 1944, over one million US troops trained at the Desert Training Center/California-Arizona Maneuver Area (DTC/C-AMA) in preparation for combat in World War II (Bischoff, 2000). Tanks maneuvered across many types of desert terrain, including geomorphically stable desert pavement where they left track scars that are still clearly visible today. This historical event provides a unique opportunity to study long term impacts of vehicle maneuvers on this particular desert landform because it permits accurate deduction of the timing of pavement alteration and its cause.

The study site for this research lies along a single interfluve on the gently sloping surface of an incised alluvial fan issuing from Butler Pass in western Arizona (Fig. 2). The pass rests at approximately 520 m in elevation. Mean temperature at Needles, CA, 208 km to the northwest, is 11°C in January and 35°C in June. Rainfall varies considerably from year to year with an annual average of 112 mm at Needles (Prose and Wilshire, 2000). Vegetative ground cover within the study area is 5-10%, consisting mostly of creosote bush (*Larrea tridentata*), triangle-leaf bursage (*Ambrosia deltoidea*), and brittle bush (*Encelia farinosa*).

The entire site is only several hundred square meters in area along a single interfluve - a sufficiently small size to offer confidence that geomorphic processes working on the site are similar. The study area is comprised of a discrete alluvial surface covered by well-developed desert pavement. Rock types are similar in the desert pavement, consisting of angular granitic, gneissic, and quartzite clasts. The slope of 1-2° remains consistent over the study area. Soil Av and B horizons in undisturbed areas display uniform characteristics across the study area. Other than a power line over 100 m away, the surface area investigated does not appear to have been subject to significant anthropogenic damage other than from relic DTC/C-AMA tank maneuvers.

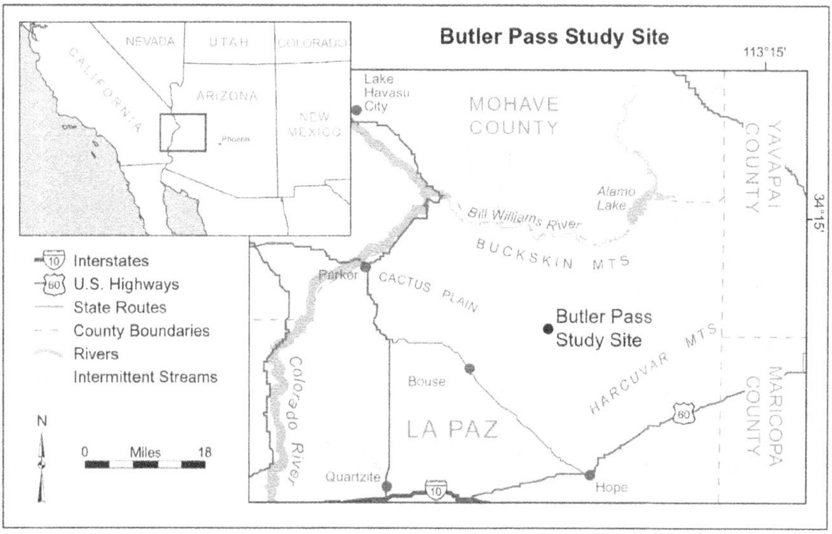

Figure 2. Location map of the Butler Pass study site in western Arizona.

3. METHODS

Measurements of track scar width and base were compared with military vehicle specifications to determine which type of vehicle created the scars. The historical context of training associated with the DTC/C-AMA isolated the approximate time of alteration. Tanks in the army inventory and maneuvering in the area in the early 1940s were the M3A1 Stuart, the M3A5 General Grant, or the M4 Sherman (Bischoff, 2000). The majority of tanks using the DTC/C-AMA in 1943-44 were M4s (Bischoff, 2000). It is possible that M60 tanks, maneuvering during a 1964 exercise designated Operation Desert Strike (Prose and Wilshire, 2000), also operated in the Butler Pass area. Tanks that could have traveled in the Butler Pass area differ in track width, track base, and weight (Table 1). The M60 tank has a much larger track width and base, and is heavier than World War II era vehicles. Measurement of each track scar pair indicates whether the tank causing the damage is more likely to have been an M60 or an earlier model tank, and therefore, isolates when alteration to the desert pavement occurred. Civilian 4x4 off-road wheel vehicle dimensions are also compared. Track scar width and track base measurements were observed at 10 cm intervals and subjected to a one-sample t-test comparing these measurements to specifications from tanks used in training at the DTC/C-AMA (Table 1).

Table 1. Characteristics of the tracked military vehicles that are thought to have operated in the Butler Pass, AZ study area. The measurement convention for track base is from the outside of one track to the outside of the other. This makes the value useful in calculations concerning track transport on rail or ship where space is a prime factor.

Vehicle	Track Base	Track Width	Vehicle Weight	Ground Pressure
M3A1 Stuart Light Tank[1]	224 cm (88 in)	29.5 cm (11.6 in)	12,900 kg (12.7 tons)	0.91 kg/cm^2 (12.9 lb/in^2)
M3A5 Grant Medium Tank[1]	272 cm (107 in)	40.6 cm (16 in)[1] or 42.1 cm (16.6 in)[2]	28,100 kg (27.7 tons)	0.89 kg/cm^2 (12.7 lb/in^2)
M4 Sherman Medium Tank[1]	262 cm (103 in)	42.1 cm (16.6 in)	30,300 kg (29.8 tons)	1 kg/cm^2 (14.3 lb/in^2)
M60A1 Main Battle Tank[3]	363 cm (132 in)	71.12 cm (28 in)	57,406 kg (56.5 tons)	0.87 kg/cm^2 (12.37 lb/in^2)
For Comparison: Civilian 4x4 Dodge Dakota[3]	(wheel base) 178 cm (70 in)	(wheel width) 22.86 cm (9 in)	2,430 kg (2.7 tons)	not available

[1] Information from http:/www.onwar.com/tanks/usa, accessed 17 September 2002.
[2] Measurements vary depending on track type.
[3] Measured by author (M60A1 Tank is located on display at Bouse, AZ).

Moisture penetration was estimated by a non-standard field method where water was artificially added to the pavement surface. A small trench was dug approximately 1.5 m long, 15 m wide, and 30 cm deep perpendicular to a track scar. Another trench of similar dimensions was dug a few decimeters away in undisturbed pavement. An evaporative cooler screen was carefully laid 5 cm away and parallel to the trenches. This distance provides a buffer area between the trench side and the area where water is added to the surface, so the presence of the trench does not affect moisture penetration. A total of 7.6 l (2 gal) of water (the equivalent of 2 cm precipitation) was slowly added to the swamp cooler screen that acted as an evaporation retardant and controlled surface water flow so runoff was minimized. By controlling water input, it was possible to minimize surface pooling, a variable that would have influenced moisture penetration depth, particularly in track scars. The trench was then dug back approximately 7 cm to reveal the depth of moisture penetration, which was measured with a ruler.

A variety of techniques compared surface properties of track-scarred pavement with undisturbed pavement. These methods included processing of digital imagery to compare surface particle albedo, particle surface area, and relative presence of rock coatings. Random sampling was used to collect specimens for comparison of surface particle mass, volume, density, and sphericity (e.g., Sneed and Polk, 1958). National Resource Conservation Service (NRCS) soil profile descriptions (cf. Schoeneberger et al., 1998; Birkeland, 1999) established the degree of soil development.

A common method of determining soil density is through the use of bulk density testing (cf. Iverson et al., 1981; Braunack, 1986b; Schoeneberger et al., 1998; Prose and Wilshire, 2000), which requires gravimetric analysis of a sediment sample of known volume (Birkeland, 1999). An accurate method can be used in the field with a nuclear soil density gauge commonly used in the construction industry as a nondestructive and accurate way to determine both density and moisture content of soils, aggregate, concrete, and asphalt. The gauge determines the density of material through the measurement of gamma radiation in either a direct transmission or backscatter mode (Troxler Electronic Laboratories, 1998). The use of a nuclear density gauge is a construction engineering industry standard method (cf. American Society for Testing and Materials; ASTM, 2002), but there does not appear to be significant use of these devices in geomorphic research. Soil density was measured in the field using a Troxler model 3430 Roadreader density gauge. Backscatter and direct transmission readings were acquired at 15 paired locations.

Backscatter electron (BSE) microscopy of soil samples permits a high resolution examination of the nature of soil porosity. Field samples of sediment under track scars and under undisturbed sediment were encased in epoxy at the site to protect and prepare them for analyses. Once epoxied, subsamples approximately 1 cm in diameter were removed from 0-2 cm, 3 cm, and 5 cm depths underneath track scarred pavement and undisturbed pavement. The fragments were placed in microprobe molds, polished and carbon coated. A JEOL 8600 Electron Microprobe using a BSE detector and energy dispersive spectrometry generated textural and chemical information on the composition of the soil plasma. Sample material with higher net atomic numbers scatter more electrons and appear bright in a BSE image (Reed, 1993). The BSE detector provides good compositional contrast and can be used to clearly distinguish sediment structure at scales of 1μm to 100μm and beyond.

4. RESULTS

One sample t-tests of track scar field measurements do not closely match the track width and base specifications for tanks under consideration because the scars have eroded over time. Field measurements strongly suggest, however, that World War II era tanks caused the pavement alteration (Fig. 3). All measured track base and widths are significantly smaller than can be expected from an M60 tank, and are larger than popular off-road wheeled vehicle 4x4 specifications. Because the Butler Pass area was not added to the DTC/C-AMA as a training site until March 1943, World War II tanks most likely created the scars between March 1943 and April 1944, the date when the training area was closed. Investigation of debris found in a gully within 100 m of the tracks examined in this study revealed 5 gal oil cans commonly used by the Army for tank maintenance, and tank parts including M4 track end connectors. The date stamped on the bottom of the oil cans is 24 May 1943. Thus, the track scars under study are approximately 60 yrs old.

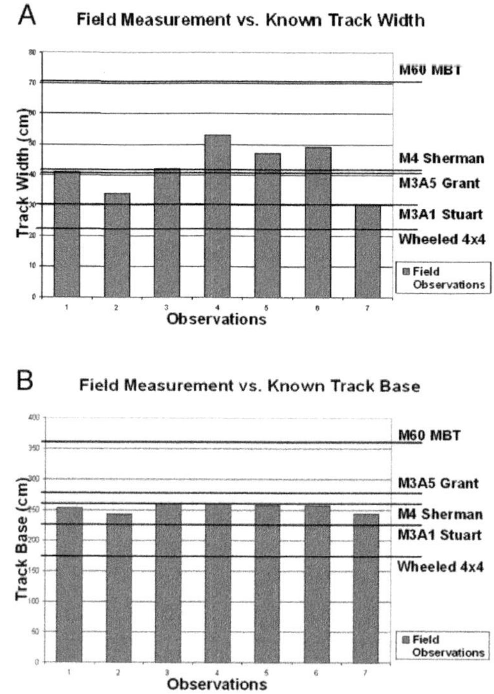

Figure 3. A. Comparison of known track base vs. track scar base, and B. known track width vs. track scar width; mostly correlate with World War II era tanks.

The introduction of water to the pavement surface in the field confirmed the anomaly of greater moisture penetration under tank track scars. The average depth of moisture penetration was 5.7 ± 0.5 cm under track scars and only 3.4 ± 0.6 cm under undisturbed pavement.

Detailed surface and subsurface field measurements confirm the basic field observation that in-track and out-of-track sediments differ substantially.

In-track sediment is lighter in color and contains more particles that do not have rock coatings (Dorn, 1998), and particles inside the tracks are smaller. The Av soil horizons inside track scars average 3.6 ± 0.6 cm in depth, whereas undisturbed areas average 2.3 ± 0.5cm. The A-B horizon boundary is rated either abrupt (0.5 to <2cm) or very abrupt (<0.5cm) in both track scars and undisturbed areas (Schoeneberger et al., 1998).

Soil is consistently denser under track scarred surfaces than adjacent, undisturbed desert pavement in all 15 paired-sample locations. Troxler gauge readings indicate that soil density averages 1825 ± 103 kg/m^3 (114 ± 6 lb/ft^3) under track scars and 1713 ± 93 kg/m^3 (107 ± 6 lb/ft^3) under undisturbed pavement from 0-20 cm in depth (Fig. 4).

BSE imagery provides compositional data that allow qualitative assessments of sediment texture. Samples at 0-2 cm depths are within the Av horizon in both undisturbed pavement and in track scars. These BSE images exhibit significant differences in the character of soil particles (Fig. 5). In track scars, the Av horizon soil structure appears fractured, whereas undisturbed pavement sediment appears comparatively solid. The trend begins to reverse at 3 cm depth, with comparatively more solid soil plasma present in track scar samples than at the 0-2 cm depth (Fig. 6). At 5 cm depth, the soil compaction is clearly opposite of the 0-2 cm samples; more compacted soil textures occur under track scars than under undisturbed pavement (Fig. 7).

5. DISCUSSION

Previous literature argues that compaction of soils by tracked vehicle or ORV passage should inhibit moisture penetration (Wilshire and Nakata, 1976; Iverson et al., 1981; Webb, 1983; Prose and Wilshire, 2000). Evidence at the Butler Pass study site indicates this characterization is not universal. Track scars created by tanks maneuvering in preparation for World War II generated a condition whereby moisture penetrates to greater depths than adjacent natural pavement. Greater moisture penetration occurs despite higher soil density within the tank tracks.

Prose and Wilshire (2000), using a surface infiltrometer, reported up to a 55% lower infiltration rate in desert pavement soils traversed by tanks maneuvering in the same exercises that created the Butler Pass tracks in the early 1940s. Moisture infiltration is directly related to soil porosity and permeability. Porosity refers to the relative amount of pore space in sediment compared to that occupied by solids. Permeability refers to movement of

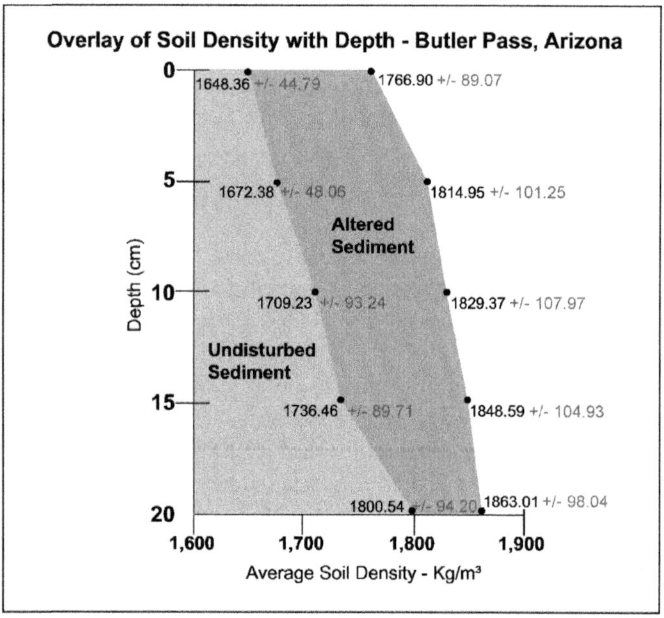

Figure 4. Soil density under track scarred pavement averages 111.17 kg/m^3 greater than under undisturbed pavement. This chart shows the difference in soil density with depth.

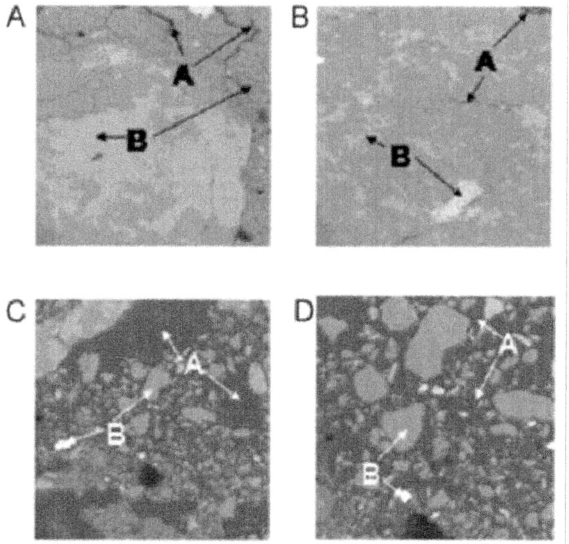

Figure 5. BSE imagery comparing the Av horizon sediment in track scars and in undisturbed pavement between 0-2 cm depth. The soil plasma in track scars is relatively well compacted with relatively few, small pore spaces. Scale: The height and width of images A, B, and C is 600 μm, compared to 300 μm for image D.

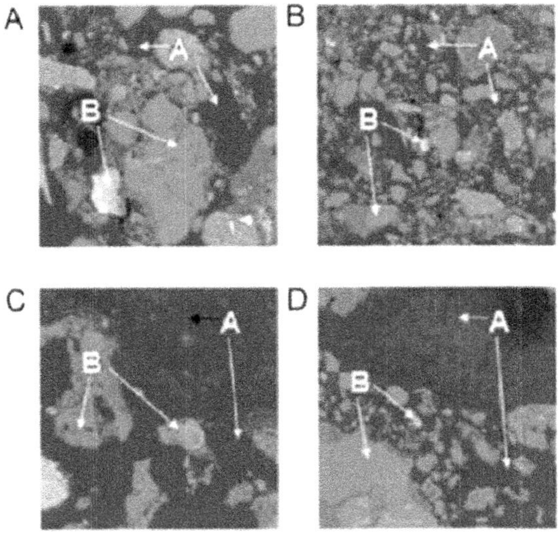

Figure 6. BSE imagery comparing the Av horizon sediment in track scars and undisturbed pavement at 3 cm depth. The soil plasma in track scars possesses relatively more pore spaces than at 2 cm depth. Scale: The height and width of image A is 2241 μm; image B, 690 μm; image C, 600 μm; and image D, 120 μm.

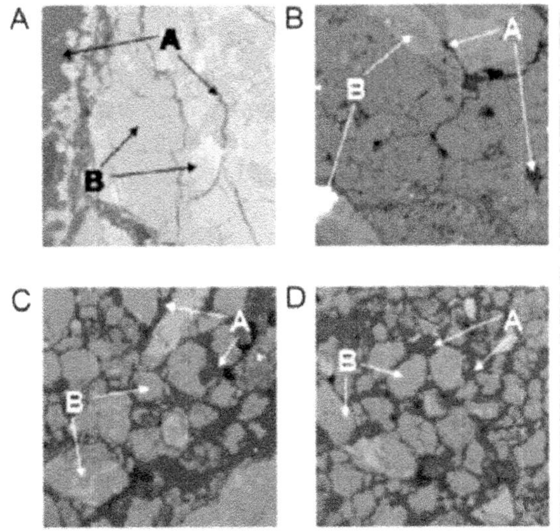

Figure 7. BSE imagery comparing Av horizon sediment in track scars and undisturbed pavement at 5 cm depth. The soil plasma in track scars possesses relatively more compaction than at 2 or 3 cm depths, but the soil plasma in undisturbed pavement is relatively less compact. Scale: The height and width is 60 μm for all four images.

water and is a function of the presence and interconnectedness of these pores. Water introduced at the surface of dry sediment fills capillaries and pore spaces; the water farthest from soil is under lower surface tension and is, therefore, able to infiltrate deeper into the soil in response to gravity and soil matrix potential (Birkeland, 1999). Larger pore spaces allow more water to migrate in this manner. Porosity and permeability are inversely related to soil density, which is susceptible to change by compaction (Thomas and Goudie, 2000).

Field observations of the upper soil profile suggest why moisture penetrates to greater depths in track scars versus undisturbed pavement at the Butler Pass study site. The NRCS measure of surface consistency reveals that the Av horizon (a layer of silt directly underneath lag) of the track scars is much more friable than peds of Av horizon material under undisturbed pavement. Indeed, the surface of track scars is easily rearranged by finger pressure, whereas the surface of undisturbed pavement adheres strongly together. The upper Av horizon in track scars was probably regenerated primarily from materials brought to the site by wind and surface water action after scar creation, so the Av sediment matrix in these locations is not well formed or compacted. The broken soil matrix permits the passage of water. Insufficient time has passed since the destruction of the desert pavement Av horizon by tank maneuvers for significant surface induration to occur. This is in direct contrast to undisturbed pavement, which displays an indurated, compact soil matrix at the surface, the major effect of which is to serve as a barrier to moisture penetration that is characteristic of undisturbed desert pavement at this study site, but not present in track scars.

BSE imagery of sediment samples provides greater explanatory detail, as electron microscope observations show differences with depth in sediment matrix porosity between track scars and undisturbed pavement soils. Tanks appear to have broken up the Av horizon, creating numerous pathways for water to penetrate to greater depths. BSE observations at the 3 cm depth reveal that undisturbed sediment appears slightly more compacted than those under track scars. The difference, however, is more subtle than at the surface, suggesting that the tank's disturbance was not as extensive. At the 5 cm depth, the situation reverses; soil matrices are more compacted under tank track scars than under undisturbed pavement. Another way of thinking of the tank effects is in terms of the trade-off between permeability and porosity. Given these BSE observations, water penetration is most rapid under track scars in the first 2 cm where the indurated crust has been crushed to a sufficient extent to generate large and interconnected pores. By the 3 cm depth, tank crushing both increased soil density and created a pathway of fractured peds; the net effect is that any moisture infiltration would be fairly similar in scarred and undisturbed pavement. Moisture penetrating to a depth of 5 cm would move far slower under the tank scars, because the soil matrix at this depth was not ground up, but simply compacted from the weight of track passage.

The implications of greater moisture penetration are profound in desert ecosystems. Increased soil compaction as a consequence of vehicle maneuver affects water infiltration and hydraulic conductivity and is therefore believed to decrease plant development in arid soils (Braunack, 1986b; Ayers et al., 1990; Watts, 1998; Milchunas et al., 1999). However, the nature of soil horizons has a critical influence on plant responses to that precipitation (Hamerlynck et al., 2000). Tank passage over desert pavement at the Butler Pass study site destroyed the Av horizon, thus removing a significant barrier to the introduction of moisture to the subsurface. The impact of the removal of surface induration on vegetation is beyond the scope of this study, but personal observations of R. I. Dorn (pers. comm., 2000) in the general study area in March of 1989 and 1993, for example, reveal enhanced germination of annuals within the tank tracks.

6. CONCLUSION

Desert pavements are ubiquitous in arid regions, occurring with regularity on weathered debris mantles, alluvial fans, and soils worldwide (Mabbutt, 1979; Cooke et al., 1993). Human encroachment on pavements will continue in the future, and military operations are certain to alter these surfaces in regions where arid zone training takes place (such as the arid southwestern United States), or where combat operations may occur in other parts of the world. While alteration to desert pavement from maneuvers is an unavoidable environmental hazard of military operations, these activities are not necessarily purely detrimental. As demonstrated in this study, increasing moisture penetration of the subsurface may encourage plant growth, which may have both positive and negative consequences to the ecological functioning of a largely barren landform.

Larger lessons of this research concern the utility of linking processes and different scales, and generalization of research on the negative consequences of anthropogenic activity. Detailed exploration of processes operating at the micron scale - in this case breaking up of soil crusts measured by electron microscopy - revealed that tank tracks had an effect opposite of those predicted in the previous literature. Articulating microscope with field experiments also explains that the consequences on soil structure and moisture penetration are limited to the upper few centimeters of the soil profile. Re-establishment of plant cover at similar sites would require vegetation that has shallow rooting depths.

Environmental management of anthropogenic activities then, benefits from research linking processes at different scales, a core theme in physical geography research (Church and Mark, 1980; Viles, 2001).

REFERENCES

American Society for Testing and Materials (ASTM). 2002. D2922-01 Standard test methods for density of soil and soil-aggregate in place by nuclear methods (shallow depth). N.P.: American Society for Testing and Materials.

Ayers, P.D. 1994. Environmental damage from tracked vehicle operation. Journal of Terramechanics 31:173-183.

Ayers, P.D., Shaw, R.B., Diersing, V.E., and Van Riper, J. 1990. Soil compaction from military vehicles. St. Joseph, MI: American Society of Agricultural Engineers, 90-1096.

Birkeland, P.W. 1999. *Soils and Geomorphology*. New York: Oxford University Press.

Bischoff, M.C. 2000. *The Desert Training Center/California-Arizona Maneuver Area, 1942-1944: Historical and Archaeological Contexts*. Tucson, AZ: Statistical Research, Inc.

Brattstrom, B.H. and Bondello, M.C. 1983. Effects of off-road vehicle noise on desert vertebrates. In *Environmental Effects of Off-Road Vehicles: Impacts and Management in Arid Regions*, R.H. Webb and H.G. Wilshire, eds., New York: Springer-Verlag, 167-206.

Braunack, M.V. 1986a. Changes in physical properties of two dry soils during tracked vehicle passage. Journal of Terramechanics 23:141-151.

Braunack, M.V. 1986b. The residual effects of tracked vehicles on soil surface properties. Journal of Terramechanics 23:37-50.

Bury, R.B. and Luckenbach, R.A. 1983. Vehicular recreation in arid land dunes: Biotic responses and management alternatives. In *Environmental Effects of Off-Road Vehicles: Impacts and Management in Arid Regions*, R.H. Webb and H.G. Wilshire, eds., New York: Springer-Verlag, 207-234.

Church, M.A. and Mark, D.M. 1980. On size and scale in geomorphology. Progress in Physical Geography 4:342-390.

Cooke, R., Warren, A., and Goudie, A.S. 1993. *Desert Geomorphology*. London: University. College Press.

Cooke, R.U. 1970. Stone pavements in deserts. Annals of the Association of American Geographers 60:560-577.

Dorn, R.I. 1998. *Rock Coatings*. Amsterdam, The Netherlands: Elsevier.

Elvidge, C.D. and Iverson, R.M. 1983. Regeneration of desert pavement and varnish. In *Environmental Effects of Off-Road Vehicles: Impacts and Management in Arid Regions*, R.H. Webb and H.G. Wilshire, eds., New York: Springer-Verlag, 225-243.

Gatziolis, D., Fried, J.S., and Ramm, C.W. 2000. Monitoring the impacts of tracked vehicle training area use at Fort Hood, TX - A GIS approach. 2nd International Conference on Geospatial Information in Agriculture and Forestry, Lake Buena Vista, FL, I542-I549.

Gillette, D.A. and Adams, J. 1983. Accelerated wind erosion and prediction of rates. In *Environmental Effects of Off-Road Vehicles: Impacts and Management in Arid Regions*, R.H. Webb and H.G. Wilshire, eds., New York: Springer-Verlag, 97-109.

Hamerlynck, E.P., Mcauliffe, J.R., and Smith, S.D. 2000. Effects of surface and sub-surface soil horizons on the seasonal performance of *Larrea Tridentata* (Creosotebush). Functional Ecology 14:596-607.

Hinckley, B.S., Iverson, R.M., and Hallet, R.M. 1983. Accelerated water erosion in ORV-use areas. In *Environmental Effects of Off-Road Vehicles: Impacts and Management in Arid Regions*, R.H. Webb and H.G. Wilshire, eds., New York: Springer-Verlag, 81-96.

Iverson, R.M., Hinckley, B.S., Webb, R.H., and Hallet, R.M. 1981. Physical effects of vehicular disturbances on arid landscapes. Science 212:915-916.

Koch, M. and El-Baz, F. 1998. Identifying the effects of the Gulf War on the geomorphic features of Kuwait by remote sensing and GIS. Photogrammetric Engineering and Remote Sensing 64:739-747.

Lathrop, E.W. 1983. The effect of vehicle use on desert vegetation. In *Environmental Effects of Off-Road Vehicles: Impacts and Management in Arid Regions*, R.H. Webb and H.G. Wilshire, eds., New York: Springer-Verlag, 153-166.

Lovich, J.E. and Bainbridge, D. 1999. Anthropogenic degradation of the Southern California desert ecosystem and prospects for natural recovery and restoration. Environmental Management 24:309-326.

Mabbutt, J.A. 1979. Pavements and patterned ground in the Australian stony deserts. Stuttgarter Geographische Studien 93:107-123.

Marston, R.A. 1986. Maneuver-caused wind erosion impacts, south central New Mexico. In Proceedings of the 17th Annual Binghamton Geomorphology Symposium, W.G. Nickling, ed., Binghamton, NY: Allen Unwin, 273-290.

McDonald, K. 2004. Military foot traffic impact on soil compaction properties. In *Studies in Military Geography and Geology*, D.R. Caldwell, J. Ehlen, and R.S. Harmon, eds., Dordrecht, The Netherlands: Kluwer Academic Publishers, 229-242.

McFadden, L.D., Wells, S.G., and Jercinovich, M.J. 1987. Influences of eolian and pedogenic processes on the origin and evolution of desert pavements. Geology 15:504-508.

Milchunas, D.G., Schulz, K.A., and Shaw, R.B. 1999. Plant community responses to disturbance by mechanized military maneuvers. Journal of Environmental Quality 28:1533-1547.

Prose, D.V. 1985. Persisting effects of armored military maneuvers on some soils of the Mojave Desert. Environmental Geology and Water Science 7:163-170.

Prose, D.V. and Wilshire, H.G. 2000. The lasting effects of tank maneuvers on desert soils and intershrub flora. US Geological Survey Open-File Report OF 00-512.

Prosser, C.W., Sedivec, K.K., and Barker, W.T. 2000. Tracked vehicle effects on vegetation and soil characteristics. Journal of Range Management 53:666-670.

Reed, S.J.B. 1993. *Electron Microprobe Analysis*. Cambridge: Cambridge University Press.

Schoeneberger, P.J., Wysocki, D.A., Benham, E.C., and Broderson, W.D. 1998. *Field Book for Describing and Sampling Soils*. Lincoln, NE: US Department of Agriculture, Natural Resources Conservation Service, National Soil Survey Center.

Sneed, E.D. and Polk, R.L. 1958. Pebbles in the lower Colorado River, Texas: A study in particle morphogenesis. Journal of Geology 66:114-150.

Thomas, D.S. and Goudie, A.S., eds. 2000. *The Dictionary of Physical Geography*. Oxford: Blackwell Publishing Company.

Thomas, D.S.G. 1989. *Arid Zone Geomorphology*. New York: Halsted Press.

Troxler Electronic Laboratories. 1998. Application Brief. Troxler Roadreader Model 3430 Moisture/Density Gauge, Research Triangle Park, NC: Troxler Electronic Laboratories, AB3430-0598.

Viles, H.A. 2001. Scale issues in weathering studies. Geomorphology 41:61-72.

Von Werlhof, J. 1987. *Spirits of the Earth: A Study of Earthen Art in the North American Deserts: 1 - The North Desert*. El Centro, CA: IVC Museum Society.

Watts, S.E. 1998. Short-term influence of tank tracks on vegetation and microphytic crusts in shrub steppe habitat. Environmental Management 22:611-616.

Webb, R.H. 1983. Compaction of desert soils by off-road vehicles. In *Environmental Effects of Off-Road Vehicles: Impacts and Management in Arid Regions*, R.H. Webb and H.G. Wilshire, eds., New York, NY: Springer-Verlag, 31-50.

Webb, R.H. and Wilshire, H.G., eds. 1983. *Environmental Effects of Off-Road Vehicles: Impacts and Management in Arid Regions*. New York: Springer-Verlag.

Wilshire, H.G., 1983, The impact of vehicles on desert soil stabilizers. In *Environmental Effects of Off-Road Vehicles: Impacts and Management in Arid Regions*, R.H. Webb and H.G. Wilshire, eds., New York: Springer-Verlag, 20-30.

Wilshire, H.G. and Nakata, J.K. 1976. Off-road vehicle effects on California's Mojave Desert. California Geology 29:123-132.

Chapter 20

DEVELOPMENT OF AN ARCHEOLOGICAL PREDICTIVE MODEL FOR MANAGEMENT OF MILITARY LANDS

Identification of Geological Variables in Desert Terrain

Eric McDonald,[1] Thomas Bullard,[1] Tad Britt,[2] and Marilyn O'Ruiz[3]
[1] *Desert Research Institute*
[2] *US Army Engineer Research and Development Center*
[3] *University of Illinois*

Abstract: A framework has been developed for an archeological predictive model based on demonstrated relations among multiple geologic variables (e.g., topography, geochronology, rock type, geomorphology) for 81 previously identified cultural resource sites across diverse desert terrain. Results indicated that the most useful variables are deposit type, piedmont setting, geometric form, deposit age, surface age, desert pavement, surface horizon, and strongest subsoil horizon. Traditional sources for geologic and soil data available to most military installations are of low resolution or are incomplete. Rapid and cost-effective methods for collecting these data in desert terrains will be required to advance and implement archeological predictive models. The new model was applied initially to desert terrain conditions at the US Army National Training Center, Fort Irwin, CA, but it has potential for application across military lands throughout the southwestern US where as much as 80% of the holdings have not been inventoried for cultural resources.

Key words: deserts, cultural resources, military geography and geology, geologic variables, arid land management

1. INTRODUCTION

The serious need for development of new tools or methods to locate cultural resource sites and ensure compliance with existing laws is highlighted by two facts: (i) nearly 70% of all Department of Defense (DoD) lands are in the arid southwestern US and (ii) 80% of these holdings have

D.R. Caldwell et al. (eds.), Studies in Military Geography and Geology, 259–270.
© 2004 *Kluwer Academic Publishers.*

not been adequately inventoried for cultural resources. Conventional archeological methods (i.e., transect walkovers by closely spaced field crews) are labor intensive and time consuming and do not meet the DoD schedule for land acquisition and commencement of training activities. All Federal landholding agencies must comply with a variety of laws, including the National Environmental Policy Act (NEPA) and the National Historic Preservation Act (NHPA), to provide protection of archeological sites and historic resources. These laws are designed to inventory, manage, monitor, and conserve natural and cultural resources.

One solution for accelerating compliance with archeological identification and evaluation requirements is development of a geology-based archeological predictive model (APM) for locating cultural resource sites. The objective of such a model is the integration geologic and cultural resource knowledge into a geographic information system (GIS) database to characterize locales for archeological site potential and to rank areas for investigation of cultural resources. Successful development of an APM will facilitate timely acquisition of property and maximize its future use.

The rationale for using geologic data to predict the location of cultural resources is that surficial geology is the foundation of all cultural landscapes. This is because surficial geology often directly and indirectly reflects human activity because the landforms and deposits have an influence on distribution and location of exploitable natural resources such as water, shelter, and food. As the landscape evolves through time, human interaction with the environment and resource exploitation practices adapt in response. Prehistoric habitation preferences, food gathering and processing areas, and lithic quarries are all predicated on their position on the landscape, proximity to suitable geologic resources, and association with underlying geology.

This paper presents an approach and initial results for applying soil and geomorphic information to determine possible relations among the location of cultural resources, soil and geomorphic attributes, and landscape history. The methodological approach employed in developing the APM facilitates meeting immediate objectives and addressing compliance issues while serving as a management tool for future planning and operations.

2. BACKGROUND AND APPROACH

2.1 GIS Advances in Developing APM

GIS are used extensively for spatially explicit models when existing point locations of events are known and areas favorable for other, not yet discovered, sites need to be delineated. In archeology, predictive models of this type have been developed mostly to support better management of cultural resources where sites are identified and ranked in terms of information potential and vulnerability (Wescott and Brandon, 2000).

Recent trends in archeological predictive modeling have focused on development of three-dimensional approaches. The Channel Hillslope Integrated Landscape Development (CHILD) model (Tucker et al., 2001) can be utilized to examine dynamic landscape evolution, effects on the environment, and site selection factors. More robust analysis tools and sophisticated, recently developed information technology systems also have contributed significantly. Collectively, such approaches allow for much larger data sets and thus larger geospatial units of analysis. Studies at these larger scales permit a better understanding of the geomorphic processes that affect archeology and the environment, improved statistical accuracy with respect to probability, and better identification of methodological biases.

Future trends in developing more statistically robust spatial models in archeology will benefit from advances in spatial analysis used in other contexts, such as the risk maps developed in epidemiology (Kitron, 2000) and models of species distribution (Stockwell and Peters, 1999). Other innovative approaches incorporate modeling of soil, climate, vegetation, human uses, and other features in a location-based decision support system (Parks et al., 2003).

2.2 Geographic Setting of the National Training Center

The National Training Center (NTC) at Fort Irwin, CA is the US Army's primary mechanized maneuver training facility, covering approximately 2600 km^2 within the Mojave Desert in southern California (Fig. 1). The pre-history of the Mojave Desert region contains a rich and diverse record of human occupation and resource exploitation. Hundreds if not thousands of archeological sites that span more than 10,000 years of human prehistory are recorded on the NTC and immediately adjacent land (Byrd and Pallett, 1994; Byrd, 1998).

The geomorphology of the NTC can be broadly classified into isolated mountain ranges and intervening basins (Peterson, 1981; Fig. 2). The spatial distribution of surficial deposits reflects the Quaternary landscape history recognized across the Mojave Desert (Wells et al., 1987; Bull, 1991; McDonald et al., 2003).

Evolution of the land surface and modification of depositional surface characteristics form the basis for differentiating relative ages of the landscape (McFadden et al., 1989; McDonald and McFadden, 1994).

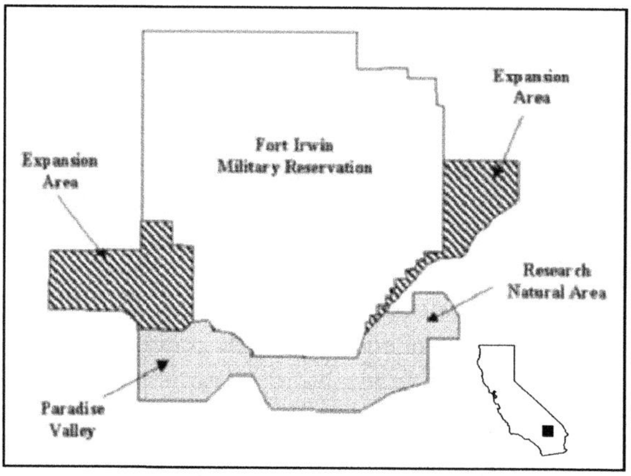

Figure 1. Location map of the US Army National Training Center at Ft. Irwin, CA.

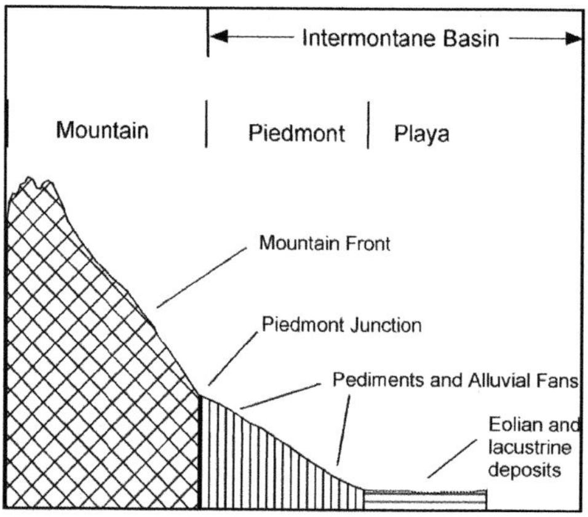

Figure 2. Generalized diagram showing common landscape components and deposits typical of the Basin and Range physiographic province that are found at the NTC.

Systematic changes in soil morphology over time (Table 1) form the basis for estimating ages and stability of desert land surfaces (Harden et al., 1991; McDonald, 1994; Birkeland, 1999; McDonald et al., 2003). The presence of well-developed soils indicates long term landscape stability and can provide important information regarding the occurrence and preservation of cultural materials.

2.3 Research Design

Soil and geomorphic attribute data were collected through field observations at 81 previously recorded cultural resource sites. Sites were selected to provide a representative sample of both principal geomorphic features and cultural resource attributes common to the NTC. At each site, 16 soil, geomorphic, geologic, and environmental parameters (Table 2) were recorded. All of the cultural resource sites visited are primarily surface sites. Results were assembled in an attribute matrix for evaluation.

Table 1. General trends in soil formation through time observed in desert soil morphology.

Approx. surface or deposit age	Soil Property							
	Av horizon[1]	Char. horizon[2]	B horizon thickness[1]	Hue[3]	Carbonate or silica stage[4]	Clay content[1]	Silt[1]	Salts[1]
Latest Holocene (<1 Ka)[5]	-	AC, Cox, C	-	10 YR	-	-	-	-
Late-Middle Holocene (1-6 Ka)	+	Bw	+	10 YR to 7.5 YR	I to I+	<+	+	+
Early Holocene (6-10 Ka)	++	Btj	++	7.5 YR	II to II+	+	++	+
Pleistocene (10-1000 Ka)	+++	Btk, Bkm, Bkqm	+++	7.5 YR to 5 YR	III to IV+	++	+++	++
Pleistocene to Tertiary (<1000 Ka)	++	Btk, Bkm, Bkqm	+ to +++	10 YR to 5 YR	I to VI	++	++	++

[1] Symbols indicate relative presence of soil property. Symbols used: - none, + weak, ++ moderate, +++ strong

[2] Horizon nomenclature follows standard Natural Resources Conservation Service terminology (USDA, 1998). A is an horizon with accumulation of organic matter and/or development of fine structure, C is unaltered soil parent material, AC is an horizon that is transitional between A and a C horizon, Bw is an incipient B horizon based on color and structure, Btj is an incipient Bt horizon, k denotes carbonate accumulation, q denotes silica accumulation, and m denotes degree of cementation.

[3] Hue according to Munsell Soil Color Charts (Munsell Color Co. Inc., Baltimore MD)

[4] Carbonate stage follows designation of Birkeland (1999)

[5] 1 Ka = 10^3 years before present

Table 2. List of 16 soil and geomorphic variables evaluated in the field.

Soil and Geomorphic Features		
Deposit type	Surface horizon	General aspect
Piedmont setting	Strongest subsoil horizon	Geomorphic stability
Basic geometric form	Dominant lithology	Potential for future burial
Deposit age	Landform morphology	Potential for buried site
Surface age	Geomorphic position	
Desert pavement	Surface texture	

3. RESULTS

Evaluation of geomorphic data for the 81 cultural resource sites indicated several general but potentially important trends (Table 3). For example, we found an association among cultural resources site type (e.g., habitation sites, lithic scatter, and lithic reduction sites), geomorphic setting within the landscape, and geologic environment. Not all 16 variables were found to provide useful information. Below, we provide results for nine of the most useful variables and relationships between the dominant site geology and site activity (Table 3).

Table 3. Summary of matrix results showing relationships between site occurrence and soil and geomorphic variables.

Deposit Type	No. [1]	Piedmont Setting	No. [1]	Basic Geometric Form[2]	No. [1]
Alluvium	47	Mountain highlands	8	Linear-concave	12
Colluvium	15	Piedmont junction	11	Linear-linear	8
Eolian	2	Pediment	11	Linear-convex	47
Playa	4	Alluvial fan	29	Concave-convex	2
Springs	6	Alluvial plain	11	Concave-linear	5
Bedrock	7	Shoreline	5	Concave-convex	1
		Playa	5	Convex-linear	1
				Convex-convex	4
	(81)[3]		(80)[3]		(80)[3]
Desert Age[4]	**No. [1]**	**Surface Age**	**No. [1]**	**Desert Pavement**	**No. [1]**
Late Holocene	12	Late Holocene	44	None	38
Middle Holocene	4	Middle Holocene	8	Lag (no soil/Av)	5
Early Holocene	9	Early Holocene	11	Weak pavement	10
Pleistocene	48	Pleistocene	17	Moderate pavement	12
Tertiary	7	Tertiary	0	Strong pavement	15
	(80)[3]		(80)[3]		(80)[3]
Surface Horizon	**No. [1]**	**Strongest Subsurface Horizon**	**No. [1]**	**Dominant Lithology[5]**	**No. [1]**
A	14	Bw/Bwk	28	H: co-grained plutonic	21
Av/Avj	43	Bt/Btk	27	H: mixed	14
B	1	C	22	LS: co-grained plutonic	7
C	22			LS: mixed	31
	(80)[3]		(80)[3]		(73)[3]

[1] Number of observed sites
[2] Convex up, concave up, 1st = down gradient, 2nd = cross gradient
[3] Total number of observed sites
[4] Age estimates defined in Table 1
[5] H: habitation sites; LS: lithic scatter and lithic reduction

3.1 Landscape and Depositional Setting

Nearly all of the cultural sites we evaluated are located within the alluvial intermontane basin and along the margin of adjacent mountain fronts. The majority of sites were associated with alluvium (58%) and colluvium (19%). The others sites (23%) were associated with eolian, playa, spring deposits, and exposed bedrock. Bedrock sites primarily consisted of tors associated with bedrock pediments or small inselbergs. Along the piedmont gradient, most of the sites (76%) were distributed throughout the intermontane basins across a range of piedmont and playa settings (Table 3; Fig. 2).

The most common piedmont setting was alluvial fan (36%) and the least common settings were shoreline (6%) and playa (6%). At the observed sites, linear slopes that parallel down-slope length or gradient were most common (84%). The predominance of linear down-gradient sites reflects the association of most sites with depositional features, primarily alluvial fans and plains, common to the intermontane basin.

3.2 Cultural Resources and Age Relations

Geomorphic data indicated clear relations between cultural sites and the approximate age of surfaces, deposits, and landforms. More than two-thirds (69%) of the evaluated sites were associated with deposits or landforms of Pleistocene or Tertiary age (features that can be readily mapped at 1:24,000 scale). This relation is important because large areas of the intermontane basin are comprised predominantly of Holocene deposits.

In contrast, we associated many of the sites with surface ages that are younger than the underlying deposit or landform age. Differences between surface and deposit ages are primarily due to some combination of surface erosion, deposition of a thin (<0.5 m) veneer of younger sediment, and/or extensive mixing of the surface by biologic processes. Surface age estimates used here, therefore, reflect the maximum age of surface stability. Surface stability is required to allow preservation at the surface of the most common types of cultural materials (e.g., lithic scatter). Soils with active surfaces are largely unstable and were assigned either late or latest Holocene surface age estimates.

Late-to-latest Holocene surface age estimates accounted for 55% of all sites, and Holocene surface age (i.e., early to latest Holocene) estimates accounted for 79% of all sites. Sites with surface ages that appeared to be younger than the deposit age strongly suggested that buried cultural features may be present if surface processes are conducive to burial and preservation of artifacts. The overall degree of soil development also suggested a relation among cultural sites and relative age as well as surface stability. The majority of sites (71%) were associated with some degree of A horizon development, denoted by the presence of an A, Av, or Avj horizon (where "v" refers to abundant vesicular pores, a common feature in Mojave desert soils; McFadden et al., 1998). The strongest subsoil horizons associated with most sites (71%) contain some degree of B horizon development. These relations indicated that most cultural features presently located at the surface are in geomorphic settings conducive to surface stability.

Development of desert pavements signifies some degree of surface stability. Sites are distributed among all of the lag/pavement descriptors ranging in distribution from no lag/pavement (47%) to lag (6%). Slightly more than half the sites (53%) had some degree of surface cover with the majority of these consisting of moderate to strongly developed desert pavement. Most of the cultural sites with evidence of habitation (campsite, rock shelter, village) are associated with a lack of either a lag or pavement surface cover. By comparison, most of the lithic sites (scatter, reduction, quarry) are associated with either a lag or pavement.

3.3 Site Activity and Lithology

Clear trends also exist between dominant site lithology and primary site activity (Table 3). Many of the habitation sites (e.g., rock shelters, campsites) were associated with coarse-grained plutonic rocks. The remaining sites were associated with deposits of mixed rock types (primarily in fan or shoreline settings) or spring deposits. In contrast, most lithic scatter and lithic reduction sites (82%) were associated with deposits of mixed rock types with only a few associated with coarse-grained plutonic rocks. Field observations indicated that mixed sediments where most of the lithic scatter and lithic reduction sites are located contained either abundant basalt cobbles or chert suitable for producing tools and projectile points.

4. DISCUSSION

Nearly all campsites, rock shelters, and similar habitation features were associated with mantled piedmont, piedmont junctions, and shoreline features. These piedmont settings are the most likely to be near water resources, especially during wetter climate periods. Many habitation sites were associated with late-to-latest Holocene surface age, reflecting either the age of the cultural materials or landscape settings subject to episodic alluvial activity (which would preclude surface preservation of older cultural resources).

One key relation is that many habitation sites are located in areas where exposed, coarse-grained plutonic bedrock is widely present. Coarse-grained plutonic rock (e.g., quartz monzonite) readily undergoes physical and chemical weathering processes that disintegrate bedrock into non-cohesive granular material (gruss), which promotes infiltration of precipitation and runoff. Quartz monzonite also is conducive to forming tafoni - cavernous weathering features with arch-shaped entrances, concave interior walls, and overhanging ledges (Mabbutt, 1977). Tafoni formation provides weathering morphology suitable for potential use as natural rock shelters, as observed on the NTC. Most rock shelters at the NTC are located in pediment settings where exposed, weathered, coarse-grained, plutonic bedrock is widely present. The general distribution of previously identified habitation sites on the NTC (not assessed in this study) also suggests a potentially strong correlation between increased frequency of habitation sites near coarse-grained plutonic bedrock, especially those areas with bedrock pediments.

4.1 Surface Stability and Lithic Cultural Sites

Evaluation of geomorphic and archeological data sets indicated that most lithic scatter, lithic reduction, and lithic quarry sites (53%) were associated with surfaces comprised of either a surface lag layer or desert pavement (Fig. 3). This association may be due to concentration and exposure of suitable resource clasts at the surface (enhancing resource collection) or may indicate that development of desert pavement promotes stability and preservation of lithics at production sites.

Figure 3. View of well-developed desert pavement showing (a) preserved prehistoric rock ring and (b) close-up photograph of lithic scatter (arrows) embedded in desert pavement.

4.2 Geomorphic Control on Core Stones

Other general trends were suggested by the overall distribution of lithic sites across the NTC. For example, most lithic sites (70%) that have cultural materials worked from basalt were located on or near alluvial fans that contain abundant basalt cobbles. Most of these fans are Pleistocene age and may be many kilometers down gradient from basalt bedrock that is exposed along high mountain ridges. Nearly all of the cryptocrystalline core stones were associated with deposits or landforms derived from, or formed on, uplifted Tertiary basin gravel.

5. CONCLUSIONS AND RECOMMENDATIONS

As indicated by this case study, the military can benefit significantly from using geologically founded predictive models that facilitate a multi-disciplinary approach to identifying, understanding, and solving complex environmental and management problems. It is important to note that the modeling concepts presented in this paper can be expanded and applied to other types of environmental and landuse issues (i.e., threatened and endangered species, wetlands management) that affect military installations.

The matrix developed for this study included a wide range of geomorphic variables (Table 2) to aid in identifying which would provide the most useful information. Many of the variables in our initial evaluation provided mixed results because of the wide range of cultural site types (each with unique ties to the desert landscape) and because desert terrain is complex with a wide range of geomorphic variables and different histories of landscape evolution. Based on the results of this preliminary survey, we determined that the number of geomorphic variables to consider could be reduced from sixteen to nine (Table 3). These nine variables should provide the greatest benefit in developing a predictive model for use in the desert terrain of the southwestern United States. Additional geoarcheological research scheduled at the NTC will be designed to test the robustness of predictive capabilities of these nine variables and to determine if further refinement is necessary or warranted.

One of the more challenging obstacles in incorporating geomorphic information into a GIS-based predictive model is lack of suitable base maps containing essential geomorphic data. Existing geology and soil maps for NTC expansion areas contain limited data. It is possible that geomorphic base maps could be prepared quickly from currently available digital photographs. Geomorphic parameters that could be readily incorporated into a new base map include deposit age, landform morphology, and overall bedrock and deposit lithology. The use of this approach is recommended to refine and reduce the number of locations and extent of cultural field surveys within NTC expansion areas and other US military installations in the desert areas of the southwestern United States, as suggested by the GIS-based predictive model results.

Future development and testing of the geomorphic attributes used here for predicting the location of cultural resources requires three additional steps. First, geomorphic data - primarily based on the nine parameters suggested above - should be collected for additional cultural resource sites. Second, additional effort is required to determine how best to link observed geomorphic data into a GIS-based model that can be used to develop favorability indices and maps for predicting the location of cultural resource features. Third, the archeological predictive model requires testing through application of the model to an area prior to conducting a cultural resources survey. Successful completion of these steps will advance development of an archeological predictive model that will greatly benefit the identification and preservation of cultural resources on military lands and help ensure compliance with existing laws (e.g., NEPA, NHPA).

REFERENCES

Birkeland, P.W. 1999. *Soils and Geomorphology*. New York: Oxford University Press.

Bull, W.B. 1991. *Geomorphic Responses to Climatic Change*. New York: Oxford University Press.

Byrd, B.F. 1998. Springs and lakes in a desert landscape: Archeological and paleoenvironmental investigations in the Silurian Valley and adjacent areas of southeastern California. Report to US Army Corps of Engineers, Los Angeles District. Davis, CA: ASM Affiliates, Inc.

Byrd, B.F. and Pallette D.M. 1994. Archaeological overview and sample survey of the Silurian Valley study area. Report to US Army Corps of Engineers, Los Angeles District. San Diego, CA: Brian F. Mooney Associates, Inc.

Harden, J.W., Taylor, E.M., Hill, C., Mark, R.L., McFadden, L.D., Reheis, M.C., Sowers, J.M., and Wells, S.G. 1991. Rates of soil development from four soil chronosequences in the southern Great Basin. Quaternary Research 35:383-399.

Kitron, U. 2000. Risk maps: Transmission and burden of vector borne diseases. Parasitology Today 16:324-325.

Mabbutt J.A. 1977. *Desert Landforms*. Cambridge, MA: MIT Press.

McDonald, E.V. 1994. The relative influences of climatic change, desert dust, and lithologic control on soil-geomorphic processes and hydrology of calcic soils formed on Quaternary alluvial-fan deposits in the Mojave Desert, California (unpublished Ph.D. dissertation). Albuquerque, NM: University of New Mexico.

McDonald, E.V., McFadden, L.D., and Wells, S.G. 2003. Regional response of alluvial fans to the Pleistocene-Holocene climatic transition, Mojave Desert, California. Boulder, CO: Geological Society of America Special Paper 368:189-205.

McDonald, E.V. and McFadden, L.D. 1994. Quaternary stratigraphy of the Providence Mountains piedmont and preliminary age estimates and regional stratigraphic correlations of Quaternary deposits in the eastern Mojave Desert, California. In *Quaternary Stratigraphy and Dating Methods: Understanding Geologic Processes and Landscape Evolution in Southern California,* S.G. Wells, J.C. Tinsley, L.D. McFadden, and N. Lancaster, eds., Boulder, CO: Geological Society of America Cordilleran Section Guidebook, *Geological Investigations of an Active Margin*, 205-210.

McFadden, L.D., Ritter, J.B, and Wells, S.G. 1989. Use of multiparameter relative-age methods for age estimations and correlation of alluvial fan surfaces on a desert piedmont, eastern Mojave Desert. California. Quaternary Research 32:276-290.

McFadden, L.D., McDonald, E.V., Wells, S.G., Anderson, K., Quade, J., and Forman, S.L. 1998. The vesicular layer of desert soils: Genesis and relationship to climate change and desert pavements based on numerical modeling, carbonate translocation behavior, and stable isotope and optical dating studies. Geomorphology 24:101-145.

Parks, B.O., Clarke, K.M., and Crane, M.P., eds. 2003. Proceedings of the 4th International Conference on Integrating Geographic Information Systems and Environmental Modeling: Problems, Prospectus, and Needs for Research. [CD-ROM, ISBN: 0-9743307-0-1]. GIS/EM4 Conf., 2-8 Sept 2000, Banff, Alberta, Canada. Jointly published by University of Colorado Cooperative Institute for Research in Environmental Sciences (Boulder, CO), US Geological Survey Center for Biological Informatics (Denver, CO), and US National Oceanic and Atmospheric Administration National Geophysical Data Center (Boulder, CO).

Peterson, F.F. 1981. Landforms of the Basin and Range Province defined for soil survey. Reno, NV: Agricultural Experiment Station Technical Bulletin 28.

Stockwell, D. and Peters, D. 1999. The GARP modelling system: Problems and solutions to automated spatial prediction. International Journal of Geographic Information Science 13:143-158.

Tucker, G., Lancaster, S., Gasparini, N., and Bras, R. 2001. The Channel-Hillslope Integrated Landscape Development Model (CHILD). In *Landscape Erosion and Evolution Modeling*, R.S. Harmon and W.W. Doe, III, eds. Dordrecht, The Netherlands: Kluwer Academic Publishers, 349-388.

US Department of Agriculture Soil Survey Staff (USDA). 1998. *Soil Survey Manual.* US Department of Agriculture Soil Conservation Service Handbook. Washington, DC: US Government Printing Office.

Wells, S.G., McFadden, L.D., and Dohrenwend, J.C. 1987. Influence of Late Quaternary climatic changes on geomorphic and pedogenic processes on a desert piedmont, eastern Mojave Desert, California. Quaternary Research 27:130-146.

Wescott, K.L. and Brandon, R.J., eds. *Practical Applications of GIS for Archaeologists: A Predictive Modeling Toolkit.* London: Taylor & Francis.

Chapter 21

THE GEOMETRY OF LINE-OF-SIGHT AND WEAPONS FAN ALGORITHMS

Peter L. Guth
US Naval Academy

Abstract: Intervisibility algorithms, when applied to digital elevation models, are used to compute line-of-sight, weapons fans or viewsheds, and accurate three-dimensional perspective views. Whereas data quality, atmospheric effects, vegetation, and buildings contribute to the final result, the geometric model has a major impact. Seven geometric parameters should be explicitly defined: Viewer and target locations, interpretation of viewshed, point interpolation, point selection along radials, viewshed creation, vertical earth curvature, and horizontal earth curvature. The importance of horizontal earth curvature - the determination of straight line distance between observer and target - has not been sufficiently appreciated. Unless Universal Transverse Mercator approximations are valid, geodetic computations should be used. Because digital elevation models available to the military typically have a geographic-based point spacing, many established procedures that implicitly assume a conformal Universal Transverse Mercator grid introduce errors. A spaced radial algorithm produces the best weapons fans.

Key words: line-of-sight, weapons fans, viewshed algorithm, digital elevation model, intervisibility

1. INTRODUCTION

Much of the military's interest in terrain analysis concerns intervisibility, because weapons, communications, and detection systems require line-of-sight (LOS) for light, radio, or radar waves. Intervisibility can be defined as a ". . . function which calculates area or line-of-sight which can be 'seen' from a specific location or locations" (Defense Mapping Agency, 1994, 131). Simple intervisibility or LOS computations determine whether two points can see each other and more complex products simultaneously display

D.R. Caldwell et al. (eds.), Studies in Military Geography and Geology, 271–285.
© 2004 *Kluwer Academic Publishers.*

a number of computations in a weapons fan or viewshed (Lee, 1991). Accurate intervisibility computations can create three-dimensional (3-D) perspective views of terrain, with graphic visualizations realistic enough to measure positions. Intervisibility also figures in many military models and simulations.

The importance of LOS (Fig. 1A) and weapons fans (Fig. 1B) has been recognized since the work of the great French military engineer Sébastien Le Prestre de Vauban (1633-1707). Military historians and geographers also recognize the importance of intervisibility analysis in understanding past operations. Published examples include the defense of West Point in the

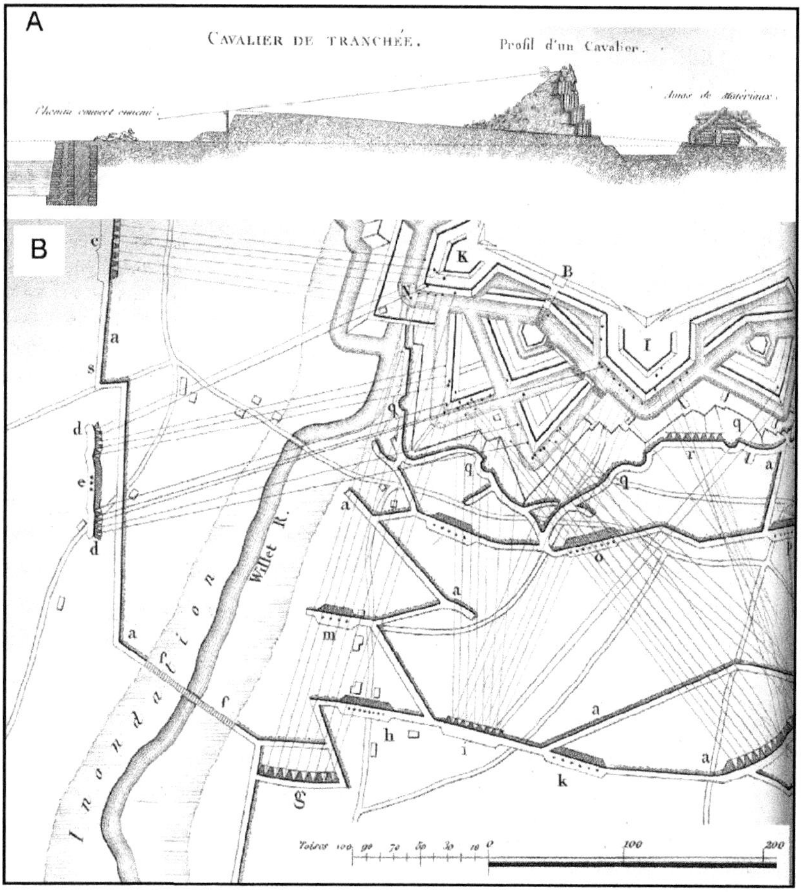

Figure 1. A. Diagrams from Vauban showing an early depiction of the military importance of line-of-sight and B. A portion of a map of the siege of Ath in 1706, showing attacking batteries and their fields of fire. From Vauban, 1828-29, Plates 6 and 10.

American War of Independence (Galgano, 2004; Palka, 2004, both this volume), artillery employment in the American Civil War (Ehlen and Abrahart, 2004, this volume), archaeological analysis of the Anglo-Zulu War of 1879 (Pollard, 2002), and German fortifications in World War II (Rose et al., 2002).

In the 21st century, the importance of intervisibility permeates Army doctrine. Operational planning incorporates METT-T (Mission, Enemy, Terrain and Weather, Troops, and Time Available; e.g., Department of the Army, 1992/2000) and OCOKA (observation and fields of fire, concealment and cover, obstacles, key terrain, and avenues of approach; e.g., Department of the Army, 1992/2001). The US Army Field Manual for topographic operations (Department of the Army, 2000, Appendix B) lists six examples of Tactical Decision Aids produced using intervisibility models. Sixteen US Army Field Manuals refer to the Digital Topographic Support System (DTSS), which is fielded to corps, divisions, and brigades; and seven refer to the TerraBase program, which is available to any unit with a computer (results from a subject search for the terms "DTSS" and "TerraBase" at General Dennis J. Reimer Training and Doctrine Digital Library, http://www.adtdl.army.mil/atdls.htm, 28 November 2003). Both the DTSS and TerraBase provide automated intervisibility analysis tools.

A complete computation of intervisibility depends on a number of factors: Characteristics of the digital elevation model (DEM), accuracy and errors in the data, the inclusion of vegetation and buildings which may not be captured (or captured inadequately), atmospheric effects ranging from refraction to obscuration, and the geometric model. The most simple intervisibility computation determines whether an observer has LOS to a target location. The binary result often is displayed with a topographic profile between the observer and target locations, allowing interpretation of intervisibility obstructions and the extent to which vegetation or data errors may affect the results. More complicated analyses use viewsheds or weapons fans, which color maps to show masked and visible terrain (e.g., Ehlen and Abrahart, 2004, this volume).

This paper focuses on the geometric considerations of LOS. By clearly establishing how geometry affects the computations, one can examine how algorithms and implementations differ. Because the digital data used for intervisibility only approximates reality, shortcuts in computation that allow speedier performance may be justified (Franklin, 2002). However, the nature of the shortcuts and their impact on the results must be clearly understood.

2. LINE-OF-SIGHT WITH DEMS

LOS requires elevation data, most commonly in the form of DEMs. Alternative data structures have not been widely implemented, although intervisibility with triangulated irregular networks is possible (Kidner et al., 2001). Older DEMs frequently use Universal Transverse Mercator (UTM) coordinates, and many intervisiblity algorithms require a UTM DEM. However, most DEMs available to the military today, such as the National Geospatial-Intelligence Agency's (NGA) Digital Terrain Elevation Data (DTED) and Shuttle Radar Topography Mission (SRTM), use geographic coordinates with 1-3 second spacing. The United States Geological Survey (USGS) has also moved to geographic coordinates for its National Elevation Dataset (NED). Any intervisibility algorithm should support DEMs in either geographic and Universal Transverse Mercator (UTM) coordinates.

This discussion will assume that the intervisibility model allows the user to enter observer and target positions, as well as their heights above the ground. Some viewshed implementations have apparently not allowed the target to be above the ground, but this will not be considered valid. The target has a specified height above the terrain, and if that point has geometric LOS to the observer, the point will be considered visible. With a glancing LOS, the target might not be detected or identified, but detection and identification are different than geometric LOS.

3. LINE-OF-SIGHT GEOMETRY

The LOS geometric algorithm requires a number of selections. Names applied to algorithms often reflect or describe only one or two of the parameters and ignore other options. Future discussion of algorithms should explicitly address all seven parameters listed in Table 1. Working implementations have used all the shortcuts and approximations described below. Fisher (1993) discussed three variables in viewshed algorithms: Point interpolation, viewpoint and target representation, and the mathematical algorithm selected and computer precision. He found the second variable to have the largest impact - an important finding highlighting the selections usually ignored in discussions of intervisibility algorithms.

3.1 Computer Precision

Computer precision is not directly under control of the program or user. Fisher (1993) and Champion et al. (1995) invoked arithmetic precision and

Table 1. Geometric selections in the intervisiblity algorithm

Selection	Options	Recommended	Justification
Viewer and target locations	Nearest grid postings, Interpolated	Interpolated	DEM models a continuous surface.
Interpretation of viewshed	Point-to-Point, Point-to-Cell, Cell-to-Cell, or Cell-to-Point	Point-to-Point	Many moderate resolution DEMs have grid postings that are not sufficiently accurate without interpolating.
Point interpolation	Nearest point, Point to SW, Average nearest points, Weighting to nearest points, Bilinear, Triangle, Polynomial, Splines	Bilinear interpolation	Reasonably smooth surface and reasonable execution times.
Point selection along radials	Nearest grid postings, All grid square sides, Grid square sides in either x or y direction, Grid triangle sides, Spacing along radial	Spacing along radial	Adjust spacing to capture peaks and ridges. With geographic grids, alternatives are not simple to compute as grid does not provide straight path. With UTM DEM and short sight-lines, other options equally good.
Creation of viewshed	Spaced radials, point to point for every grid posting, Concentric squares	Spaced radials	Fast execution, guaranteed to show either every masked or every visible point at any display scale, and ability to model sub-pixel regions.
Vertical earth curvature	None, Earth shape, Earth shape plus refraction	Earth shape plus refraction	For visible light, use integrated correction from TM-551 or Yoeli (1985). Can ignore under 1 km, must consider over 5 km.
Horizontal earth curvature	Geodesic Line, UTM, or geographic	Geodesic Line	Works for both UTM and geographic DEMs, and across UTM zones. Could use UTM for short sight-lines in one UTM zone.

rounding errors to explain differences between implementations, but it is likely other decisions in the algorithm have more impact on the results. Until the other choices are tested individually and their impact on the results assessed, this remains an unknown factor. Confirmation that computer precision has an impact on intervisibility will require running the same code on different computers to insure that all other options have the same settings.

3.2 Observer and Target on Grid Nodes and Viewshed Interpretation

Forcing the observer or target locations to lie on grid nodes in the DEM, although not explicitly stated, appears to be standard in many intervisibility computations. Algorithms that pre-compute intervisibility for an entire DEM use only the grid points, and enlargements of viewsheds from those algorithms may show a pixelated appearance. With high resolution DEMs this would not be a problem, but with typical military DEMs having 1 or 3 second spacing, this may not allow for optimal level of detail.

Fisher (1993) discussed the issue of observer and target location, and found that it had the largest impact on the final analysis. He noted that viewsheds could be considered Point-to-Point, Point-to-Cell, Cell-to-Cell, or Cell-to-Point. In the case of Cell-to-Cell, Fisher advocated going from each corner of the viewer's cell to each corner of the target cell. Potentially 16 separate calculations would be required, although the algorithm could stop with the first visible finding. Of the geographic information systems (GIS) implementations used by Fisher (1993), five used a grid-cell location for the viewpoint, one used an interpolated vector point, and one allowed either representation. If points can be interpolated within a DEM, positions for intervisibility should not have to be restricted to the grid nodes.

3.3 Point Interpolation

Kidner et al. (1999) described ten algorithms, including several variants of a quadratic equation, for point estimation within a DEM grid. Yoeli (1985) used an additional algorithm, weighting by distance to the four grid postings, which produces circular artifacts along the sides of the grid squares. Schneider (2001) points out artifacts in the linear plane and bilinear interpolation. Removing these artifacts requires a higher order surface and the estimation of derivatives for the surrounding elevation points, but can lead to other artifacts when the terrain surface undulates or overshoots the control points (Schneider, 2001). Schneider (2001) suggests Bezier splines, B-splines, or Coons patches for interpolation. Although Kidner et al. (1999)

advocate more complex algorithms, they provided only a ranked list with no quantification of the differences among the algorithms. A bilinear interpolation is simpler and probably sufficiently accurate for most DEMs. Effects of the interpolation method will decrease as the data spacing decreases, and the four surrounding values become closer in value, although elevations quantized as integer meters will introduce stairsteps in some grid cells. Interpolation differences will have most impact along these features, because ridge and hilltop features control the viewshed (Rana, 2003),

3.4 Point Selection along Profiles or Radials

Champion et al. (1995) illustrated in their figures (Figs. A1, A3, A4, and A6) four classes of algorithm used for point selection along the intervisibility radial, and named three for military modeling and simulation programs. The program names will not be familiar to users from the terrain analysis or GIS communities. Champion et al. (1995) did not provide original references to the programs, whose descriptions tend to be buried in the gray literature. Here, descriptive names are used for the point selection decision, because an LOS algorithm contains a number of choices, which should be labeled with meaningful, descriptive names.

The nearest posting algorithm, inspired by Bresenham's computer graphics line drawing algorithm, selects only points that lie on DEM grid nodes, and uses integer arithmetic for fast operation. Yoeli (1985) claimed that the grid square sides algorithm would find the extreme points along the profile, but Fig. 2, 3, and 4 of Kidner et al. (1999) demonstrate this would not always be the case. The grid triangle sides algorithm also suffers ambiguity because two sets of triangles can be chosen. Unlike the other choices, which depend on the grid spacing to determine the number of points used in the determination, the radial interpolation algorithm can use a number of points specified by the user, with a trade-off in speed versus accuracy. The point spacing could be set to the data spacing, half the data spacing (inspired by the Nyquist sampling limit), or any other desired spacing. The point spacing can become as small as the interpolation method allows.

An implicit assumption in the grid square and grid triangle side algorithms is that the intervisibility or radial line can be easily computed in grid coordinates. If horizontal earth curvature invalidates this assumption, much of the advantage of these algorithms disappears.

3.5 Vertical Earth Curvature

Vertical earth curvature applies to visible, radio, and radar portions of the spectrum differently because of varying effects of refraction. It consists of two components, a geometrical parameter depending on profile length and earth radius, and a refraction component. Equations for visible light from the Department of the Army (1970) and Yoeli (1985) combine both factors into a single correction. These algorithms differ slightly due to the small number of significant figures. The selected equation would not significantly affect computations except at very long ranges. As with horizontal earth curvature, for short distances, vertical curvature can be ignored with little impact. Radio refraction becomes much more complicated (Freeman, 1998) and cannot be considered part of the geometric model

3.6 Horizontal Earth Curvature

Horizontal earth curvature affects the "straight line" path between the observer and the target. For simplicity, this is generally computed in the DEM grid coordinates, as shown in published figures (Yoeli, 1985; Champion et al., 1995). For intervisibility over short distances this should make no difference, but over long distances it will compute a different sight line for a UTM or a geographic DEM. This effect does not appear to have been previously appreciated.

Sample line-of-sight profiles were used to test this effect. The end points were selected graphically, and converted to both geographic and UTM coordinates. For the geographic case, this assumes that the DEM grid in latitude and longitude approximates a rectangular Cartesian reference system. The DEMs have nearly identical posting resolution and come from the same producer and underlying data. From the end-point coordinates, intermediate points along the profiles were interpolated within the DEM data grid. Fig. 2 shows the resulting profiles. At the two ends, the profiles are identical, because both ends are constrained to the coordinates selected by the user. At the center of the profile, the coordinates differ substantially because of the method used to handle curvature of the earth. Fig. 3 shows the deviation between points used to compute the profiles as a function of profile length. At the center of the profile in Fig. 2, points along the profile are offset horizontally by over 750 m. In this case, the maximum vertical difference does not occur, because elevations around this point are similar. However, the center point has the potential for creating a significant difference, because it has the largest offset along the profile.

These results suggest three ways to compute the straight line required for intervisibility: Accurate geodetic computation using a geodesic line

(Vincenty, 1975), UTM coordinates, or geographic coordinates. Experiments show that UTM and coordinates along a geodesic line produce nearly identical results, but geographic coordinates differ substantially. UTM coordinates can be used for short sight lines, until the deviations for the correct geodesic line exceed the DEM grid spacing.

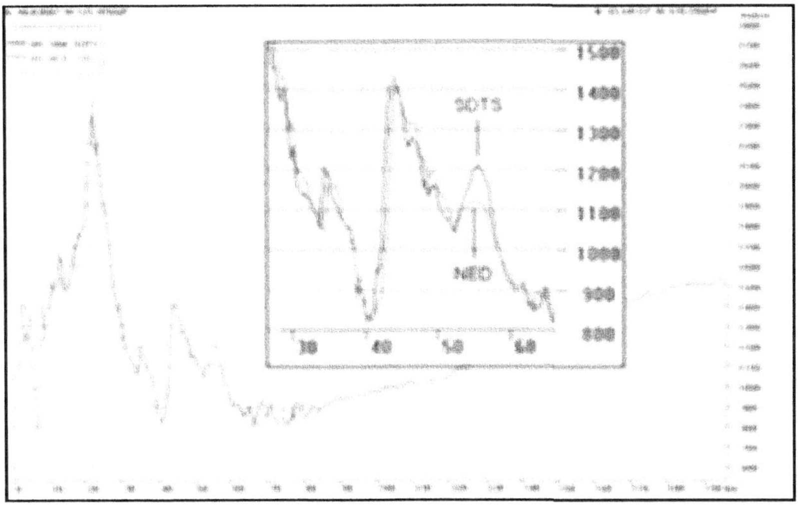

Figure 2. Profile computed from two USGS DEMs of the same area: A merge of 30 m Spatial Data Transfer Standard (SDTS) DEMs, and a 1 second NED DEM. The DEMs are based on the same data, and should differ only in the reprojected format. The selected end points were converted to UTM for the SDTS DEM and geographic for the NED DEM, and then points interpolated between those coordinates in the data grids. The enlargement shows a portion of the profile with the greatest differences in elevation. The two profiles coincide at the endpoints, whereas in the center, there are significant differences, including a 100 m difference created by missing a tall hill.

Geographic coordinates can only be used for very short sight lines - meaning that DEMs currently available for military use, DTED and SRTM, cannot use the coordinate scheme depicted in simple diagrams showing the intervisibility algorithm (e.g., Yoeli, 1985; Champion et al., 1995). These show a rectangular DEM grid that assumes the coordinates are equally spaced in the x and y directions, which is not the case for geographic DEMs. Fig. 3 shows how the difference between UTM and geographic coordinates increases with profile length. Furthermore, point interpolation schemes that use the sides of grid squares and triangles will not be easy to code, because the desired path curves in the geographic coordinate space. Finding the intersections along the sides of the grid squares and triangles is not simple. For any DEM, a geodesic line algorithm can calculate the straight line and coordinates of evenly spaced points along the profile in any desired coordinates. Correct geometry, however, comes with computational costs.

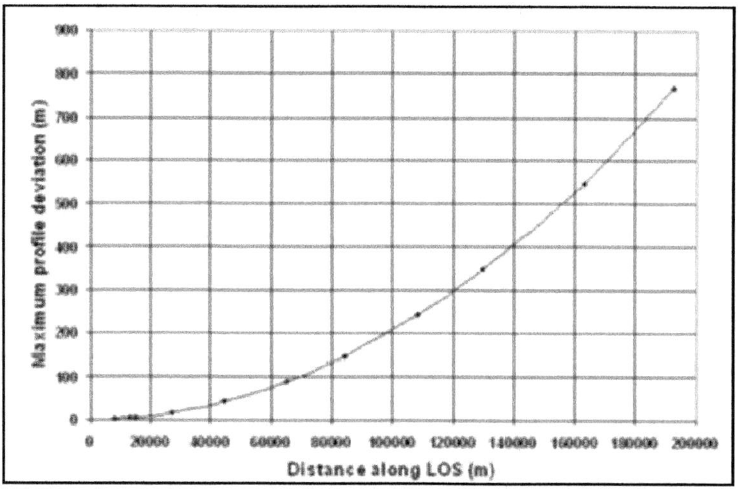

Figure 3. Maximum profile divergence between UTM and geographic coordinates as a function of profile length. For profiles less than 15 km long, the maximum deviation is under 5 m and the horizontal earth curvature can be ignored for many applications. Beyond 50 km the deviation will exceed 50 m and increases exponentially. At this distance, the horizontal earth curvature should not be ignored. This graph is for a profile constrained at both ends and the maximum deviation occurs in the middle of the profile; if the profile used a direction and azimuth, distances should be divided by two to estimate the divergence, and the maximum divergence would occur at the end of the profile.

3.7 Combining Radials for Fan

Early weapons fan solutions relied on a series of radial profiles, and drew each point along the profile as either masked or visible. Increasing the angular profile spacing reduced processing time for a quick initial assessment of intervisibility. This approach possesses speed advantages, because intervisibility at one point depends on the calculations at every other point between it and the observer. Increased computing power allows computation of a continuous fan, which can be done in two ways (Fig. 4): (1) brute force point to point calculations, or (2) modified radials, with the spacing set so that angular radial spacing is one screen pixel at the maximum range. Whereas the radial method produces far too many points close to the viewer, it exploits the fact that intervisibility is cumulative: Each point along the profile must consider all other points before it. For point-to-point calculations without shortcut approximations, multiple grid points will not lie on radials, and each point will be an independent computation.

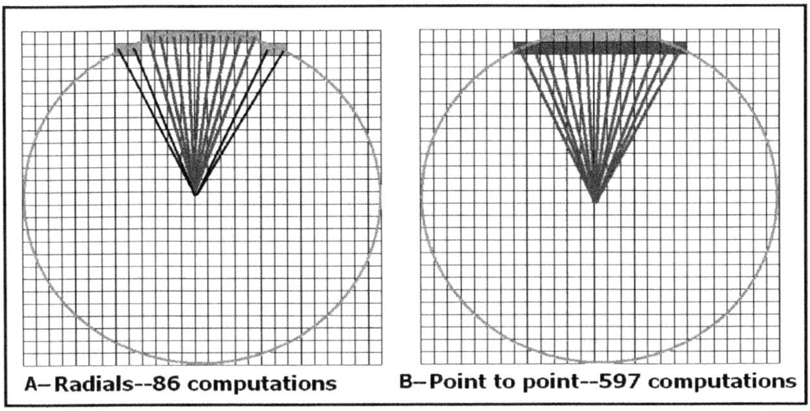

Figure 4. Comparison of viewshed construction algorithms. A. In the radial method, only points on the circumference are computed and all interior points are filled in as a byproduct. B. In the point-to-point method, each point is computed separately. In this case, the difference is a factor of six in the number of profiles. As the region size increases, the radial method increases with the radius and the point-to-point method with ½ (radius)^2.

The radial method can show all masked or hidden terrain, even if the computer screen or printout cannot display the DEM data at full resolution. High resolution DEMs often contain more data points than can be displayed, and thinning to fit the data on screen may omit critical points. A small region of either visible or masked points may or may not materially affect the analysis, but the algorithm should not miss critical points. All points of interest should be displayed on the screen and not overdrawn or omitted by thinning. Results of the radial fan method are directly linked to screen display resolution, and if saved, cannot be restored at another resolution. However, rapid display allows preservation of the fan parameters to recreate the fan at another resolution or with a different base map. Point-to-point algorithms can only guarantee displaying all masked or all visible points if they perform a raster-to-vector conversion and display the vector viewshed, but they cannot show subpixel visibility. If it is assumed that meaningful elevations can be interpolated between grid nodes in the DEM, then intervisibility can also be interpolated. Sorenson and Lanter (1993) discuss sub-cell viewshed analysis.

4. DISCUSSION

Intervisibility computations prove to be a complicated issue. Even when restricting the discussion to the geometric model, issues of data quality, format, and accuracy versus computational speed, complicate the issue. This paper has considered seven aspects of a geometric intervisibility model. Although most have been previously discussed in the literature, the issue of horizontal earth curvature - the selection of points on the straight line between observer and target - does not appear to have been fully appreciated. Horizontal curvature, like vertical curvature, does not affect very short paths, but as higher resolution DEMs become available, even small differences will lead to the selection of different grid postings to estimate intermediate elevations. The horizontal curvature algorithm probably becomes important when the deviations between algorithms exceed the DEM data spacing, and is particularly critical for military DEMs like DTED or SRTM because the geographic grid does not make it easy to compute the straight line path.

Table 1 shows recommended initial choices for geometric parameters in the intervisibility algorithm. The choice assumes the DEMs have geographic coordinates with a spacing of 1-3 second, the anticipated spacing for military DEMs for the foreseeable future.

Algorithm validation (Champion et al., 1995; Swanson, 2003) requires careful design. Commercial vendors should be encouraged to categorize their algorithms, and where possible, algorithms should allow varying parameters. Testing should proceed on two fronts: Comparing viewsheds among algorithms to determine terrain characteristics that cause algorithms to predict differing results, and comparing predicted results to ground truth. Fisher (1993) reported up to 50% variability in viewshed results using different algorithms, although Swanson (2003) reported much closer agreements. Since existing ground truth tends to follow radials and report individual points (Champion et al., 1995; Swanson, 2003), the comparisons must use the radials or individual points, extracted from viewsheds. This proves difficult with software not designed to supply output in such a fashion. Rounding errors, not from computer precision but from the depiction of results at the edges of visibility regions, probably account for many of the variations between algorithms.

Fast viewshed algorithms employ approximations that must be explicitly listed and understood. For instance, Wang et al. (2000) make no mention of earth curvature or how to incorporate it. Izraelevitz (2003) requires (1) reprojection of geographic DEMs to UTM, and (2) the calculation of earth curvature as a preprocessing step. Reprojection of a DEM can be easily accomplished, but end users might question whether they should be performing this operation. Seamless operation over large areas, difficult with UTM data, has led to the "best available" DEMs from both USGS and NGA using geographic coordinates. Gains in intervisibility computation speed from reprojection to UTM coordinates may well be offset by decreased efficiency in other operations. DEMs should remain in their native coordinates, and if an intervisibility algorithm requires a UTM DEM, the cost in storage of having two versions of the DEM, or the cost in time of reprojecting on-the-fly when required, should be clearly attributed to the intervisibility algorithm. In addition, the effects of reprojecting geographically spaced DEMs to a UTM grid need to be assessed. While the interpolation algorithm will affect the reprojection, the reprojected grid will likely differ along the ridges that preferentially determine intervisibility. Preprocessing for vertical earth curvature will be different for every potential observer's location (rings of increasing correction radiating outward from the observer) and thus will have to be performed before every computation. Again, this preprocessing must be factored into timing results. Current shortcuts, explicit and implicit, limit the effective length for LOS computations.

5. CONCLUSION

A number intervisibility algorithm parameter choices and methods of combining profiles into a weapons fan affect intervisibility results. Many of the algorithms in use, especially in commercial programs, do not clearly list these choices. Intervisibility paths should be computed using geodesic computations, unless the approximations of the UTM coordinates are considered and validated. A spaced radial algorithm produces the best weapons fans in the shortest time, assures that all desired points, either masked or visible are shown, computes sub-pixel intervisibility, and works equally well with UTM or geographic DEMs. Simplified algorithms that require reprojection of the DEM or preprocessing must be carefully analyzed to assess their full costs in accuracy, storage, and speed.

ACKNOWLEDGMENTS

This paper draws heavily on discussions at the LOS Technical Working Group Meeting organized by Ray Caputo, Danny Champion, and Lou Fatale at the USA Engineer Research and Development Center, Topographic Engineering Center, Alexandria, VA on 5 June 2003. I thank Mark Adams for his work during the past six years with TerraBase II, the military version of MICRODEM. Without his testing efforts the program would never have reached its present state of development. Discussions with Randy Swanson encouraged me to add the intervisibility algorithm testing and comparison functions. The current version of the GIS program MICRODEM (Guth et al., 1987) can be downloaded at:

http://www.usna.edu/Users/oceano/pguth/website/microdem.htm.

REFERENCES

Champion, D.C., Pankratz, K.G., and Fatale, L.A. 1995. The effects of different line-of-sight algorithms and terrain elevation representations on combat simulations. US Army Training and Doctrine Command Analysis Center - White Sands Missile Range Technical Report TRAC-WSMR-TR-95-032(R).

Defense Mapping Agency. 1994. Glossary of mapping, charting, and geodetic terms. Military Handbook MIL-HDBK-850. (Online Version: http://www.adtdl.army.mil/cgi-bin/atdl.dll/ fm/ 7-20/toc.htm)

Department of the Army. 1970. Topographic surveying. Technical Manual 5-441.

Department of the Army. 1992/2000. The infantry battalion. Field Manual 7-20. 6 April 1992, with Change 1 dated 29 Dec 2000. (Online Version: http://www.adtdl.army.mil/cgi-bin/atdl.dll/fm/7-20/toc.htm)

Department of the Army. 1992/2001. Infantry rifle platoon and squad. Field Manual 7-8. 22 April 1992, with Change 1 dated 1 March 2001. (Online Version: http://www.adtdl.army.mil/cgi-bin/atdl.dll/fm/7-8/toc.htm)

Department of the Army. 2000. Topographic operations. Field Manual 3-34.230, 3 August 2000. (Online Version: http://www.adtdl.army.mil/cgi-bin/atdl.dll/fm/3-34.230/toc.htm)

Ehlen, J. and Abrahart, R.J. 2004. Terrain and its affects on the use of artillery in the American Civil War - The Battle of Perryville 8 October 1862. In *Studies in Military Geography and Geology*, D.R. Caldwell, J. Ehlen, and R.S. Harmon, eds., Dordrecht, The Netherlands: Kluwer Academic Publishers, 155-172.

Fisher, P.F. 1993. Algorithm and implementation uncertainty in viewshed analysis. International Journal of Geographical Information Systems 7:331-347.

Freeman, R.L. 1998. *Telecommunication Transmission Handbook*, 4th edition. New York: John Wiley and Sons.

Galgano, F.A., 2004. Decisive terrain - A military geography of Fortress West Point, 1775-1797. In *Studies in Military Geography and Geology*, D.R. Caldwell, J. Ehlen, and R.S. Harmon, eds., Dordrecht, The Netherlands: Kluwer Academic Publishers, 105-120.

Guth, P.L., Ressler, E.K., and Bacastow, T.S. 1987. Microcomputer program for manipulating large digital terrain models. Computers & Geosciences 13:209-213.

Izraelevitz, D. 2003. A fast algorithm for approximate viewshed computation. Photogrammetric Engineering and Remote Sensing 69:767-774.

Kidner, D., Dorey, M., and Smith, D. 1999. What's the point? Interpolation and extrapolation with a regular grid DEM. Geocomputation 99: Proceedings of the 4th International Conference on GeoComputation, Fredericksburg, VA, USA, 25-28 July, 1999, Diaz, J., Tynes, R., Caldwell, D., and Ehlen, J., eds., CD-ROM ISBN 0-9533477-1-0.

Kidner, D.B., Sparkes, A.J, Dorey, M.I., Ware, J.M., and Jones, C.B. 2001. Visibility analysis with the multiscale implicit TIN. Transactions in GIS 5:19-37.

Lee, J. 1991. Analyses of visibility sites on topographic surfaces. International Journal of Geographical Information Systems 4:413-429.

Maloy, M.A. and Dean, D.J. 2001. An accuracy assessment of various GIS-based viewshed delineation techniques. Photogrammetric Engineering and Remote Sensing 67:1293-1298.

Palka, E.J., 2004. A military geography of the Hudson Highlands - Focal point in the American War of Independence. In *Studies in Military Geography and Geology*, D.R.

Caldwell, J. Ehlen, and R.S. Harmon, eds., Dordrecht, The Netherlands: Kluwer Academic Publishers, 89-103.

Pollard, T. 2002. The mountain is their monument: An archaeological approach to the landscape of the Anglo-Zulu War of 1879. In *Fields of Battle - Terrain in Military History*, P. Doyle and M.R. Bennett, eds., Rotterdam, The Netherlands: Kluwer Academic Publishers, 117-135.

Rana, S. 2003. Fast approximation of visibility dominance using topographic features as targets and the associated uncertainty. Photogrammetric Engineering and Remote Sensing 69:881-888.

Rose, E.F.P., Ginns, W.M., and Renouf, J.T. 2002. Fortification of island terrain: Second World War German military engineering on the Channel Island of Jersey, a classic area of British geology. In *Fields of Battle - Terrain in Military History*, P. Doyle and M.R. Bennett, eds., Rotterdam, The Netherlands: Kluwer Academic Publishers, 265-309.

Schneider, B. 2001. Phenomenon-base specification of the digital representation of terrain surfaces. Transactions in GIS 5:39-52.

Sorensen, P.A. and Lante, D.P. 1993. Two algorithms for determining partial visibility and reducing data structure induced errors in viewshed analysis. Photogrammetric Engineering and Remote Sensing 59:1149-1160.

Swanson, R. 2003. Multi-vendor line-of-sight application validation against field-collected data. Proceedings of the International Conference on Military Geology and Geography, 15-18 June 2003, West Point, NY. CD-ROM.

Vauban, S.L.P. 1828-1829. *De l'Attaque et de la Défense des Places*. Paris: Anselin. Atlas accompanying two-volume book.

Vincenty, T. 1975. Direct and inverse solutions of geodesics on the ellipsoid with application of nested equations. Survey Review 23:88-93.

Wang, J., Robinson, G.J., and White, K. 2000. Generating viewsheds without using sight-lines. Photogrammetric Engineering and Remote Sensing 66:87-90.

Yoeli, P. 1985. The making of intervisibility maps with computer and plotter. Cartographica 22:88-103.

Chapter 22

A GIS-BASED SPATIAL ANALYSIS OF CAVES AND SOLUTION CAVITIES

Application to Predicting Cave Occurrence in Limestone Terrain

Michael R. Gross, Kajari Ghosh, Alex K. Manda, and Dean Whitman
Florida International University

Abstract: Solution cavities and caves at a variety of scales were analyzed within a GIS to quantify their spatial heterogeneity and to evaluate potential geologic controls on their distribution. Techniques were first established on robust datasets of small pores taken from photographs of limestone in core and outcrop. Pore density maps and histograms of pore area proved highly effective in identifying horizons of enhanced dissolution that fall along linear trends parallel to stratigraphic layering. The technique proved equally effective for a complex network of man-made caves, suggesting that geospatial analysis can identify specific geologic horizons of localized cave development. When combined with lithologic maps and structural interpretation, these trends of high cave frequency may serve as potential targets for locating undiscovered caves in rugged terrain, an important mission confronting coalition forces in Afghanistan.

Key words: caves, GIS, spatial analysis, limestone, Afghanistan, stratigraphy, fractures

1. INTRODUCTION

For millennia caves in mountainous terrain have served as hideouts and military bases for irregular troops confronting larger, more conventional forces (Yadin, 1971; Holzimmer, 1995; Grau and Jalali, 2001; Day, 2004, this volume; Eastler, 2004, this volume). The military significance of caves re-emerged in the wake of the September 11, 2001 terror attacks against the United States and the consequent military activity in Afghanistan. One main goal of US-led coalition forces is to dislodge the enemy from cave networks scattered throughout the rugged landscape, most notably in the eastern Afghan provinces along the Pakistani border (Wines, 2001; Chang, 2001;

D.R. Caldwell et al. (eds.), Studies in Military Geography and Geology, 287–306.
© 2004 *Kluwer Academic Publishers.*

Richter, 2001). Initial activity focused on searching previously known caves as well as caves identified directly from field reconnaissance and high resolution imagery. However, numerous caves may remain undetected using these procedures. Thus, in an effort to uncover additional hideouts, a potentially valuable strategy may involve identifying specific geologic horizons where cave development is most prominent.

Caves can form by natural processes, human excavation, or a combination of both types of activity. The spatial distribution of caves and solution cavities reflects the underlying spatial distribution of the controlling geologic factors such as lithology, stratigraphy, localization of ore minerals, and planar discontinuities (White, 1988; Palmer, 1991; Knez, 1998). These geologic factors, in turn, follow systematic, non-random trends that can be quantified and analyzed within a Geographic Information System (GIS) framework. The ultimate goal of our research is to develop a GIS-based strategy for predicting the occurrence and location of undiscovered caves in limestone based on geospatial analysis of imagery acquired at a variety of scales. To this end, specific goals include (1) quantifying and characterizing the spatial distribution of solution cavity density in limestone, (2) evaluating geometric attributes of solution cavity populations, and (3) identifying high density trends that represent enhanced development of solution cavities and caves, which serve as the mappable units for the predictive strategy.

This contribution does not cover methods of direct cave detection, but rather develops a predictive strategy based on cave openings that have already been identified. Regardless of the detection method used, a significant number of caves will always remain undiscovered. Thus our strategy involves analyzing the population of known caves, which represents a segment of the total cave population, in an effort to identify regions where undiscovered caves are most likely to occur.

2. BACKGROUND

Caves can develop by means of natural processes in a variety of rock types, including carbonates, weathered intrusive rocks, lava flows, and evaporites (White, 1988). However, the most extensive cave networks and solution cavities occur in limestone. Dissolution of limestone results from the chemical interaction between circulating groundwater and calcite, the predominant mineral of limestone (Drever, 1997). Carbonic acid within groundwater converts the solid calcite into dissolved calcium and bicarbonate ions, leaving behind solution cavities within the limestone. Extensive limestone dissolution may result in the formation of enormous interconnected caverns in the subsurface, and a unique surface expression referred to as karst (White, 1988; Ford and Williams, 1989).

The extent of cave development in limestone depends upon a combination of factors, including CO_2 concentration in groundwater, position of the water table through time, rate of discharge, mineralogical composition, and age of the rock (White, 1988; Palmer, 1991; Gabrovsek and Dreybrodt, 2001). Thick sequences of limestone are composed of interbedded lithologies of varying bed thicknesses, compositions and grain sizes. Dissolution rates are strongly dependent on mineralogical composition and grain size (Rauch and White, 1977), thus a typical limestone sequence may display stratigraphic intervals with enhanced cave development alternating with layers of minor cave development and layers where caves are entirely absent. In order for solution cavities to form, the dissolving fluids must be placed in direct contact with the limestone. This is most efficiently accomplished through a network of interconnected planar discontinuities such as bed partings, fractures, and faults (Gabrovsek and Dreybrodt, 2001; Osborne, 2001). As a result of their high conductivity, dissolution occurs preferentially along these fracture planes and at fracture intersections.

3. METHODS OF ANALYSIS AND RESULTS

A GIS provides the means to organize, visualize, and merge spatial datasets from different sources and facilitates quantitative spatial analysis and predictive modeling of these data (Openshaw, 1991). Further, it provides a framework for exploring the spatial interrelationships between two or more map layers or themes (Bonham-Carter, 1994; Whitman et al., 1999).

A variety of GIS-based techniques are applied in order to evaluate the distribution of solution cavities (i.e., "microcaves") in limestone. The methodology is first established by analyzing hand specimens and close-range outcrop photographs, where abundant mm-cm scale solution cavities provide robust datasets. Once established, the techniques are then upscaled to analog cave complexes using photographs from the published literature and the internet. Digital photographs are first rectified in ERDAS IMAGINE®, then imported into the ESRI ArcView®GIS software package where solution cavities and caves are digitized as vector objects. All subsequent analysis, modeling, and map generation are performed in ArcView. The following sections describe the methods and results for each technique.

3.1 Solution Cavity Density

The locations of 405 small (mm scale) solution cavities were digitized using a core photograph taken from a water well drilled into the Biscayne aquifer of Miami-Dade County, FL (Fig. 1a). The area of analysis was divided into a grid composed of 0.1 mm x 0.1 mm square cells, for a total of ~500,000 cells. A solution cavity density was calculated for each cell in the map area using the "calculate density" function of the spatial analyst extension (ESRI, 1996a), which sums the total number of solution cavities

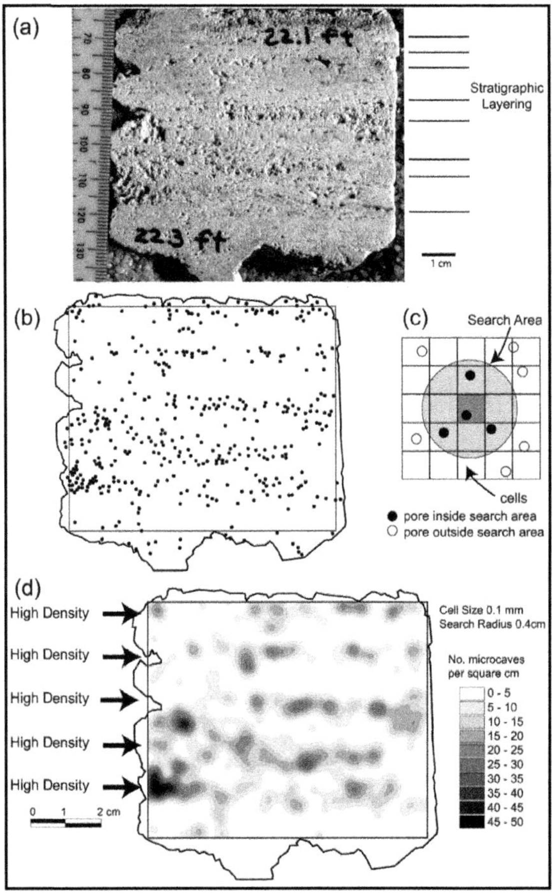

Figure 1. (a) Photograph of limestone core; (b) Map of points representing digitized solution cavities; (c) Method used for calculating density of solution cavities in the GIS; (d) Contoured map of solution cavity density.

(defined as digitized points, Fig. 1b) within the prescribed search area (defined by the search radius emanating from the cell center, Fig. 1c). The results were contoured to generate maps of solution cavity density (Fig. 1d).

The distribution of solution cavity density is heterogeneous, with regions of high solution cavity density (>35 cavities/cm^2) interspersed with regions of low density (<5 cavities/cm^2; Fig. 1d). Further, values of high density reveal a tendency to cluster in linear trends parallel to the subtle stratigraphic layering. The resulting pattern consists of high density trends alternating with trends of low density (Fig. 1d). In light of these results, we suggest that analyzing the spatial distribution of cave density is a technique that may help identify stratigraphic intervals of enhanced cave development, especially in limestone sections of interbedded lithologies. The technique is most useful where solution cavities are abundant and relatively uniform in size. Results from the analysis are automatically extracted into an exportable table, and presented as summary statistics for equal-area intervals of 0.25 cm height (Table 1). These results can be incorporated into digital files that outline boundaries of stratigraphic layers, thus enabling direct correlation between trends of high cave density and discrete stratigraphic intervals (Fig. 2).

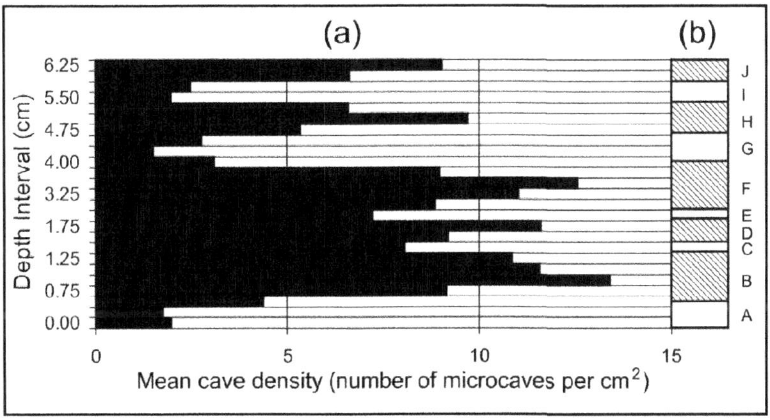

Figure 2. (a) Plot of mean solution cavity density versus depth for core sample in Fig. 1a using results compiled in Table 1; (b) Division of stratigraphy into layers of high microcave occurrence (shaded) and layers of low microcave occurrence (white) based on results from density analysis. Letters refer to assigned stratigraphic intervals.

3.2 Geometric Attributes of Solution Cavity Populations

For this application we evaluated an outcrop photograph of the Gerofit Formation (Fig. 3a), a limestone of Late Cretaceous age exposed in southern Israel (Eyal et al., 2001). The solution cavities were digitized as vector objects, and then the ArcView extension "XTools" was used to calculate the

Table 1. Quantitative data on microcave density extracted automatically using GIS. Data were collected in 0.25 cm intervals from the core sample (Figure 1a). The unit no/cm^2 refers to number of microcaves per square centimeter. Each interval contains 20,952 cells.

Interval Cm	Area cm^2	Min no/cm^2	Max no/cm^2	Range no/cm^2	Mean no/cm^2	St. Dev. no/cm^2	Sum no/cm^2
0 - 0.25	2.095	0	16.53	16.53	2.00	3.45	41899
0.25 –0.5	2.095	0	13.66	13.66	1.76	2.60	281074
0.5 – 0.75	2.095	0	18.62	18.62	4.37	3.94	242045
0.75 – 1.0	2.095	0	35.18	35.18	9.15	5.37	230634
1.0 – 1.25	2.095	0	49.91	49.91	13.42	11.07	31601
1.25 – 1.5	2.095	0	33.13	33.13	11.55	6.69	138095
1.5 – 1.75	2.095	0	49.82	49.82	10.85	11.36	189086
1.75 – 2.0	2.095	0	44.77	44.77	8.07	8.24	36902
2.0 – 2.25	2.095	0	23.47	23.47	9.18	5.32	191732
2.25 – 2.5	2.095	0	35.14	35.14	11.57	7.99	227253
2.5 - 2.75	2.095	0	25.83	25.83	7.22	5.67	168986
2.75 - 3.0	2.095	0	42.81	42.81	8.85	8.51	242513
3.0 - 3.25	2.095	0	42.81	42.81	11.01	7.80	151239
3.25 - 3.5	2.095	0	30.35	30.35	12.56	7.40	185367
3.5 - 3.75	2.095	0	29.31	29.31	8.97	6.35	263238
3.75 - 4.0	2.095	0	17.52	17.52	3.11	3.34	187905
4.0 - 4.25	2.095	0	11.22	11.22	1.51	2.34	65213
4.25 - 4.5	2.095	0	28.01	28.01	2.79	4.37	58355
4.5 - 4.75	2.095	0	28.70	28.70	5.33	5.68	111745
4.75 - 5.0	2.095	0	25.65	25.65	9.69	6.62	202963
5.0 - 5.25	2.095	0	24.20	24.20	6.59	5.03	41742
5.25 - 5.5	2.095	0	15.95	15.95	1.99	2.60	52452
5.5 - 5.75	2.095	0	17.27	17.27	2.50	3.75	138443
5.75 - 6.0	2.095	0	22.88	22.88	6.61	5.42	192379
6.0 - 6.25	2.095	0	22.88	22.88	9.02	6.54	91524

area and perimeter for each of the 334 cavities. The digitized solution cavities were then converted to raster objects in order to evaluate the centroid function of the zonal geometry script, which yielded the lengths of major and minor axes of the best-fit ellipse and its orientation (ESRI, 1996b).

The histogram of solution cavity area (i.e., the size of "microcave" openings) is positively skewed and resembles a negative exponential or power-law distribution (Fig. 3b). Although cavity openings range in area from 0.01-2.18 cm^2, the overwhelming majority of cavity openings are small (mean = 0.12 cm^2; median = 0.07 cm^2), with only three solution cavities greater than 1 cm^2 in area. The ellipticity of each solution cavity was calculated as the major axis divided by the minor axis, thus yielding a value of 1 for a circle. A plot of ellipticity versus area shows considerable scatter (Fig. 3c). Most solution cavities are elliptical in shape (mean ellipticity = 1.95). Although ellipticity does not appear to correlate with area, circular shapes are found only for small pores, whereas all pores greater than 0.5 cm^2

are elliptical. Solution cavity geometries display preferred orientations, with major axes preferentially aligned vertically and horizontally (Fig. 3d). This may reflect preferential flow along the two planar fabrics at this outcrop, horizontal stratigraphic layering and vertical joints (Eyal et al., 2001).

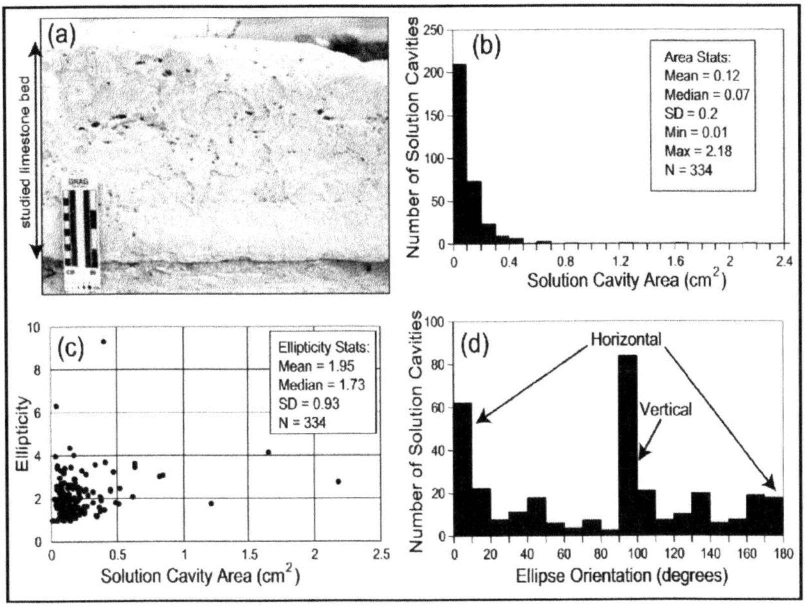

Figure 3. (a) Photograph of limestone bed imported into GIS. Note abundant small solution cavities and contact with underlying marly bed; (b) Histogram of solution cavity areas; (c) Plot of ellipticity (major axis/minor axis) versus area for solution cavities; (d) Histogram of ellipse orientation.

3.3 Porosity Profiles

The third GIS-based technique for analyzing cave development generates profile histograms of two-dimensional porosity, defined as the percentage of surface area occupied by solution cavities. The analysis was performed on the vector file of digitized pores derived from the outcrop photograph used in the previous example. The vector objects were converted to raster format, and then a grid composed of identical rectangular polygons was generated to overlay the map area. The summed area of all solution cavities within each rectangular zone was calculated and plotted as a function of vertical height.

Rectangular heights of 1 cm and 4 cm were selected for the 28 cm thick limestone bed, yielding two sets of equal area rectangles composed of 28 and 7 polygons, respectively (Fig. 4). The results identify a zone of enhanced limestone dissolution at a height of 16-20 cm above the base of the limestone bed (Fig. 4a, b). As expected, greater detail is observed using rectangles of 1 cm height, revealing subsidiary trends at bed heights of 7-9 cm and 24-27 cm (Fig. 4a). The advantages of this technique are that it can be performed rapidly, results are not affected by variability in cave dimensions, and dominant trends of enhanced cavity development can be easily recognized on the histograms. However, this technique describes spatial variability of dissolution in only one dimension, and does not characterize the distribution in the dimension parallel to the trends.

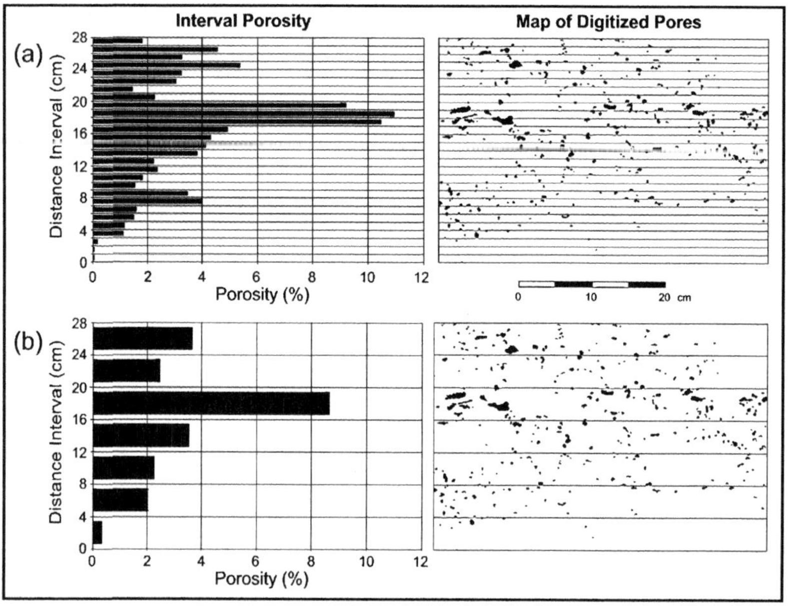

Figure 4. Porosity profile histograms for photograph in Figure 3a. Interval widths are 1 cm (a) and 4 cm (b). Histogram peaks indicate horizons of enhanced dissolution.

3.4 Upscaling to Analog Cave Complexes

As a demonstration of potential, these techniques have been applied to an analog cave complex at a scale relevant to military activity. A spectacular network of caves is exposed on the limestone cliff faces of Matala Bay along the southern coast of central Crete. With over 75 cave entrances, the Matala site is an ideal analog for this study; the caves are formed by a combination of natural processes and human excavation; they are located on a cliff face as are many caves in rugged terrain; and their abundance provides a robust dataset for analysis.

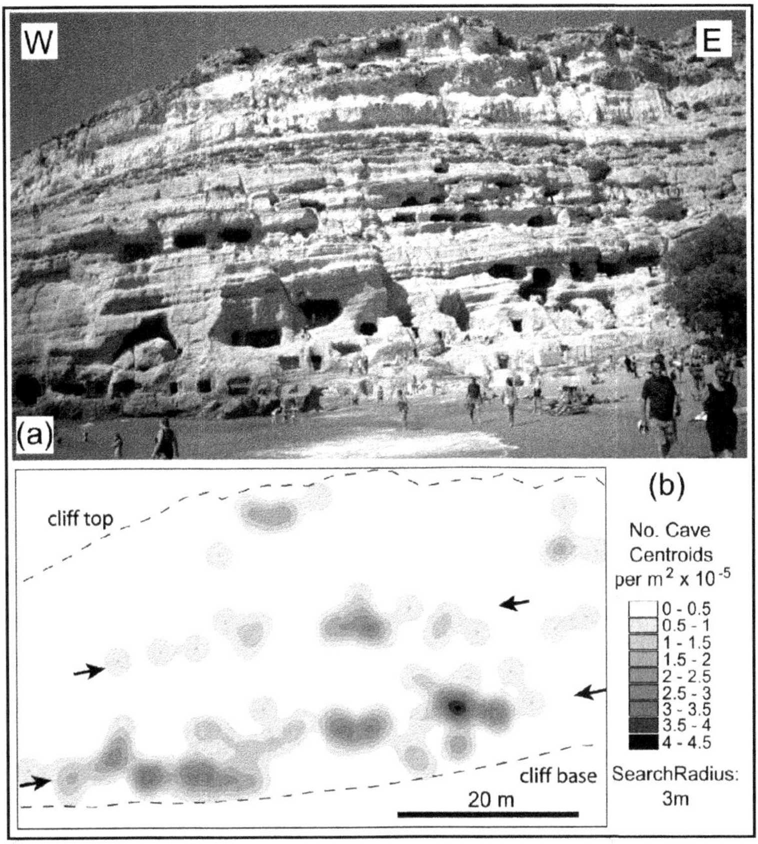

Figure 5. Results of geospatial analysis of Matala caves, Crete. (a) Photograph of cave complex exposed on cliff face (© Grisel Gonzalez and Jeff Prosise). Stratigraphic layering dips to the left (west); (b) Map of cave density; note linear trends of high density indicated by arrows, aligned parallel to dipping beds.

Digital images of caves from Matala were downloaded from the internet, thus rectification was not accurate due to the lack of surveyed control points. Nonetheless, effects of stratigraphic layering on cave distribution are dramatic; caves are found in abundance in the marly limestone beds, whereas they are absent in the prominent resistant beds (Fig. 5a). Maps of cave density were generated using a cell size of 0.05 m and search radii of 2, 2.5 and 3 m (the photograph was scaled by approximating the height of people standing against the outcrop face; the units are highly approximate and for relative comparison only). Linear trends of high cave density dip slightly to the left (west), reflecting the regional dip of the limestone strata (Fig. 5b). The profile histograms show three dominant peaks of greater than 10% cave area (Fig. 5c, d) corresponding to three main horizons of cave development (Fig. 5a). Interlayered regions of low percentages identify the resistant beds where caves are absent (Fig. 5a, c). Cave openings are mostly elliptical (mean ellipticity ~ 2) with major axes mostly horizontal (Fig. 6b), reflecting the influence of stratigraphy on cave geometry. Most cave entrances are relatively small in area (<1 m^2), with only one very large cave and a few intermediate size cave entrances (Fig. 6a, c).

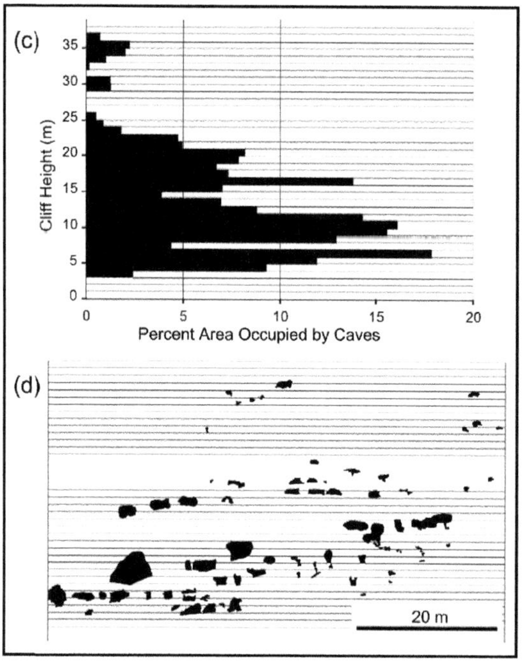

Figure 5 (cont). Results of geospatial analysis of Matala caves, Crete. (c) Profile histograms of percent cave area. Note three main peaks corresponding to horizons of enhanced cave development; (d) Cave map divided into intervals used in porosity analysis. Scale is highly approximate and should be used for relative comparisons only.

3.5 Size-Frequency Analysis of Cave Area Distributions

Populations of geologic features often adhere to power-law frequency distributions, expressed as:

$$N = aU^{-b}$$

where U is a measure of size, a is a constant, b is the power-law exponent and N is the cumulative number of values greater than or equal to U (Pickering et al., 1995). Natural phenomena that follow power-law

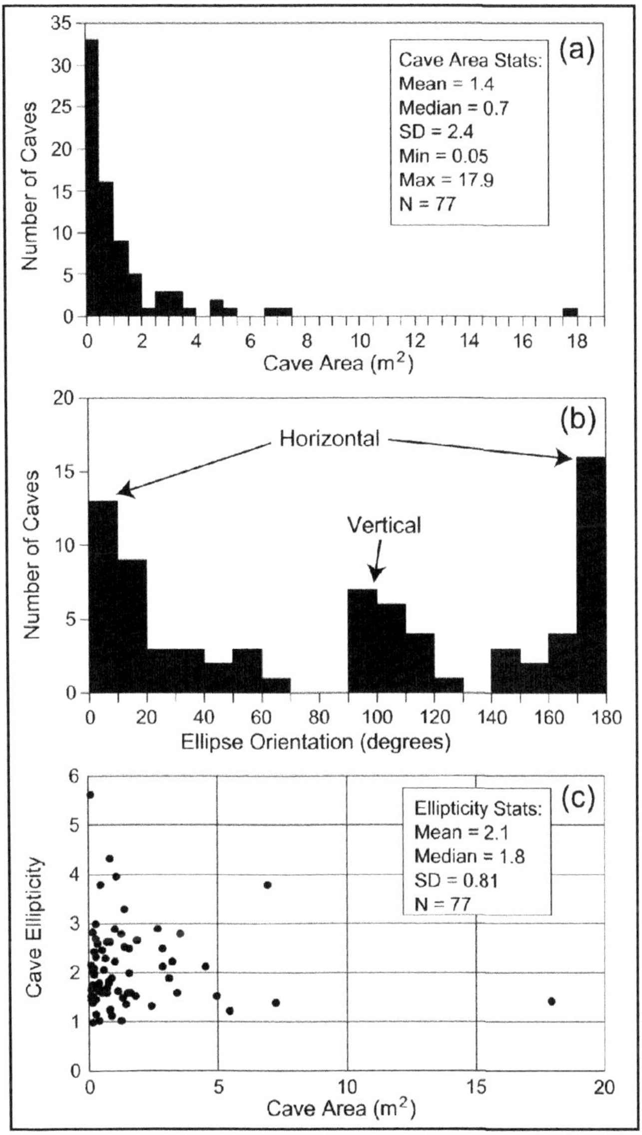

Figure 6. Compilation of Matala cave attributes. (a) Histogram of cave entrance areas; (b) Histogram of ellipse orientations; (c) Plot of ellipticity versus cave area.

distributions are described as self similar over a range of scales. When plotted in log-log space the power-law frequency curve is a straight line whose negative slope equals the power-law exponent (Fig. 7a). Many geologic datasets are undersampled, either due to limits of resolution (a typical problem for the smallest features) or to large recurrence intervals (a typical problem for the largest features). The power-law relation for such a dataset is determined from the linear segment of the population (Fig. 7a). This theoretical relationship can then be used to estimate the frequency and dimensions of under-sampled features within the overall population. A size-frequency analysis was performed on cave area values extracted by the GIS from the limestone bed (Fig. 3) and the Matala cave network (Fig. 5). Both datasets reveal strong linear correlations in cumulative frequency plots (Figs. 7b, c), implying that populations of cave openings adhere to power-law distributions in terms of their areas. Note that data for the largest caves follow the best-fit regression line, indicating that the largest features within both populations are identified. However, the smallest caves are undersampled by virtue of their deviation from the theoretical curve (Fig. 7b, c).

4. DISCUSSION

Our analysis identified several promising GIS techniques that are effective in evaluating the spatial distribution of caves in limestone. Numerous small solution cavities in hand specimens and outcrop exposures provided robust datasets for establishing quantitative relationships. The results identified trends of high cave occurrence as maps of contoured cave density, as well as peaks in histograms of porosity. The trends and peaks are prominent and easily recognized, despite subtle changes in lithologic composition and texture within the small sample areas. Application of these techniques to analog cave complexes at Matala, where differences in stratigraphic layering are more pronounced, yielded similar results. Trends of high cave density are aligned in linear swaths parallel to stratigraphic layering, and histograms identify specific horizons within the cliff face where cave development is most intense. Small solution cavities and large caves display similarities in geometry, with a tendency for elliptical openings and positively skewed distributions for entrance area (Figs. 3 and 6). The most significant result from this study is that GIS techniques can identify specific stratigraphic intervals where caves are most likely to occur, thus providing the basis for the prediction of undiscovered caves.

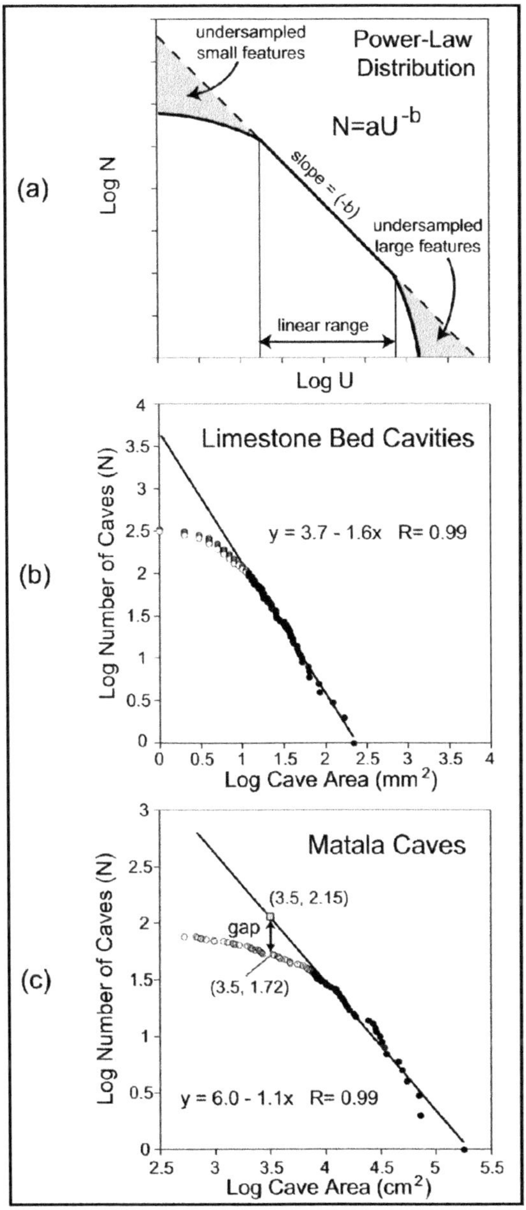

Figure 7. (a) Generalized log-log cumulative frequency graph illustrating a range-limited power-law population; (b) Cumulative frequency graph of solution cavity areas measured from the limestone bed photograph in Fig. 3a; (c) Cumulative frequency graph of cave areas measured from the photograph of Matala caves in Fig. 5a.

4.1 The Need for Developing a Predictive Strategy

Each technique - mapping cave density, generating area histograms, and compiling population attributes - has advantages and limitations, thus the overall analysis is most powerful when all three methods are used together. The combined analysis can be applied to high resolution imagery at a variety of scales, including field reconnaissance photographs, air photos, and satellite images. Image analysis techniques can be used to map the known cave openings, which will eliminate the time-consuming process of digitizing cave openings by hand. Once the known caves are converted to vector features in the GIS, the entire process from computation to map/figure generation takes less than one hour.

The search for undiscovered caves and enemy forces hiding within them can only be accomplished by troops on the ground (Richter, 2001; Schafer, 2002a), thus output from the analysis should be geared toward the tactical military scale. Predicting the locations of undiscovered caves became a major factor in Afghanistan, where once on the ground, coalition forces found cave networks much larger and more complex than previous intelligence had indicated (Schafer, 2002b). When asked about the Zhawar Kili complex in eastern Afghanistan, Rear Admiral John Stufflebeem noted with surprise "we just didn't know how extensive it was", noting the camp contains more than 50 caves, many high on the cliff walls (Schrader and Slater, 2002). In the same interview, Stufflebeem makes a further case for predictive modeling by stating the surrounding region is "riddled with hillsides and valleys of caves ... therefore there are likely other valleys with other complexes, and they, in fact, may very well have [Al Qaeda militants]." The limitations of relying solely on the direct interpretation of imagery were echoed by Joint Chiefs Chairman General Richard Myers, who noted that "other types of surveillance or reconnaissance" did not reveal the huge extent of cave complexes in the Zhawar Kili area (Schafer, 2002b).

4.2 Predictive Strategy for Interpolating Between Known Cave Openings

The power-law relationship established for dimensions of cave openings is a powerful tool for predicting the size and number of undetected cave openings within a surveyed area. Specific stratigraphic horizons may be identified by GIS as zones of enhanced cave development on the basis of known cave openings; however, numerous caves belonging to the same population may nonetheless remain undetected. For example, the size-frequency distribution of Matala cave openings implies that we were unable to detect many small caves from direct inspection of the photograph (Fig.

7c). The number of undetected caves can be estimated by evaluating the gap between the data and the theoretical best-fit curve (Fig. 7c). Direct analysis of the Matala photograph yields 52 (log N = 1.72) cave openings with diameters greater than or equal to 65 cm (log cave area = 3.5 cm^2), large enough for a person to enter. However, the power-law model predicts 141 (log N = 2.15) caves in this range. Thus, an estimated 89 caves with diameters ranging from 65-100 cm (the minimum cave diameter that adheres to the power-law fit) were undetected by visual inspection of the photograph. A search for these undetected caves within the area covered by the photograph should concentrate on the horizons identified by GIS analysis as regions of preferential cave development. For Matala, these correspond to stratigraphic horizons at heights of 6, 10, 17, and 20 m above the cliff base (Fig. 5).

4.3 Predictive Strategy for Extrapolating to Surrounding Regions

Combining results from the geospatial analysis with geologic factors that control cave development may provide the means to predict cave occurrence in surrounding areas (Hung et al., 2002). To illustrate this point, consider a sequence of sedimentary rocks that has been folded and faulted, a common occurrence in regions of compressional tectonics (Fig. 8). Stratigraphically controlled caves are identified in one limestone unit (Fig. 8a), and quantitative spatial analysis further identifies specific horizons within the limestone characterized by enhanced cave development (e.g., Fig. 5). Fracture traces mapped from air photos fall into two main trends, one parallel and the other perpendicular to the regional fold axes (Fig. 8b). The fracture traces, geomorphic expressions of fracture zones in the underlying bedrock, are preferential pathways for dissolving fluids. Thus, buffer zones are generated around the fracture traces in order to identify the major fluid flow conduits in the bedrock (Fig. 8c).

A map predicting regions of likely cave occurrence can then be generated by combining the lithologic (limestone outcroppings) and structural factors (fracture zones) that control cave development (Fig. 8d). A low probability of cave occurrence is assigned to regions underlain by non-carbonaceous lithologies, whereas a high probability is assigned to the limestone outcrop belt. The highest probability for cave occurrence is found in regions of overlap between the limestone and the buffered fracture zones (Fig. 8d). These are the areas where the rocks most suitable for dissolution are placed in direct contact with the largest volumes of dissolving fluids. Further, the linear trends of overlap zones may represent preferred orientations for subsurface cave systems.

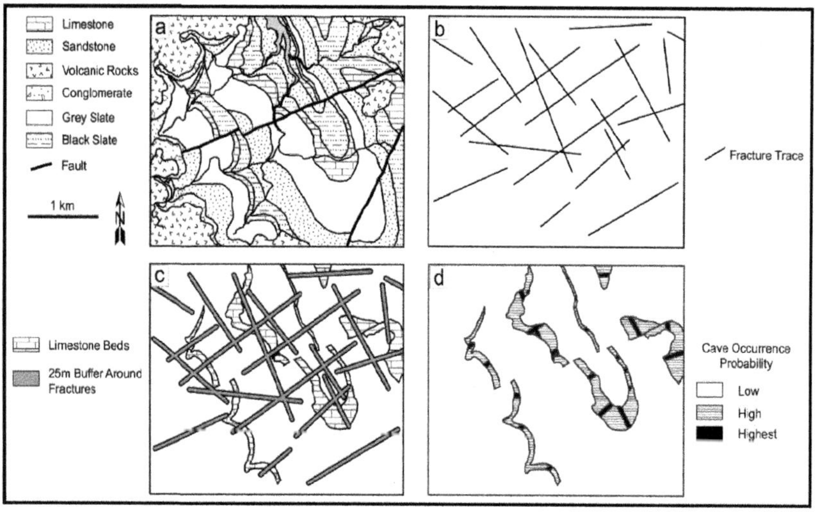

Figure 8. Schematic example for developing a cave probability map. (a) Geologic map (from Powell, 1992); (b) Fracture trace map; (c) Map of limestone outcrops with 25 m buffer around fracture traces; (d) Probability map of cave occurrence. See text for more details.

Military reconnaissance units can be provided with a cave probability map (e.g., Fig. 8d) along with results from the quantitative analysis of known caves in the region (e.g., Fig. 5). The method thereby incorporates three geologic factors that control cave distribution, namely lithology, planar discontinuities, and most importantly, stratigraphic layering. Regions of highest cave probability constitute a small percentage of the total area (1.3% in the example in Fig. 8d), thus personnel in the field can focus initially on a systematic and efficient search for possible cave openings. Because all maps and analyses are prepared within a GIS, regions of predicted cave occurrence are georeferenced and thus are navigable via GPS.

4.4 An Example from Afghanistan

The US Department of Defense released a pair of intriguing photographs showing effects of a 1998 air strike on cave openings in one of the Zhawar Kili cave complexes (US Department of Defense, 1998). Zhawar Kili is located in the eastern Afghan province of Paktia close to the border with Pakistan, and was the scene of intense coalition bombing and special operations activity in 2001-02 (Schafer, 2002a; Schrader and Slater 2002). We emphasize the limitations inherent in using this example - image resolution is low, the scale is unknown, the actual site among the many camps at Zhawar Kili is unknown, and the lithology is uncertain.

The photograph was filtered to enhance cave openings and deformed in order to restore geomorphic features to appropriate inclinations (Fig. 9a). A coarse geologic interpretation identifies three subvertical cliff faces with horizontal lineations that may signify stratigraphic layering (Fig. 9b). Ten cave openings on cliff face #1 were initially detected (Fig. 9c). The observed cave openings were digitized into the GIS and analyzed for spatial relationships and population attributes. Maps of cave density and area histograms identify the horizontal trend of cave openings near the base of the cliff in terms of a linear trend of high density and elevated cave areas, respectively (Fig. 9d, e). These trends delineate a significant "cave horizon" that may be traced across the outcrop area in the hope of locating undiscovered caves. In fact, an eleventh cave opening may exist along the continuation of this trend to the south, although it cannot be confirmed due to poor image quality (Fig. 9c). Geometries of the Zhawar Kili cave openings can be analyzed individually and collectively. The distribution of entrance areas at Zhawar Kili is similar to previous examples, with two large cave openings (# 9, 10) and numerous small cave openings (#1-8). Cave shapes are mostly elliptical, with major axes of the larger caves aligned subhorizontally.

5. CONCLUSIONS

(1) The spatial heterogeneity of solution cavities and caves can be effectively characterized using GIS. The GIS provides for automated extraction of both geometric attributes and results of quantitative spatial analysis. (2) Regions of enhanced dissolution can be identified and quantified by mapping cavity/cave density and generating profile histograms to stratigraphic layering. (3) Attributes of cavity/cave geometry indicate that most openings are elliptical in shape with major axes preferentially aligned parallel to stratigraphic layering. Areas of cave openings adhere to power-law distributions, providing the means to estimate the number and sizes of undetected cave openings within a sample population. (4) Cave probability maps can be generated by combining lithologic information and fracture interpretation. Within high probability areas, searches should focus on those of cavity area. These regions frequently follow linear trends that correspond to stratigraphic horizons identified by geospatial analysis as zones of enhanced cave development. (5) Surveillance of caves via remote sensing does not provide a complete picture of cave networks in rugged terrain. However, the geologic interpretation of spatial relationships may provide additional, and potentially valuable, information on the nature of cave populations and their distributions.

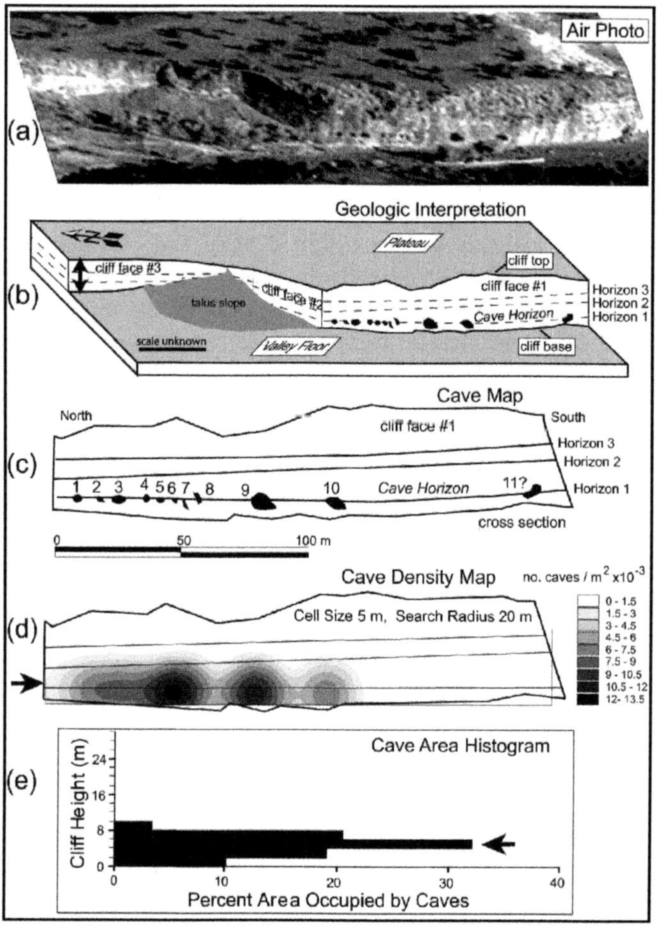

Figure 9. (a) Oblique air photo released by US Department of Defense (August, 1998) showing caves in Zhawar Kili region, eastern Afghanistan; (b) Geologic interpretation of (a) identifying cave openings in the cliff face; (c) Digitized map of caves; (d) Cave density map; (e) Profile histogram of cave area as a function of cliff height. Arrows indicate stratigraphic position of maximum cave development. Scale is relative and does not represent true length.

ACKNOWLEDGEMENTS

We thank Kevin Cunningham and Michael Wacker of the US Geological Survey and Ram Weinberger of the Israel Geological Survey for access to core and outcrops. Thanks to Grisel Gonzalez for granting us permission to use her Matala photograph. Florida International University students Sumanjit Aich, Tammy Eisner, and Pedro Alvarez provided valuable assistance. Comments and suggestions from Yehuda Eyal, Gren Draper, and Andrew Macfarlane are appreciated. Insightful reviews by Judy Ehlen, Doug Caldwell, and Irene Boland greatly improved the paper. This material is based upon work supported by the US Army Research Office under contract/ grant number DAAD19-99-1-0306.

REFERENCES

Bonham-Carter, G.C. 1994. *Geographic Information Systems for Geoscientists*. Tarrytown, NY: Pergamon Press.

Chang, K. 2001. "Nature made the perfect hiding place," The New York Times, 26 November 2001.

Day, M.J. 2004. Military campaigns in tropical karst - The Maroon Wars of Jamaica. In *Studies in Military Geography and Geology*, D.R. Caldwell, J. Ehlen, and R.S. Harmon, eds., Dordrecht, The Netherlands: Kluwer Academic Publishers, 79-88.

Drever, J.I. 1997. *The Geochemistry of Natural Waters: Surface and Groundwater Environments,* 3rd edition. Upper Saddle River, NJ: Prentice Hall.

Eastler, T.E. 2004. Military use of underground terrain - A brief historical perspective. In *Studies in Military Geography and Geology*, D.R. Caldwell, J. Ehlen, and R.S. Harmon, eds., Dordrecht, The Netherlands: Kluwer Academic Publishers, 21-37.

Environmental Systems Research Institute, Inc. (ESRI). 1996a. *Using the ArcView® Spatial Analyst*. Redlands, CA: ESRI, Inc.

Environmental Systems Research Institute, Inc. (ESRI). 1996b. *Using ArcView®GIS*. Redlands, CA: ESRI, Inc.

ERDAS, Inc. 2001. *ERDAS IMAGINE® Tour Guides (V8.5)*. Atlanta, GA: ERDAS, Inc.

Eyal, Y., Gross, M.R., Engelder, T., and Becker, A. 2001. Joint development during fluctuation of the regional stress field in southern Israel. Journal of Structural Geology 23:279-96.

Ford, D. and Williams, P. 1989. *Karst Geomorphology and Hydrology*. London: Unwin Hyman.

Gabrovsek, F. and Dreybrodt, W. 2001. A model of the early evolution of karst aquifers in limestone in the dimensions of length and depth. Journal of Hydrology 240:206-24.

Grau, L.W. and Jalali, A.A. 2001. The campaign for the caves: the battles for Zhawar in the Soviet-Afghan war. The Journal of Slavic Military Studies 14:69-92.

Holzimmer, K.C. 1995. Walter Krueger, Douglas MacArthur, and the Pacific War: The Wakde-Sarmi Campaign as a Case Study. The Journal of Military History 59:661-85.

Hung, L.Q., Dinh, N.Q., Batelaan, O., Tam, V.T., and Lagrou, D. 2002. Remote sensing and GIS-based analysis of cave development in the Suoimuoi Catchment (Son La-NW Vietnam). Journal of Cave and Karst Studies 64:23-33.

Knez, M. 1998. The influence of bedding-planes on the development of karst caves. Carbonates and Evaporites 13:121-31.

Openshaw, S. 1991. Developing appropriate spatial analysis methods for GIS. In *Geographical Information Systems, Principles and Applications*, D.J. Maguire, M.F. Goodchild, and D.W. Rhind, eds., New York: Longman Scientific.

Osborne, R.A.L. 2001. Halls and narrows: network caves in dipping limestone, examples from eastern Australia. Cave and Karst Science 28:3-14.

Palmer, A.N. 1991. Origin and morphology of limestone caves. Geological Society of America Bulletin 103:1-21.

Pickering, G., Bull, J.M., and Sanderson, D.J. 1991. Sampling power-law distributions. Tectonophysics 248:1-20.

Powell, D. 1992. *Interpretation of Geological Structures through Maps*. Essex, UK: Longman House.

Rauch, H.W. and White, W.B. 1977. Dissolution kinetics of carbonate rocks. 1. Effects of lithology on dissolution rates. Water Resources Research 13:381-94.

Richter, P. 2001. "Hundreds of U.S. troops to search Al Qaeda caves," The Los Angeles Times, 21 December 2001.

Schafer, S.M. 2002a. "U.S. striking eastern Afghan targets," Associated Press, 7 January 2002.

Schafer, S.M. 2002b. "U.S. forces seize 2 Al Qaeda, equipment," The Los Angeles Times, 8 January 2002.

Schrader, E. and Slater, E. 2002. "Afghan caves daunt the U.S.," The Los Angeles Times, 15 January 2002.

US Department of Defense. 1998. Zhawar Kili cave complex, Afghanistan, pre-strike photograph. (Online Version: http://www.defenselink.mil/photos/Aug1998)

White, W.B. 1988. *Geomorphology and Hydrology of Karst Terrains*. New York: Oxford University Press.

Whitman D., Gubbels T.L., and Powell, L. 1999. Spatial interrelationships between lake elevations, water tables, and sinkhole occurrence in Central Florida: A GIS approach. Photogrammetric Engineering and Remote Sensing 65:1169-78.

Wines, M. 2001. "Heavily fortified 'ant farms' deter Bin Laden's pursuers," The New York Times, 26 November, 2001.

Yadin, Y. 1971. Bar-Kokhba: *The Rediscovery of the Legendary Hero of the Second Jewish Revolt against Rome*. London: Weidenfeld and Nicolson.

Chapter 23

GROUNDWATER

Past, Present, and Future Uses in Military Operations

Christopher A. Gellasch
US Military Academy

Abstract: During the 20th century, armies conducted many major operations in areas without enough surface water to meet operational requirements. Geologists were used in World War I to locate and develop groundwater resources by deep drilling. More recently, locating groundwater sources in the Saudi desert during the Gulf War was made possible by using geological expertise. In the Balkans, geologists were employed to locate and develop groundwater for peacekeeping operation base camps. Groundwater is often ideal for these semi-permanent installations, housing thousands of troops, because wells require little maintenance and well yields generally remain constant over time. When comparing drinking water sources, groundwater has advantages over both surface water, which requires treatment, and bottled water, which is prohibitive in terms of both cost and logistical requirements. Groundwater will continue to be a critical source of water for military operations in the future.

Key words: drinking water, groundwater, military geology, military operations, well drilling

1. INTRODUCTION

Many factors influence the outcome of a military operation. Although logistics are not always perceived as critical to the success of an operation, history has proven otherwise. Even Napoleon acknowledged that an army travels on its stomach. Without adequate food and water, the effectiveness of any military unit is degraded. Obtaining adequate supplies of water from nearby sources can be a challenge in terrain with little or no surface water. In other cases, water quality may be the primary issue when local supplies have been contaminated. In either case, finding ample quantities of clean water from a local source can greatly benefit a military force. An army that must

D.R. Caldwell et al. (eds.), Studies in Military Geography and Geology, 307–319.
© 2004 *Kluwer Academic Publishers.*

transport large quantities of water great distances to soldiers in the field places a significant burden on its logistical system.

During the last several centuries especially, soldiers have used groundwater to provide a safe, reliable source of water. Fortified localities during medieval times commonly obtained drinking water from groundwater wells (Eastler, 2004, this volume; Rose, 2004). As troops moved into an area, private wells that were established by residents would be commandeered for use by the military. However, the dedicated use of military geologists and well drilling units to locate and produce water from subterranean sources has a much shorter history.

During the first half of the 20th century, armies began to use groundwater on a large scale and so required the use of geologists and specialized well drilling units. By the end of the century, the US military employed a combination of active duty and reserve forces drilling units in conjunction with civilian-contracted drillers to develop groundwater sources. Geologists, both military and civilian, contribute to locating and evaluating potential aquifers worldwide.

In this short overview, only a few historical examples have been selected to illustrate the importance of groundwater in military operations. After summarizing the importance of groundwater usage in both World Wars, more current examples are given for balance. This background gives perspective to the US military's current principles and practices of locating and developing groundwater sources and the advantages of using groundwater for future military operations.

2. HISTORICAL EXAMPLES

2.1 World War I (1914-18)

The German army developed the use of geology in warfare in 1914. The initial focus of German military geology was concerned with locating and developing groundwater supplies (Brooks, 1920). By November 1918 the German army had employed approximately 250 geologists and typically over 50% of their work was concerned with either water supply or drainage (Rose et al., 2000). German geologists were used to find groundwater sources in many locations, notably the Lorraine front and the Moselle basin (Brooks, 1920).

The British also realized the importance of locating water sources during the near-static trench warfare along the Western Front. In 1915, as British troop concentrations increased, serious water supply problems emerged (Rose, 2004). The concentration of 300,000 British troops plus 100,000 horses and pack animals in a relatively small area in the lowlands of northern France and Belgium caused a significant demand for water (Pittman, 1998). Both the British and French estimated that when troops were massed for an advance, at least 568 m^3 of water per day were required for each 50 km^2 occupied (Brooks, 1920). This area lacked adequate surface water sources so armies on both sides turned to geologists to determine the availability of groundwater. The other alternative was to bring water forward from rear areas in containers or by pipeline, but this was not always practical. In the Ypres area of Flanders several aquifers were discovered above and below the Ypres Clay (Doyle et al., 2000). The British army then utilized specialized well drilling units to develop usable water sources.

When the United States entered the war in 1917, American military leaders learned some lessons about military geology and groundwater from their allies. Geologists were employed at many levels within the US Army (Brooks, 1920). The 26th Engineer Regiment was formed as a special water supply unit. The regiment consisted of 1434 men (primarily oil well drillers) and 51 officers organized into seven water supply companies, although no geologists were assigned to the unit (Pittman, 1998). The workload was so great that an additional regiment was being formed when the war ended. By the end of the war it was clear that groundwater was a key resource for military operations in many parts of the world.

2.2 World War II (1939-45)

After World War I, military geology diminished in importance and most military geologists were demobilized from active military service. In 1937, the German army re-established its military geology organization and began evaluating potential combat zones for several geologic factors including groundwater sources (Rose et al., 2000). When World War II began in 1939, the German geologists were organized and prepared to support military operations. In preparation for Operation Sealion, the proposed invasion of England in 1940, German military geologists prepared numerous water supply maps of southeastern England (Rose et al., 2002; Rose and Willig, 2004, this volume).

After Britain declared war on Germany in September 1939, W.B.R. King (a World War I military geologist) was recalled to active duty and deployed with the British Expeditionary Force to France (Rose, 2004). Later, British geologists played a significant role in evaluating the groundwater resources of North Africa and behind the beaches of Normandy before the Operation Overlord invasion in 1944. After D+5, water from existing wells in Normandy was supplemented by Royal Engineers drilled wells (Rose and Pareyn, 1998).

The North Africa campaign is an excellent example of how geological expertise was important for locating groundwater sources. In the British army, geologists advised commanders as to the depth and amount of water that could be obtained from various locations along geologic structures. The 42nd Geological Section of the South African Engineer Corps used electrical resistivity traverses in selected areas to assist in finding water sources (Rose and Rosenbaum, 1998). Although geological methods of locating groundwater were becoming more reliable, some high levels of military command initially relied on more primitive methods. One example is the British 8th Army in the Middle East and North Africa, which initially used dowsing as a primary means to locate groundwater supplies (Moseley, 1973). Once dowsing fell out of favor and geology became the primary method of locating water in desert regions, commanders had a reliable, scientific tool to assist in planning the movement of troops through inhospitable lands. In the Fuka Basin of western Egypt, geologists assisted Royal Engineers well drilling units in the installation and development of shallow wells that yielded up to 23 m^3 of water per hour from 0.25 m diameter holes (Rose, 2004).

The Americans formed the Military Geology Unit (MGU) in 1942 as part of the US Geological Survey to conduct similar studies to benefit allied operations. The MGU consisted of 100 personnel by 1945 and was responsible for producing 313 area reports called Strategic Engineering Studies (Terman, 1998). Many of these reports dealt with the quality and quantity of both surface and subsurface water sources.

Geologists used increasingly sophisticated means of locating groundwater during World War II. Development of groundwater resources was conducted proactively instead of in response to an army arriving in theater and realizing surface water supplies were not adequate. This advanced research conducted by both the Axis and Allied armies indicates that the lessons of World War I were not forgotten or were quickly relearned. A proactive approach eased the demand on supply lines stretched thin by mechanized warfare and set a precedent for future military operations.

3. RECENT OPERATIONS

Since the end of World War II, the use of groundwater in military operations has continued to be an important part of logistical planning. Whereas contributions by German and other forces in operations worldwide are significant, two recent examples well documented in English serve to demonstrate groundwater's continuing importance to military operations.

3.1 Gulf War (1990-91)

During the Gulf War of 1990-91, large numbers of coalition forces were gathered in the northern Saudi Arabian desert before the liberation of Kuwait. Hundreds of thousands of troops lived in an area that receives approximately 100 mm of rain per year and has almost no surface water. Many of the units spent more than six months living in the desert and required substantial amounts of water. Fortunately, there is a large amount of water contained beneath the arid terrain in nine major aquifers. The presence of one of the most productive aquifers in the Arabian Peninsula (the Umm er Radhuma aquifer) allowed the build-up of troops with very little impact on groundwater resources (Knowles and Wedge, 1998).

Military well drilling units installed some water supply wells in rear areas and rehabilitated existing wells. Civilian contractors installed large, municipal-type wells deeper than 1000 m (L.C. Dwyer, US Army Corps of Engineers (USACE), Engineer Research and Development Center (ERDC), pers. comm., 1999). Once the ground war began, military planners needed to secure water for forces moving into Iraq. Geoscientists used remote sensing techniques to locate existing wells in locations where coalition forces were expected to operate (Knowles and Wedge, 1998). Without the aquifers in the Saudi Arabian desert, it would not have been feasible to supply the assembled coalition forces with enough water to conduct the build-up and subsequent liberation of Kuwait.

3.2 Bosnia and Kosovo

Armed conflict in the Balkans led to United Nations intervention in Bosnia in the early 1990s and the North Atlantic Treaty Organization (NATO) eventually assumed this peacekeeping role. Escalating instability in Kosovo resulted in a NATO air war and occupation of that region by ground forces in 1999. In both Bosnia and Kosovo, several thousand peacekeeping troops were deployed for extended periods of time and operated from base camps. Due to the poor environmental conditions and lack of adequate infrastructure, clean sources of water became a concern. A well drilling team from the British army deployed to Bosnia in 1993 to construct wells for base camps (Nathanail, 1998). When US military forces deployed to Bosnia in 1996, USACE evaluated sites for hydrogeologic conditions and the eventual installation of wells by military well drilling units. Due to cost issues, civilian contractors were used instead of military units to drill the wells (L.C. Dwyer, ERDC, pers. comm., 1999).

After reviewing lessons learned from Bosnia, base camp selection criteria were developed for Kosovo. One of these criteria, water availability, was a critical factor used to evaluate ten sites as potential base camps for US forces (L.C. Dwyer, ERDC, pers. comm., 1999). One location, Camp Bondsteel, was thought to have bedrock suitable for well development. The limestone bedrock in this area has undergone dissolution and forms a karst aquifer. Although groundwater was recognized as important for establishing base camps, the difficulty of locating wells in karst aquifers was not initially considered. As with Bosnia, local contractors were used instead of military well drillers as a cost saving measure (Fig. 1).

Figure 1. Local contractors in Kosovo drilling a water supply well near Camp Bondsteel. Source: USACE, ERDC, Topographic Engineering Center.

After contractors encountered difficulty drilling wells with an adequate yield, LTC Kenneth W. McDonald, a US Army Engineer officer deployed to Kosovo, used geologic data and field methods to determine three new well locations. These wells produced over 2000 m^3 of water per day and contributed to the success of the operation in Kosovo. Although using contractors was supposed to be less expensive than using military well drilling teams and geologists, the operation cost several hundred thousand dollars more than had been expected (K.W. McDonald, US Military Academy (USMA), pers. comm., 2003).

4. US MILITARY RESOURCES FOR FINDING AND DEVELOPING GROUNDWATER SOURCES

4.1 Military Well Drilling Units

All three US military services have the capability to deploy well drilling units. The US Army has twelve well drilling detachments within the Corps of Engineers, although all of them are in the reserve components. Two additional detachments will be created in 2004 (T. Scarbrough, US Army 416th Engineer Command (ENCOM), pers. comm., 2003). An army well drilling detachment is normally lead by a sergeant first class and has 10 personnel, one truck-mounted drill rig, and well completion kits (Scarbrough and Lang, 2001; P.A. Lang, US Army Reserve Command (USARC), pers. comm., 2003). These kits allow the unit to install wells of either 180 m (600 ft) of PVC or 455 m (1500 ft) of steel pipe (US Army, 1994; Baehr, 2001). Both the US Air Force and Navy have active duty well drilling units. The Air Force has three Rapid Engineer Deployable, Heavy Operational Repair Squadron (REDHORSE) drilling teams on active duty and one in the reserve component (P.A. Lang, USARC, pers. comm., 2003). These units typically consist of 12 personnel and have the ability to drill to depths of 610 m using their own equipment. The Navy has eight well drilling teams, each capable of 24 hour drilling operations (T. Scarbrough, 416th ENCOM, pers. comm., 2003). Each service uses its well drilling units for training and operational missions around the world. None of the three services has an officer assigned to well drilling units nor do they have a requirement for a geologist.

Well drilling equipment is not standardized across the services. The Army uses the LP-12 drill rig mounted on a 6 x 6 truck. The Air Force REDHORSE teams do not have organic well drilling equipment but normally use the commercially available Schramm T450MII Rotadrill for well installation. Navy personnel use the ISO/Air Transportable Water-Well Drill that has the advantage of high maneuverability due to its 5.9 m length (Scarbrough and Lang, 2001).

4.2 Water Detection Response Team

Although US military units may have a need for groundwater to sustain operations, the required geological expertise may not exist at the Division or Corps level. The Water Detection Response Team (WDRT) consists of government geoscientists and is tasked with locating groundwater sources on short notice to assist military well drilling operations (US Army, 1994). The WDRT was established in 1985 and most manpower comes from two

USACE agencies: ERDC and the Mobile District; with additional assistance from the US Geological Survey. Without adequate geologic knowledge, some drilling units must resort to a trial-and-error method of locating water supplies. This results in dry holes, which wastes valuable time and money. Since 1994, the more than 100 military-drilled water wells have surpassed a 95% success rate due to WDRT assistance (Dwyer and Markley, 1998). In cases where insufficient data are available, the WDRT may deploy personnel to the theater to assist with site selection and well design. The difficulty in locating high yield wells at Camp Bondsteel, Kosovo is a good example of where the use of the WDRT could have made a difference.

4.3 Tactical and HCA Wells

Most US military well drilling teams focus on the installation of tactical wells. Typically, these wells are drilled quickly under adverse conditions, require additional treatment by a Reverse Osmosis Water Purification Unit (ROWPU; Fig. 2) to ensure potability, and are only intended for short term missions. Tactical wells have a yield of less than 11 m^3/day (Baehr, 2001).

Figure 2. A trailer-mounted Reverse Osmosis Water Purification Unit (ROWPU) in operation. Source: US Army.

Although installation of tactical wells is a valuable skill, these temporary wells are not used in most current operations. The establishment of a base camp or other semi-permanent facility housing thousands of personnel requires a well with higher yields and better water quality. The USACE developed Humanitarian Civic Assistance (HCA) wells as an alternative to tactical wells. The HCA well is larger and designed to be a permanent water source that will not require treatment before consumption (Baehr, 2001). Table 1 gives the specifications for HCA and tactical wells. Military drillers have deployed to many areas of Central America as part of the New

Horizons Program to install HCA wells for local communities in need of reliable, clean drinking water (Fig. 3). These missions are excellent training for the drilling teams that install HCA wells similar to those appropriate for base camp operations (Baehr, 1998).

Table 1. A comparison of HCA and tactical wells.

Parameter	HCA Well	Tactical Well
Yield per day	~125 m^3	<11 m^3
Water Quality	Must be free of contaminants	Less important; water treated with ROWPU
Well Grouting	Required to be sanitary	Inadequate seal
Sand Content	<5 parts per million	Not a concern due to short well life

Figure 3. US Army Engineer soldier drills a Humanitarian Civic Assistance (HCA) well in El Salvador. Source: USACE Mobile District.

5. DISCUSSION

Although use of groundwater is a viable option for many military operations, it must be compared to other water sources such as bottled water and surface water. It is important to determine if the financial and logistical benefits of using a groundwater source outweigh the other options. Beginning with the Gulf War, bottled water has grown in popularity with both military commanders and troops. Bottled water has the appeal of convenient packaging and can rapidly supply small numbers of troops during the initial stages of an operation (US Army, 2003). The drawbacks of bottled water include generation of solid waste from empty bottles and the large amount of transportation and in-theater storage required.

Surface water is available in many areas and does not require geological expertise or well drilling to locate and develop. The drawbacks of surface water are that it requires substantial treatment before consumption and may have seasonable variations in flow. Groundwater normally has a low chemical or biological threat of contamination and does not usually experience a large seasonal variation in quantity. If accessibility and cost factors are equal, groundwater is the preferred choice for a water supply (Roebuck, 2003). After Preventive Medicine personnel test and approve a groundwater source, treatment other than chlorination usually is not required. However, the initial transport requirement for a well drilling unit is considerable. For example, an Army well drilling detachment requires eight C-130, five C-17, or two C-5A aircraft to transport all equipment into theater (Baehr, 2001; P.A. Lang, USARC, pers. comm., 2003). A switch to more modern drilling equipment that is smaller and lighter can reduce this burden.

In terms of cost, groundwater generally has an advantage over both bottled and surface water. Bottled water requires the purchase, shipment, and storage of thousands of plastic bottles. A ROWPU must normally treat surface water before consumption. An estimate of costs was conducted by the US Army Center for Health Promotion and Preventive Medicine in 2002 based on a three month period for operations in Afghanistan. The cost to supply 294,840 gal (1,114,495 l) of bottled water to soldiers during this period was approximately $1.4 million whereas the estimated cost to provide the same amount of ROWPU-treated water was $315,000 (US Army, 2003). Whereas the initial fixed costs to drill and develop a water supply well are variable depending on depth to water and local geology, long term costs may be significantly lower than bottled water or ROWPU-treated surface water (US Army, 1987). Although there are many variables in calculating costs of groundwater usage, based on the author's experience with installing groundwater wells, the cost of using groundwater can be much less than the cost of using ROWPU-treated surface water. This cost savings becomes even greater over a period of several years due to the low amount of maintenance required to keep a water supply well functioning.

6. CONCLUSIONS

Water is an essential logistical component to the success of military operations. Finding ample supplies of clean water has been a concern of military leaders since ancient times. As armies grew larger and logistics became more complex, the use of groundwater as a safe and reliable water source became more important. Beginning with World War I, armies fielded special well drilling units and employed geologists to locate and develop groundwater supplies to supplement surface water sources. More recent military operations requiring the establishment of semi-permanent base camps with populations of several thousand soldiers highlight the benefits of using groundwater as a primary water supply. Whether drilled by military personnel or civilian contractors, these wells have proven important to the success of operations. When easily accessible, a clean and high yield groundwater water source is an ideal option for current and future military operations. The cost to provide groundwater to troops can be much less than using either bottled or surface water. The assets to locate, drill, and develop groundwater sources exist within the US military. As demonstrated in the Balkans, these military resources have not always been used effectively and should be fully utilized in future operations.

ACKNOWLEDGEMENTS

Ms. Laura C. Dwyer at the USACE ERDC Topographic Engineering Center and Ms. Laura Waite Roebuck at the USACE Mobile District provided information on US military well drilling operations. CPT Paula Lang, USARC, and CPT Tim Scarbrough, 416th ENCOM, gave me updated information on the number and composition of US military well drilling units. LTC Kenneth W. McDonald, USMA, shared his Kosovo deployment experiences to give me a better understanding of that operation. Dr. Edward P.F. Rose of Royal Holloway, University of London, provided invaluable comments on the history of military groundwater usage and suggested many additional references to the author.

REFERENCES

Baehr, J.N. 1998. Potable water well design for Humanitarian Civic Action well drilling missions. In *Military Geology in War and Peace*, J.R. Underwood, Jr. and P.L. Guth, eds., Boulder, CO: Geological Society of America Reviews in Engineering Geology XIII: 227-231.

Baehr, J.N. 2001. Operational guidelines for Humanitarian Civic Action water well drilling, 2nd edition. Mobile, AL: US Army Corps Engineer District.

Brooks, A.H. 1920. The use of geology on the Western Front. US Geological Survey Professional Paper 128-D, 85-124.

Doyle, P., Bennett, M.R., and Cocks, F.M. 2000. Geology and warfare on the British sector of the Western Front 1914-18. In *Geology and Warfare: Examples of the Influence of Terrain and Geologists on Military Operations,* E.P.F. Rose and C.P. Nathanail, eds., London: The Geological Society, 179-235.

Dwyer, L.C. and Markley, B. 1998. Topo team finds water in Haiti. Engineer Update 22(11):8-9.

Eastler, T.E. 2004. Military use of underground terrain - A brief historical perspective. In *Studies in Military Geography and Geology*, D.R. Caldwell, J. Ehlen, and R.S. Harmon, eds., Dordrecht, The Netherlands: Kluwer Academic Publishers, 21-37.

Knowles, R.B. and Wedge, W.K. 1998. Military geology and the Gulf War. In *Military Geology in War and Peace*, J.R. Underwood, Jr. and P.L. Guth, eds., Boulder, CO: Geological Society of America Reviews in Engineering Geology XIII:117-124.

Moseley, F. 1973. Desert waters of the Middle East and the role of the Royal Engineers. Royal Engineers Journal 87:175-186.

Nathanail, C.P. 1998. Hydrogeological assessments of United Nations bases in Bosnia Hercegovina. In *Military Geology in War and Peace*, J.R. Underwood, Jr. and P.L. Guth, eds., Boulder, CO: Geological Society of America Reviews in Engineering Geology XIII:211-215.

Pittman, W.E. 1998. American geologists at war: World War I. In *Military Geology in War and Peace*, J.R. Underwood, Jr. and P.L. Guth, eds., Boulder, CO: Geological Society of America Reviews in Engineering Geology XIII:41-47.

Roebuck, L.W. 2003, Operational guidelines for Humanitarian Civic Assistance water supply systems. Mobile, AL: US Army Corps Engineer District.

Rose, E.P.F. 2004. The contribution of geologists to the development of emergency groundwater supplies by the British army. In *200 years of British Hydrogeology,* J.D. Mather, ed., London: Geological Society Special Publications, in press.

Rose, E.P.F. and Pareyn, C. 1998. British applications of military geology for 'Operation Overlord' and the battle in Normandy, France, 1944. In *Military Geology in War and Peace,* J.R. Underwood, Jr. and P.L. Guth, eds., Boulder, CO: Geological Society of America Reviews in Engineering Geology XIII:55-66.

Rose, E.P.F. and Rosenbaum, M.S. 1998. British military geologists through war and peace in the 19th and 20th centuries. In *Military Geology in War and Peace*, J.R. Underwood, Jr. and P.L. Guth, eds., Boulder, CO: Geological Society of America Reviews in Engineering Geology XIII:29-39.

Rose, E.P.F., Häusler, H., and Willig, D. 2000. A comparison of British and German military applications of geology in world war. In *Geology and Warfare: Examples of the Influence of Terrain and Geologists on Military Operations,* E.P.F. Rose and C.P. Nathanail, eds., London: The Geological Society, 107-140.

Rose, E.P.F., Mather, J.D., and Willig, D. 2002. German hydrogeological maps prepared for Operation 'Sealion': the proposed invasion of England in 1940. Proceedings of the Geologists' Association 113:363-379.

Rose, E.P.F. and Willig, D. 2004. German military geologists and geographers in World War II - Roles in Planning for Operation Sealion - The Invasion of England Scheduled for September 1940. In *Studies in Military Geography and Geology*, D.R. Caldwell, J. Ehlen, and R.S. Harmon, eds., Dordrecht, The Netherlands: Kluwer Academic Publishers, 199-219.

Scarbrough, T. and Lang, P. 2001. Military Water-Well Drilling. Engineer 31 PB 5-01-3:4-9.

Terman, M.J. 1998. Military Geology Unit of the US Geological Survey during World War II. In *Military Geology in War and Peace*, J.R. Underwood, Jr. and P.L. Guth, eds., Boulder, CO: Geological Society of America Reviews in Engineering Geology XIII:49-54.

US Army. 1987. Water supply sources and general considerations. Washington DC: Departments of the Army and Air Force TM 5-813-1.

US Army. 1994. Multiservice procedures for well-drilling operations. Fort Leonard Wood, MO: US Army Engineer School FM 5-484.

US Army. 2003. Use of bottled water for deployment support. Aberdeen Proving Ground, MD: US Army Center for Health Promotion and Preventive Medicine Water Quality Information Paper 31-034.

Chapter 24

MANAGING GROUNDWATER RESOURCES AT CAMP SHELBY TRAINING SITE, MS (USA)

David M. Patrick,[1] Kai M. Roth,[1] and Robert A. Lemire[2]
[1] *The University of Southern Mississippi*
[2] *Mississippi Army National Guard*

Abstract: A descriptive groundwater model was constructed for Camp Shelby and its environs to understand the relationships between surface hydrology, military training, water supply aquifers, and civilian encroachment around the site. The model was based upon data from over 1200 permitted water wells. Main results were that: (i) most of the site is underlain by a clay-rich buffer having low hydraulic conductivity that would prevent significant aquifer contamination by surface activities; (ii) wetland water tables are controlled by this buffer and are not interconnected with water supply aquifers; (iii) most water wells are screened in confined aquifers at depths of 100-750 ft below ground surface; (iv) these laterally discontinuous aquifers are separated by clayey units and are not interconnected; and (v) the water pressures in the deep aquifer under the eastern part of the site, determined from potentiometric surface difference maps, declined as much as 150 ft over the last 40 years.

Key words: groundwater modeling, natural resource management, military training

1. INTRODUCTION

This paper presents the results of a groundwater study conducted as a part of natural resource management activities at Camp Shelby, a Mississippi Army National Guard (MSARNG) training and maneuver site in southern Mississippi. The study was motivated by the need to collect information for the training site's Integrated Natural Resource Management Plan (INRMP) and to determine the effects of encroachment of civilian population and increased civilian water usage. The objectives of the study were to identify the nature and distribution of the sediments underlying the training site in terms of their buffering capability, water-bearing capacity,

D.R. Caldwell et al. (eds.), Studies in Military Geography and Geology, 321–332.
© 2004 *Kluwer Academic Publishers*

and their control of wetland locations, aquifer and aquiclude stratigraphy and correlation, and the effects of historic pumping on aquifer pressures.

Camp Shelby consists of about 134,000 acres and is the largest contiguous Reserve Component training site in the United States. It is located in southern Mississippi approximately 100 mi southeast of Jackson, 10 mi south of Hattiesburg and 55 mi north of Gulfport and the Mississippi Gulf Coast (Fig. 1). The training site is situated within the Gulf Coastal Plain Physiographic Province, which consists of gently rolling uplands and broad stream valleys developed upon coarse- and fine-grained Tertiary sediments of continental origin. Deep, confined aquifers within this upper Tertiary (Neogene) sequence are the exclusive sources of municipal and industrial water supply throughout this region.

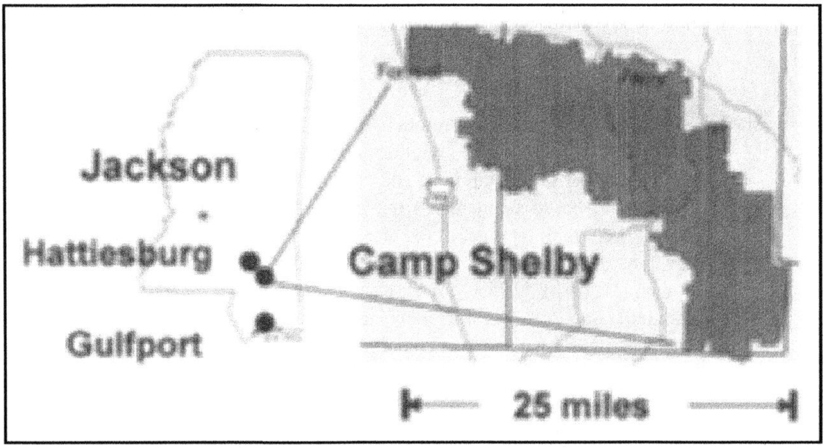

Figure 1. Outline map of Mississippi (left) showing the location of Camp Shelby (right) in the southeastern part of the state.

The facilities and training areas are used by the Army, Navy, and Air Force. As many as 100,000 troops are trained each year, by approximately 1000 full-time employees. The training areas include: Impact area, air-to-ground range, airfield, drop zone, artillery and multiple-launch rocket system firing points, and a combined arms area supporting platoon-level maneuver training. A multiple-purpose range complex (heavy) is nearing completion and a C-17 landing strip is in the planning stages.

2. MILITARY LAND MANAGEMENT

2.1 Integrated Natural Resource Management

Military land is considered a resource consisting of three functional areas that include biological, earth, and cultural resources. Biological resources include the flora and fauna; earth resources pertain to air, soils, and water (surface and subsurface); and cultural resources include archaeological and historical sites (see McDonald et al., 2004, this volume). Furthermore, these functional areas are not to be managed independently; rather, the goal is for integrated management which requires that the relationships between the three functional areas be understood (Patrick et al., 1994; US Army, 1998). The geographic information system (GIS) provides the mechanics of integration; however, determining interrelationships is not an easy task. For this reason, installations have prepared an INRMP to achieve this goal. Natural and cultural resource activities are supported by the Army through the Integrated Training Area Management (ITAM) Program. This program provides guidance and funding to assist Army installations to attain the best possible use of training lands, develop a decision-making process which integrates training requirements with the management of natural and cultural resources, promote proactive land management, and align land management with training, testing, and readiness priorities.

2.2 Land Management Environmental Tasks

In order to understand, manage, and integrate the three functional areas, a military installation natural and cultural resource manager must: (i) ensure that the installation is in compliance with all applicable State and Federal statutes and guidelines; (ii) conduct sufficient applied research to fully understand and integrate natural and cultural resources; and (iii) educate troop trainers, trainees, and the public regarding the natural and cultural resource ground rules of that installation.

2.3 Encroachment

Encroachment, i.e., the gradual build up of civilian population around the installation, often results in additional resource management issues and challenges. Noise, air pollution, flight space, traffic, recreational activity, intrusion, and unexploded ordinance are common issues. Surface and subsurface waters are additional issues because, like air and noise, these waters are not confined to the installation.

2.4 Land Management at Camp Shelby

Critical habitat management is an on-going activity; the major habitat issues include wetlands, the gopher tortoise, the Louisiana quillwort, and the red cockaded woodpecker. There are over 15,000 acres of riverine and slope wetlands on Camp Shelby (Minkin et al., 1998) whose locations are controlled by the training site geology and hydrology. There have been no environmental issues related to groundwater in this region to date.

Although Camp Shelby is relatively isolated along most of its southern and eastern periphery, the western, particularly the northwestern, side is only 2.5 mi from the Hattiesburg city limits (Fig. 2). Although Hattiesburg has grown toward Camp Shelby, most of the population growth has occurred to the west of Hattiesburg, some distance from the training site. Even so, this area of Mississippi is one of the fastest growing areas in the state and encroachment may become an issue in the future. The MSARNG coordinates military training activities and environmental/encroachment issues with the Hattiesburg Area Development Partnership (HADP) and other local agencies and organizations. One of the missions of the HADP is to promote economic development and this organization realizes Camp Shelby's local importance as the third largest employer in the area.

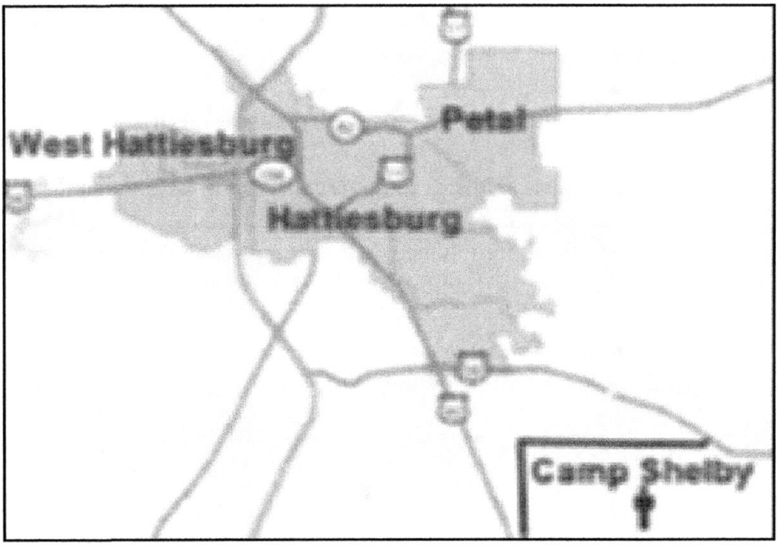

Figure 2. Map showing the northwest corner of Camp Shelby and its proximity to the southeastern city limits of Hattiesburg. The distance is approximately 2.5 mi.

3. REGIONAL HYDROGEOLOGY

The Miocene and younger sediments are several thousand feet thick and exhibit a south or southwest dip toward the Gulf of Mexico. Prior to European development, the groundwater flow direction was similar to the direction of dip. The regional stratigraphy is shown in Table 1. The oldest units of interest relative to groundwater are the subsurface Miocene Catahoula Fm. and the overlying Miocene Hattiesburg Fm. In outcrop, the Catahoula Fm. is sand or sandstone; whereas the Hattiesburg Fm. is predominantly fine grained. In the subsurface, confined aquifers occur in the sandy facies or beds of both formations; the interbedded clays and silts are the confining beds. Both formations are overlain by a few tens of feet of surficial, Plio-Pleistocene, coarse clastics of the Upland Complex also called the Citronelle Fm. (Brown, 1944; Bicker, 1969; Dockery, 1981; Autin et al., 1991; Meylan and Li, 1995; Patrick and Boyd, 2001).

Table 1. Generalized stratigraphic column for the Camp Shelby region. Modified after Dockery (1981).

Era	System	Series	Formation
Cenozoic	Quaternary	Holocene	Undifferentiated alluvium
		Pleistocene	Upland Complex[1]
	Tertiary	Miocene	Hattiesburg Fm.
			Catahoula Fm.[2]

[1] Also called Citronelle Formation
[2] In subsurface

Geologic mapping has shown that most of the northwestern part of Camp Shelby is underlain by the Hattiesburg Fm. (Patrick and Boyd, 2001). Previous investigations have shown that the mineral smectite, an expansive clay, is present in the clay fraction of the Hattiesburg Fm. (Adamczak, 1986; Meylan and Li, 1995). The presence of smectite would further decrease the permeability of the formation and contribute to absorption and retardation of contaminants and buffer the training site.

Stratigraphic correlation of aquifers is usually one of the goals of a regional hydrogeologic study and it was a goal of this investigation; however, previous experience in nearby counties demonstrated that correlation beyond a distance of a few miles was difficult (Patrick and Zhao, 1989; Sturdivant and Patrick, 1990). In these previous studies, contacts between the Catahoula Fm. and Hattiesburg Fm. could not be identified, nor could the principal aquifers be traced any significant distance within a county. The inability to include this correlation may be explained by facies changes in these continental sediments.

4. METHODOLOGY

The basic data elements of the study came from driller's logs and geophysical logs for State-permitted, water and petroleum test wells obtained from the Mississippi Department of Environmental Quality (MDEQ), Oil and Gas Board, and US Geological Survey (USGS). Currently, the MDEQ requires water wells 12 in or greater in diameter to be permitted; thus, wells smaller than this size would not be included. It was apparent from the outset of this study that there were relatively few water wells on the training site and much of the required data would have to be developed based upon wells from outlying areas. For this reason, data was collected for all of Perry County and parts of adjacent counties; however, only those data relative to the general area around Camp Shelby are reported here. The inclusion of an area significantly larger than that of Camp Shelby resulted in two additional advantages besides more and better stratigraphic information; the effects of civilian pumping in nearby parts of Hattiesburg on aquifers underlying Camp Shelby could be documented, and the accuracy of potentiometric surface contour maps would be improved. The well data was imported into a Microsoft Excel database that included over 1200 wells.

Well locations, originally identified by latitude-longitude, were converted to Universal Transverse Mercator grid using the USGS program J-380 (GEOUTM). Selected well data were then imported into ArcView GIS for map presentations, and into RockWorks 99 for construction of north-south (dip) and east-west (strike) cross-sections. Water elevations for selected time periods were imported into ArcView from which potentiometric surface contour maps were constructed. Collectively, these data and data elements would comprise a descriptive model of hydrogeology of the region. The inclusion of hydraulic conductivity and storativity information with these data would permit the development of a numerical model of the training site in the event such a model were required (Fetter, 1994; Dingman, 2002).

Water level changes (drawdown) for a given well in a given aquifer could not be determined because of the absence of continuous monitoring. Therefore, changes in water level in a given aquifer had to be based upon changes in water levels within nearby wells completed at different times in this same aquifer. Furthermore, rather than year-to-year comparisons, changes had to be based upon differences between periods of time. For example, water level comparisons were made for the period 1975-79 versus the period 1980-84. The change in the potentiometric surface could be determined by subtracting (in ArcView) one surface from the other (Roth, 2002).

5. RESULTS AND DISCUSSION

5.1 Overview

Initially, dip and strike cross-sections were prepared on a county-wide basis to determine the geographic extent of the aquifer units. As expected, these data indicated that correlation could not be accomplished either along the dip or the strike. Therefore, the county was subdivided, more or less along the dip direction, into three areas, north, middle and south. Dip and strike cross-sections, as well as potentiometric surface and difference maps, were prepared for each of these areas. Most of Camp Shelby lies in the middle area, and the data for this area are described in this paper.

5.2 Strike and Dip Sections

The locations of the strike and dip cross-sections are shown in Fig. 3; and the strike section is shown in Fig. 4. The east and west dip cross-sections are illustrated in Figs. 5 and 6, respectively. The lithology shown in the sections was based primarily on the drillers logs which were, in turn, based upon

Figure 3. Location map showing water wells and strike and dip cross-section locations. The shaded area in the southwest part of the training site is the impact area.

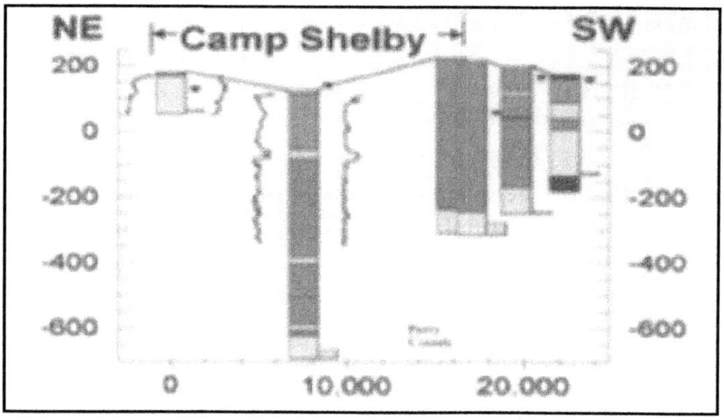

Figure 4. Strike cross-section (C-C'). Aquifers and sands are dotted (light tone), gravels are dotted (dark tone), and confining beds are dashed. The self-potential (SP) log is on the left, and the electrical resistivity log is on the right of the lithologic log. The screened intervals of the aquifers are shown by the small, cross-hatched rectangles on the right side; the water levels (elevations when drilled) are shown by the triangles. The boundary of Camp Shelby is shown by vertical lines at the top of each section. The scale is in feet. These notations also apply to the dip cross-sections shown in Figs. 5 and 6.

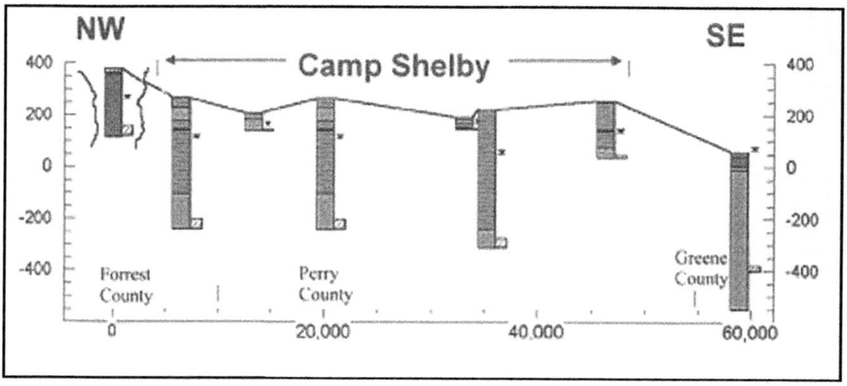

Figure 5. East dip cross-section (B - B'). The direction of dip is to the southwest (right).

cuttings brought to the surface by the circulating drilling mud. For those wells that had geophysical logs run in them, there is a general agreement between the geophysical data and cutting data; however, there may be an underestimation of the amount of sand in the lithologic logs.

The cross-sections in Figs. 5 and 6 show that, by and large, aquifers lie several hundred feet below the surface and are overlain by significant thickness of fine-grained confining beds. Usually, a well is not drilled deeper than the target aquifer and, therefore, there is no information on the nature of the strata underlying the target sand. Most pressure heads in these wells

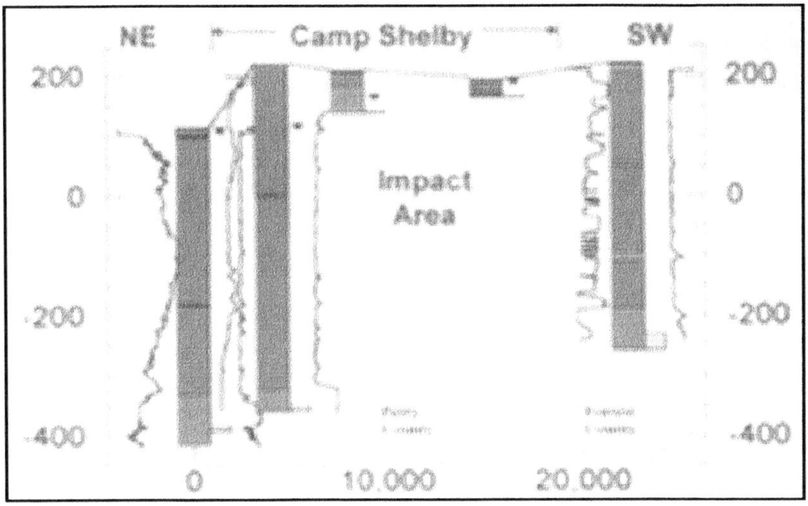

Figure 6. West dip cross-section (A-A'). The direction of dip is to the southwest (right).

(when drilled) were scores to hundreds of feet above screen elevation, which is also an indicator of a confined aquifer. The sections also show that most of Camp Shelby is directly underlain by fine-grained and impermeable strata that control the locations of most the wetlands and serve to "buffer" the training site.

Stratigraphic correlation across Camp Shelby was found to be no easier than regional correlation. Most likely the deeper strata would be considered to be a part of the Catahoula Fm., whereas surface or near-surface ones would be a part of the Hattiesburg Fm. The cross-sections also support evidence elsewhere that lateral and vertical facies changes are an important part of this stratigraphic sequence. For the reasons given above, the individual aquifers were correlated on the basis of depth as shallow, middle, and deep. These designations were based upon a generalized statistical distribution of well elevations rather than any stratigraphic relationship. On this basis the aquifers were classified as upper, middle, and deep with the following approximate elevations. Upper: surface to 0 ft; middle: 0 ft to ~300 ft; and deep: below ~300 ft.

5.3 Potentiometric Surface Change Maps

Potentiometric surface change maps were made for each aquifer. Fig. 7 is an example of one such a map for the deep aquifer. This map shows that the potentiometric surface elevation on the Hattiesburg side of Camp Shelby had

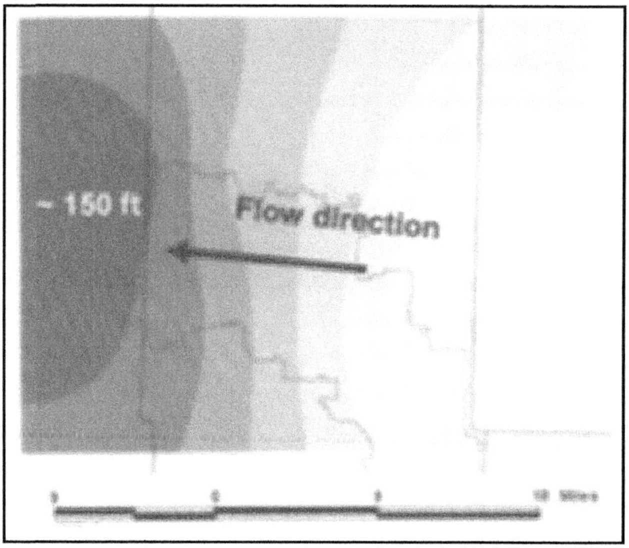

Figure 7. Potentiometric surface change map for the deep aquifer showing a decrease of approximately 150 ft between the periods 1960-69 and 1990-2000.

decreased by as much as 150 ft between the years 1960-69 and 1990-2000. Potentiometric surface change maps for the middle and upper aquifers for similar time periods indicated elevation decreases of 75-100 ft. These decreases in potentiometric surface elevation have occurred due to population growth and increased pumping in the Hattiesburg area. The potentiometric surface and potentiometric surface difference maps generally confirm the decline in water pressure that has been observed in other south Mississippi counties in which there has been no deterioration in the quality and quantity of available groundwater. What is of more importance in regard to Camp Shelby is the direction of groundwater flow that, for the three aquifers, is mainly away from the training site and towards the pumping centers in the Hattiesburg area. This being the case, a contaminant in the aquifers underlying Camp Shelby would be pumped toward these centers. Considering that most of the training site is buffered by the Hattiesburg Fm., the occurrence of surface-derived contamination would be unlikely. Even so, the possibility exists that a contaminant could reach these aquifers, either inadvertently or intentionally, through unprotected well heads on the training site. This means that unused wells on Camp Shelby including those drilled for training purposes by military well drillers should be properly plugged and capped prior to abandonment.

6. CONCLUSIONS

In general, the strike and dip sections indicated that most of the training site was found to be underlain by the Hattiesburg Fm. which would prevent significant aquifer contamination by surface activities. These data also support the notion that this formation also controls the locations of most of the wetlands on Camp Shelby.

Most water supply wells are screened in confined aquifers that were classified as shallow, medium, or deep at approximate depths of 100-750 ft below ground surface. These laterally discontinuous aquifers are separated by clayey units and are not interconnected.

Pumping center, radii of influence of pumped wells, changes in flow direction, and historic drawdown could be identified from potentiometric surface maps. They showed that the historic reduction in water pressures in the deep aquifer had declined approximately 150 ft during the last 40 years, and as much as 100 ft for the middle and upper aquifers. Furthermore, these maps indicated that, currently, groundwater flow under Camp Shelby is away from the training site and towards major population centers.

ACKNOWLEDGMENTS

The studies presented in this paper were funded by the US National Guard Bureau (NGB) and ITAM program through the MSARNG. R.A. Lee, Environmental Program Manager, at Headquarters MSARNG and the GIS, Environmental and Range Control staffs at Camp Shelby are acknowledged for their assistance and encouragement. S. Jennings (MDEQ) and L. Slack (USGS) provided helpful advice and access to their respective organization's groundwater data. M. Meylan and G. Russell, geology faculty at the University of Southern Mississippi (USM), also provided administrative and technical assistance. The authors acknowledge the helpful critiques of K. McDonald and R. Harmon.

DISCLAIMER

The views and opinions stated in this paper are exclusively those of the authors and have not been endorsed by the US Department of Defense, NGB, the MSARNG, or USM. Reference to commercial brand names is made for informational purposes only and does not constitute an endorsement or approval of these products by the above organizations.

REFERENCES

Adamczak, D.L. 1986. The petrology of the Hattiesburg Lutite (Miocene), Northern Forrest County, Mississippi (unpublished Master's thesis). Hattiesburg, MS: The University Southern of Mississippi.

Autin, W.J., Burns, S.F., Miller, B.J., Saucier, R.T., and Snead, J.I. 1991. Quaternary geology of the Lower Mississippi Valley. In *Quaternary Nonglacial Geology,* R.B. Morrison, ed., Boulder, CO: Geological Society of America, Geology of North America, K-2:547-582.

Bicker, A.R. 1969. Geologic map of Mississippi (scale 1:50,000). Jackson, MS: Mississippi Geological Survey.

Brown, G.F. 1944. Geology and ground-water resources of the Camp Shelby area. Mississippi State Geological Survey Bulletin 58.

Dingman, S.L. 2002. *Physical Hydrology*. Upper Saddle River, NJ: Prentice-Hall, Inc.

Dockery, D.T. 1981 Stratigraphic column of Mississippi. Jackson, MS: Mississippi Bureau Geology.

Fetter, C.W. 1994. *Applied Hydrology*. Upper Saddle River, NJ: Prentice-Hall, Inc.

McDonald, E., Bullard, T., Britt, T., and O'Ruiz, M. 2004. Development of an archeological predictive model for management of military land - Identification of geologic variables in desert terrain. In *Studies in Military Geography and Geology*, D.R. Caldwell, J. Ehlen, and R.S. Harmon, eds., Dordrecht, The Netherlands: Kluwer Academic Publishers, 259-270.

Meylan, M.A. and Li, Z. 1995. Geologic mapping of south-central Mississippi - A model for the distribution of Neogene and Quaternary sediments. Transactions of the Gulf Coast Association Geological Society, 435-440.

Minkin, P., Packer, W., Gravid, D., Bishop, M., and Bishop, A. 1998. Delineation of wetlands and other regulated waters, Camp Shelby. Report to the National Guard Bureau, Vicksburg, MS: US Army Engineer Waterways Experiment Station.

Patrick, D.M. and Boyd, S.A. 2001. Wetlands and erosion studies in support of military training, Camp Shelby Training Site, Mississippi, USA. In *The Environmental Legacy of Military Operations*, J. Ehlen and R.S. Harmon, eds., Geological Society of America Reviews in Engineering Geology XIV:137-149.

Patrick, D.M. and Zhao, S. 1989. Groundwater extraction trends in Forrest and Lamar Counties, Mississippi. Proceedings of the Mississippi Water Resources Conference, Starkville, MS: Mississippi Water Resources Research Institute, 85-89.

Patrick, D.M., Corcoran, M.K., Albertson, P.E., and Smith, L.M. 1994. Earth resources stewardship at Department of Defense installations. Interim Report. Earth Resources Task Area, Legacy Resources Management Program, Vicksburg, MS: US Army Engineer Waterways Experiment Station, Technical Report GL-94-9.

Roth, K. 2002. The geohydrology of the Neogene aquifer system of Perry County and environs (unpublished Master's thesis). Hattiesburg, MS: The University of Southern Mississippi.

Sturdivant, R., Jr. and Patrick, D.M. 1990. Groundwater models of the Miocene aquifer system in Jones County, Mississippi. Starkville, MS: Proceedings of the Mississippi Water Resources Research Institute, 136-146.

US Army. 1998. Integrated Training Area Management (ITAM). Washington, DC: US Government Printing Office, Army Regulation 350-4.

Chapter 25

WATER AND ENVIRONMENTAL SECURITY IN THE MIDDLE EAST

J. David Rogers
University of Missouri-Rolla

Abstract: Israel, Jordan, and the Palestinian Authority are linked by common aquifers that have been subject to over-drafting, contamination, and negotiation. The Israelis used military force to secure the Jordan River watershed in 1964 and 1967. The environmental security of the region's scarce water resources has become a contentious issue and an impediment to regional development because the three countries have been unable to develop protocols for managing their water. The expanding populations of Israel, Jordan, and the Palestinian Authority have created an unprecedented demand for additional potable water needed to sustain life, maintain sanitation, and irrigation for agriculture. Desalination of seawater using reverse osmosis will likely emerge as a major source, but it requires significant capital outlay and energy. Management of water resources has emerged as a national priority upon which each country sees its survival depending. Armed conflicts and sporadic clashes will continue if the issue is not resolved.

Key words: water wars, environmental security, water resources, over-drafting, desalination

1. INTRODUCTION

Israel and the Palestinian Territories are separated from Jordan by the Syrian-African Rift, the longest valley in the world (Figs. 1 and 2). The three nations are linked by common aquifers, but their groundwater reserves have been over-utilized for crop production, even when many foodstuffs might be imported at lower unit prices. Since the mid-1980s, the expanding populations of Israel, Jordan, and the Palestinian Authority have competed for dwindling water resources. Middle East population growth is currently

D.R. Caldwell et al. (eds.), Studies in Military Geography and Geology, 333–343.
© 2004 *Kluwer Academic Publishers*

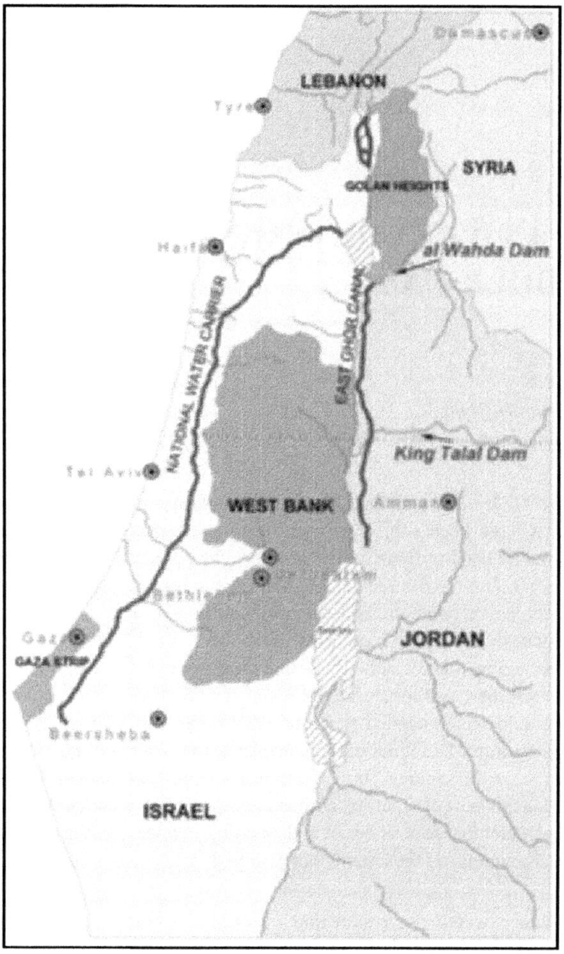

Figure 1. Map of border region between Israel, Jordan, and the Palestinian Authority, showing the Golan Heights, the National Water Carrier, the East Gohr Canal, and the proposed al Wahda Dam site.

averaging a staggering 3% annually. The population of Israel and the Palestinian Territories now stands at approximately eight million. The population of the Palestinian Authority occupying Gaza and the West Bank recently exceeded one and a half million. Israel's growth has come about in large part through three waves of immigration: The first following establishment of the British mandate at end of World War I (1918); a second pulse after World War II, particularly 1948, when the State of Israel was created; and a more recent influx of immigrants between 1987-2002, when two million Russian Jews settled in Israel. By 2020 the population of Israel is expected to grow to nine million and that of the Palestinian Authority to exceed three million.

According to the World Bank, the Middle East has the highest median cost of water supply and sanitation in the world, reaching $300 per capita in 1985, about double what it costs in the United States and about five times the cost of water in Southeast Asia (Starr, 1991). Israel, the Palestinian lands (West Bank and Gaza), and Jordan are jointly facing a combined water deficit of at least 300-400 m^3 per year, and as much as 500-600 million m^3, depending on weather patterns and consumption (Casa, 1991). Jordan's population is increasing at a rate of 3.8% per year, one of the world's highest growth rates. Like Israel, Jordan has exhausted the country's natural water resources and must begin looking seriously at desalination and water import schemes to meet future demands.

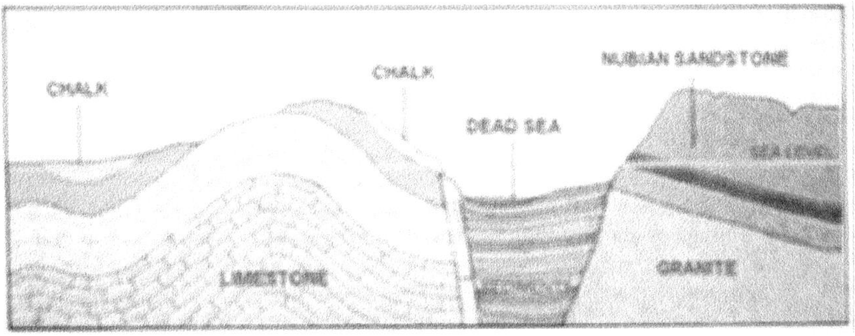

Figure 2. Generalized geologic cross-section from west to east through the Syrian-African Rift, between Israel and the West Bank territory on the left and Jordan on the right. The Dead Sea occupies the lowest point in a massive graben formed by a pull-apart basin between the Mediterranean and Arabian tectonic plates. Modified from Beitzel (1985).

Security concerns have been an impediment to cooperative resource development by Israel and her neighbor Arab states since the creation of Israel in 1948. The Israelis are concerned that their water resources infrastructure is secure from intervention and terrorism. Armed conflict erupted between Israel and Syria in 1964-65 when Syria attempted to divert flow from the upper Jordan River (Gleick, 1993). Israeli concerns about environmental security manifested themselves again in the June 1967 war, when Israel doubled her water resources by occupying the Golan Heights and south Mt. Herman, where the waters of the Jordan River coalesce. Israel has refused to relinquish any of this valuable watershed.

The Palestinian Authority has the greatest need for developing additional sources of water, but no formal agreement has been reached on this issue in the ongoing negotiations with Israel because of the overpowering concerns about security, which erupted when the Infantada began in 1999.

Development of new water sources has become a bargaining chip for Israel and the Palestinian Authority in their ongoing negotiations. The region's Arab neighbors see the United States as the only peace broker capable of funding key infrastructure improvements on the West Bank that the Israelis would refrain from targeting in any retaliatory strikes. Water has emerged as the key component in the region's environmental security. Exploitation of water resources will reign supreme in any economic development in this volatile region in the foreseeable future.

2. DEVELOPMENT OF WATER RESOURCES

The first modern water conveyance system in the region was initiated by Jewish settlers in 1935 to bring well water from the Jezreel Valley southward through Palestine, when it was a British mandate, to the northern Negev Desert. Jewish settlers constructed three experimental settlements in the Negev in 1943, followed by 16 more in 1946-47. This first water pipeline leading to the northwestern Negev was only 0.15 m in diameter, but stretched 190 km. It was completed in 1947. The first large scale supply system was a 1.68 m diameter pipeline extending 130 km from the Yarkon River to the Negev completed by the Israelis in 1948. It was capable of supplying 100 million m^3/yr.

In the late 1950s Jordan and Israel embarked on a race to collect, convey, and disperse the free-flowing waters of the Jordan River below the Sea of Galilee. In 1955 the Johnston Unified Water Plan was adopted by both countries as a non-ratified treaty of allocation rights, which was more or less successfully implemented until the June 1967 war between Israel and her Arab neighbors (Efrati, 2000). By 1961 the Jordanians completed their 110 km long East Ghor Canal in the Jordan Valley below the Sea of Galilee (the name was changed to the King Abdullah Canal in 2001). During the decade of the 1960s, the Israelis constructed their National Water Carrier, an 85 km long system of pipelines, open channels, tunnels, re-regulation pools, and distribution reservoirs (Kantor, 2001).

In 1964 the Arab League countries tried to sabotage the Israeli system by diverting water from the Jordan River just downstream of the Sea of Galilee. The Israelis responded by moving their intake to the northwestern shore of Galilee, near Tabgha, further from the border. The Israelis changed their plan so a 372 m lift would be required at the intake, from whence the flow would be taken by gravity to Israel's coastal plain. When they realized that the Israelis were proceeding with construction of the Tabgha intake, the Syrians began diverting the headwaters of the Jordan River during the winter of 1964-65. The Israelis responded with a series of air strikes and commando raids on Syria's diversion works (Gleick, 1993). This military action contributed to the tensions that led to the June 1967 war, when Israel secured the Golan Heights and Mt. Herman, effectively doubling its domestic water supply (shown as the Golan Heights occupation zone in Fig. 1). The National Water Carrier is often referred to as the Kinneret-Negev Conduit, and was constructed between 1964 and 1969. Subsequent extensions and crossovers have been added to the conduit on a near-continuous basis, and water is pumped, in stages, to the kibbutzim in the northwestern Negev.

Since completion of the National Water Carrier distribution system, Israel has augmented its conveyance capacity by drilling hundreds of wells to tap groundwater resources along the route, so it presently conveys around 400 million m^3/yr, which supplies about 25% of Israel's needs. In January 1990 and the summer of 1991 the water carrier was shut down by drought and deliveries to agricultural users were slashed 50%. This sudden cessation of water deliveries brought international attention to how tenuous the water situation had become in the face of Israel's commitment to allow two million Russian Jews to immigrate during the succeeding decade.

For the Jordanians, water quality has slowly but continuously declined in the Jordan River Valley. The majority of Jordan's fresh water for Amman comes from the Yarmuk River and wells drilled in nahals and wadis in upper catchments east of the Jordan Valley. Natural springs occur along the rift's bounding faults. The Zarqa River supplies about 25% of the annual supply and the remainder is collected from side wadis. After the 1994 Treaty of Peace was signed between Jordan and Israel, fresh water began moving through a pipeline from the Sea of Galilee to the East Ghor Canal.

3. MANAGEMENT OF WATER RESOURCES

The Israelis have tapped two major aquifers since 1948. The Yarkon/Taninim or "Mountain" Aquifer lies beneath north-central Israel and the West Bank territory of the Palestinian Authority (Fig. 3). 70-80% of the Mountain Aquifer theoretically lies beneath the West Bank region claimed by the Palestinian Authority, as well as 70-80% of the effective recharge area. But the recharged waters flow westward, toward the coastal plain, which belongs to Israel. Since the mid-1960s the Israelis have tapped 25-45% of their agricultural water from this aquifer, causing a gradual but sustained depletion.

The Coastal Aquifer underlies the coastal plain, along the Mediterranean Sea in west-central Israel. It is comprised of Plio-Pleistocene age sands and calcareous sandstone. Although the coastal aquifer contributes about 250 million m^3/yr, sea water intrusion has become a nagging problem, obviating withdrawals within 40-80 m of the ground surface. The Coastal Aquifer does not extend beneath the West Bank but does lie beneath the entirety of the Gaza Strip.

During the first half century of development (1948-98), the Israelis succeeded in over-drafting the country's water resources between 15 and 20% beyond the recharge capacity (see lower water table in Fig. 3). Although recharge efforts increased significantly each decade, so did consumption. An additional headache for all three countries has been increasing levels of groundwater pollution, mostly from pesticides, fertilizers, and untreated sewage disposal. In the highly concentrated Gaza Strip (population just over one million), the Coastal Aquifer has become seriously contaminated, requiring additional water and sewage treatment infrastructure to be constructed (Committee on Sustainable Water Supplies for the Middle East, 1999). Loans for these improvements, however, have not been forthcoming from foreign sources, which are fearful of the region's historic instability.

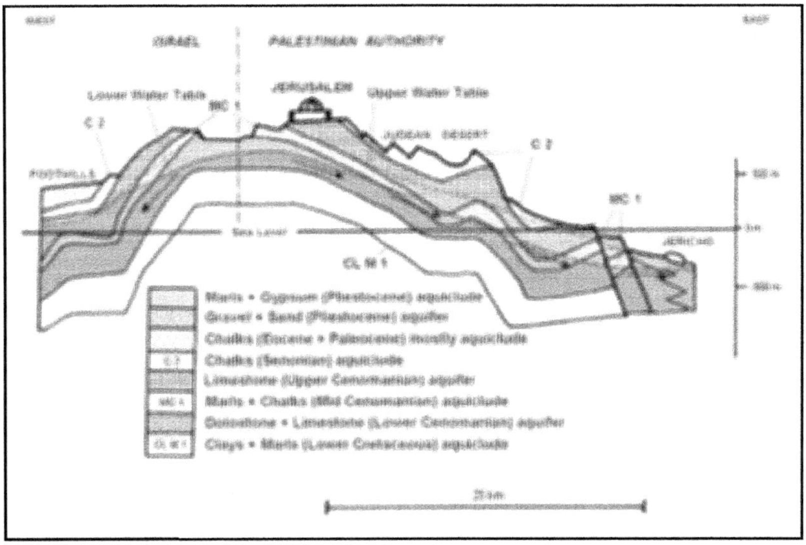

Figure 3. Cross-section through the Yarkon-Taninim, or "Mountain," Aquifer in central Israel and the West Bank. This aquifer was steadily reduced between 1948 and 1991. The Israelis have not been able to recharge this aquifer as readily as the younger Coastal Aquifer underlying the western coastal lowland.

The Israelis have developed a multi-faceted approach to solve their short and long term problems with over-utilization of groundwater. Over the past 25 years they have instituted sweeping conservation measures by employing drip irrigation, later adding fertilizers to create a dual irrigation process they termed "fertigation." Another avenue of research was focused on developing salt-tolerant species of their agricultural crops. In the 1990s, they reported some surprising successes raising crops irrigated with brackish water.

In 1995 the Israelis began recycling 120 million m^3/yr from the Shafdan Sewage Treatment Plant in Tel Aviv to recharge their Coastal Aquifer (Sitton, 2003). The effluent undergoes secondary treatment before being discharged into spreading basins for one day, then is allowed to dry for two to four days. Filtration occurs naturally via percolation through the Coastal Aquifer. The Coastal Aquifer is then tapped to provide agricultural water for the western Negev through the Third Negev Pipeline, which presently conveys about 115 million m^3/yr.

Israel's Arab neighbors have spearheaded the employment of desalination for domestic water supply. Until recently, 60% of the world's desalination capacity lay in Persian Gulf states, with Saudi Arabia accounting for almost one-third of annual world production (Starr, 1991). Kuwait, Saudi Arabia, and the United Arab Emirates are almost totally dependent on desalination for their fresh water supply. The Saudis have repeatedly expressed concern over the security of their immense power generation/desalination plants, which are strategically vulnerable to attack or sabotage.

The Israelis recently decided to pursue desalination. By 2004 they hope to complete a reverse-osmosis desalinization plant at Ashkelon, which will produce 50 million m^3/yr of fresh water. Israel is in process of designing a second desalination plant with equal capacity. They estimate that the reverse-osmosis process can produce one cubic meter of water for just US \$0.57, making it an attractive alternative (the average price their neighbor states pay is about US \$1/m^3). If these pilot plants are successful, the Israelis hope to expand their desalinization capacity to between 500-600 million m^3/yr by 2012. This could allow additional expansion of their agricultural holdings into irrigable parts of the Negev or compensate them for groundwater withdrawal concessions to the Palestinian Authority.

4. DISCUSSION

4.1 Palestinian Concerns

When the Declaration of Principles between Israel and the Palestinian Liberation Organization was signed in 1993, a provisional government was established over the West Bank and Gaza Strip known as The Palestinian Authority. West Bank Palestinians obtain their water from pre-1967 wells, but were not permitted to drill any additional wells until the 1993 treaty. Palestinians have always been able to purchase water from Israel's National

Water Carrier for a charge. Whereas the Palestinians acknowledge that Israel provides requisite water to the West Bank settlements for domestic and industrial use, Israel refuses to increase the volume of deliveries, which has become a major obstacle in ongoing negotiations. The Palestinians want to draft 200 million m^3 per year from the Mountain Aquifer for 10 years (until 2010), and gradually increase these withdrawals to 400 million m^3 per year over the succeeding 10 years (2010-20). They believe that this water is needed for new agricultural crops and their expanding population, which is growing at a rate of 3% per annum.

Both Jews and Arabs living in Israel use more water per capita than Palestinians in the West Bank or Gaza. Domestic and industrial consumption only account for about 30% of Israel's water consumption: 70% of Israel's water is used to support their expansive agricultural enterprise, including water-loving crops such as cotton and watermelons, which target the lucrative European market. But the agricultural sector supplies just 5% of Israel's Gross National Product. Despite this disparity, there is a bipartisan feeling among Israel's leadership that food production should be given a high priority, to better insure self-reliance in a world filled with potential enemies.

4.2 Agreements Forged Between Jordan and Israel

In 1994 a new water allocation plan was reached between Israel and Jordan as part of the Arava peace treaty, which has not been popular with the Jordanian people (Scham and Lucas, 2001). The treaty promised crucial allocations of water from Israel, cooperative efforts aimed at finding additional resources, establishing increased storage within Jordan, water quality and protection measures and protection of shared groundwater resources, and exchange of hydrologic information between the two countries (Scham and Lucas, 2001).

Jordan's water management policies have evolved as reactions to increasing Israeli development and utilization of the region's extractable water. The Jordanians have employed incremental measures, including dams, canals, deep withdrawal wells, and drip irrigation technology. They have also investigated solar-powered pumping and desalination of brackish groundwater in the northern Arava Valley (south of the Dead Sea) and pumped storage schemes aimed at establishing massive desalination plants near Aqaba.

An example of the frustrations felt by Israel's neighbors is the long-awaited al-Wahda Dam on the Yarmuk River, between Syria and Jordan (Fig. 1). Originally conceived in 1955, it was not until 1987 that engineering plans were completed and the search began for funding. The World Bank turned the project down because Israel, Jordan, and Syria could not agree on apportionment, even though Israel only controls 3% of the watershed. The Bank's established policy is that water must be managed to meet national objectives, including social, security, and economic concerns, and that water is an international resource whose apportionment and distribution requires extensive research and international cooperation (Klump, 2002). The Bank insists that all entities owning a portion of any watershed must agree on a protocol for development. To date, this has been impossible to achieve in the Jordan River watershed because of the historic enmity between Israel, Jordan, Lebanon, Syria, and the Palestinian Authority.

In July 1999 relations between Jordan and Syria had warmed sufficiently that the two countries agreed to build the al-Wahda Dam by circumventing the World Bank, using money from the Jordanian national treasury, the Arab Fund for Economic and Social Development, and the Islamic Development Bank. The project has since stalled because of a contractual requirement that 40% of the project funds be directed to Jordanian and Syrian subcontractors.

5. CONCLUSIONS

Israel, Jordan, and the Palestinian Authority are linked by common aquifers that have been exploited to the maximum since the mid-1930s and have recently been subject to over-drafting and contamination. The region's geologic and climatologic settings limit future development by conventional methods. Formal protocols need to be developed for managing these shared water resources. Assistance from international organizations and donor countries has historically been withheld because of perceived instability of the region.

Security concerns have also been an impediment to cooperative resource development by Israel and her neighbor Arab states since 1948. The Israelis are concerned that any above-ground pipeline could serve as a magnet for terrorists, similar to the Los Angeles-Owens River Aqueduct in the 1920s and the repeated attacks on a multinational oil pipeline in Colombia, which has accounted for almost half of the world's terrorist attacks the past few years (152 attacks in 2000 and 178 in 2001).

The Palestinian Authority has the greatest need for developing additional sources of water, but no formal agreement has been reached on this issue in the ongoing negotiations with Israel because of the overpowering concerns about security during the Infantada that began in 1999. Development of new water sources will be a fundamental bargaining chip for both Israel and the Palestinian Authority in future negotiations.

The only viable alternatives currently being discussed involve US Aid for International Development funded programs, similar to those granted to the Egyptians in 1979 and the Jordanians in 1994, when those countries signed peace treaties brokered by the US which formally recognized Israel (Starr, 1995). Since 1994 the Palestinians have sought Arab funding for desalination plants, but without success. The Israelis have gone forward with construction of desalination plants, which could replace 200 million m^3 in groundwater losses from the Mountain Aquifer beneath the Palestinian Authority by 2008, at a cost to Israel of at least US $114 million/ yr. This figure could triple by 2012 if the Israelis are successful in bringing more desalination plants online. The 3% annual growth of the Palestinian population demands that something be done to provide additional sources of water or armed conflict will become almost certain (Darwish, 2003).

Over the next century, survivability in the harsh climate of the Middle East will be driven by economic sustainability. The most challenging aspect of expanding populations and infrastructure in Israel, Jordan, and the Palestinian Authority will be potable water. Water infrastructure needs to be constructed ahead of domestic, industrial, and political demands. Water is fundamental to life, both for bodily intake and maintenance of sanitation. Past experience with Middle Eastern countries has shown that sustenance will most likely rely on reverse-osmosis desalination of seawater, which requires significant electrical power.

After signing the 1979 peace treaty with Israel, Egyptian President Anwar Sadat said the only reason his country would go ever go to war again would be to protect its water resources (Darwish, 2003). In the 21st century, arid and semi-arid Middle Eastern nations will be supporting vastly increasing populations, demanding more water than is currently available. These nations will need to employ redundant water supply systems, so the loss of one or two supply lines can be obviated by tapping into parallel or alternative sources, similar to a power grid. The United States will likely continue to serve a key role as the region's treaty broker, facilitating foreign investment in critical engineering infrastructure. Much of that infrastructure will be vulnerable to interdiction by terrorist attacks or overt acts of war. But, these nations' water resources are inevitably intertwined, so that it will be in their mutual interests to cooperate in both the development and common defense of such infrastructure.

REFERENCES

Beitzel, B.J. 1985. *The Moody Atlas of the Bible Lands.* Chicago, IL: Moody Press.

Casa, K. 1991. Water: The real reason behind Israeli occupations: Washington Report on Middle Eastern Affairs. (Online Version: http://www/Washington-report.org/backissues/0791/9107626.htm)

Committee on Sustainable Water Supplies for the Middle East. 1999. *Water for the Future: The West Bank and Gaza Strip, Israel, and Jordan.* Washington, DC: National Academy Press.

Darwish, A. 2003. Analysis: Middle East water wars. (Online Version: http://newsvote.bbc.co.uk/mpapps/pagetools/print/news.bbc.co.uk/2/hi/middle_east/294976...)

Efrati, C. 2000. The successful failure: The Johnston water negotiations. (Online Version: http://www.maxwell.syr.edu/maxpages/faculty/gmbonham/2000-Fall-IR-rojects/Website-Group/ICN/Copy%20of%20ICNweb/efrati1.htm)

Gleick, P., ed. 1993. *Water in Crisis: A Guide to the World's Fresh Water Resources*: New York: Oxford University Press.

Kantor, S. 2001. The National Water Carrier (Ha' Movil Ha' Artsi). (Online Version: http://research.haifa.ac.il/~eshkol/kantorb.html)

Klump, V. 2002. Hydro-politics along the Jordan River: Science, technology and international affairs. Georgetown University, School of Foreign Service. (Online Version: http://www.georgetown.edu/sfs/programs/stia/students/vol.02/klumpv.htm)

Niemi, T.M., Ben-Avraham, Z., and Gat, J.R. 1997. Dead Sea research - An introduction. In *The Dead Sea: The Lake and Its Setting,* T.M. Niemi, Z. Ben-Avraham, and J.R. Gat, eds., New York: Oxford University Press, 3-7.

Scham, P.L. and Lucas, R.E. 2001. "Normalization" and "anti-normalization" in Jordan: The public debate. Middle East Review of International Affairs Journal 5:1-20.

Sitton, D. 2003. Advanced agriculture as a tool against desertification. (Online Version: http:// www.mfa.gov.il/mfa/go.asp?MFAH00u70)

Starr, J.S. 1991. Water wars. Foreign Policy 82:17-36.

Starr, J.S. 1995. *Covenant over Middle Eastern Waters.* New York: Henry Holt and Co.

Index

GPSR Compliance

The European Union's (EU) General Product Safety Regulation (GPSR) is a set of rules that requires consumer products to be safe and our obligations to ensure this.

If you have any concerns about our products, you can contact us on ProductSafety@springernature.com

In case Publisher is established outside the EU, the EU authorized representative is:

Springer Nature Customer Service Center GmbH
Europaplatz 3
69115 Heidelberg, Germany

Batch number: 10755284

Printed by Printforce, the Netherlands